Jackfruit

Botany, Production and Uses

Jackfruit

Botany, Production and Uses

Edited by

Sisir Mitra

Former Professor, Faculty of Horticulture, and Dean, Postgraduate Studies, Bidhan Chandra Krishi Viswavidyalaya (State Agricultural University), Mohanpur, West Bengal, India

CABI

CABI is a trading name of CAB International

CABI
Nosworthy Way
Wallingford
Oxfordshire OX10 8DE
UK

Tel: +44 (0)1491 832111
E-mail: info@cabi.org
Website: www.cabi.org

CABI
200 Portland Street
Boston
MA 02114
USA

Tel: +1 (617)682-9015
E-mail: cabi-nao@cabi.org

A catalogue record for this book is available from the British Library, London, UK.

Library of Congress Control Number: 2022950107

ISBN-13: 9781800622296 (hardback)
9781800622302 (ePDF)
9781800622319 (ePub)

DOI: 10.1079/9781800622319.0000

Commissioning Editor: Rebecca Stubbs
Editorial Assistant: Emma McCann
Production Editor: Shankari Wilford

Typeset by Exeter Premedia Services Pvt Ltd, Chennai, India
Printed and bound in the UK by Severn, Gloucester

Contents

Contributors

Manivannan Arivalagan, ICAR-Indian Institute of Horticultural Research, Bengaluru, Karnataka 560 089, India, E-mail: arivalagan2100@gmail.com

Gundappa Baradevanal, ICAR-Central Institute for Subtropical Horticulture, Rehmankhera, Lucknow-226101, Uttar Pradesh, India. E-mail: Gundappa@icar.gov.in

Jigzcel Divine F. Basoy, College of Human Ecology and Food Sciences, University of Southern Mindanao, Kabacan, Cotabato, The Philippines. E-mail: jdfbasoy@usm.edu.ph

Chalinda K. Beneragama, Department of Crop Science, Faculty of Agriculture, University of Peradeniya, Peradeniya, Sri Lanka. E-mail: chalindab@agri.pdn.ac.lk

Anup Kumar Bhattacherjee, ICAR-Central Institute for Subtropical Horticulture, Rehmankhera, Lucknow-226101, India. E-mail: dadabhatu@gmail.com

R.M.S. Ruwan Chamara, Department of Crop Science, Faculty of Agriculture, University of Peradeniya, Peradeniya, Sri Lanka. E-mail: ruwanchamara034@gmail.com

Michel Chauvet, 15 Avenue du Quartier Bas, 34230 Vendemian, France. E-mail: Michel.chauvet34@orange.fr

Vasudevamurthy Dileep, Department of Biotechnology, University of Agricultural Sciences, GKVK, Bengaluru, India. E-mail: Dileepgautam16@gmail.com

Phebe Ding, Department of Crop Science, Faculty of Agriculture, Universiti Putra Malaysia, Malaysia, 43400 UPM, Serdang, Selangor, Malaysia. E-mail: phebe@upm.edu.my

D.M. Samantha B. Dissanayaka, Department of Crop Science, Faculty of Agriculture, University of Peradeniya, Peradeniya, Sri Lanka. E-mail: samanthad@agri.pdn.ac.lk

David Guest, School of Life and Environmental Sciences, University of Sydney, Sydney, NSW, Australia. E-mail: david.guest@sydney.edu.au

Mathala J. Gupta, ICAR-Central Coastal Agricultural Research Institute, Ela, Goa 403402, India. E-mail: mathalajuliet@gmail.com

Azimah Hamidon, Department of Crop Science, Faculty of Technical and Vocational, Universiti Sultan Idris, 35900 Muallim, Perak, Malaysia. E-mail: azmiah.hamidon@ftv.upsi.edy.my

Chikkavalli K. Harshitha, Department of Biotechnology, University of Agricultural Sciences, GKVK, Bengaluru, India. E-mail: ckharshitha@gmail.com

Hamizah Hassan, Department of Crop Science, Faculty of Agriculture, Universiti Putra Malaysia, 43400 UPM, Serdang, Selangor, Malaysia. E-mail: miss_sah@yahoo.com

Ganesan Karunakaran, Division of Fruit Crops, ICAR-Indian Institute of Horticultural Research, Bengaluru, Karnataka 560 089, India. E-mail: ganesan.karunakaran@icar.gov.in

Kundan Kishore, Central Horticultural Experiment Station (ICAR-Indian Institute of Horticultural Research), Bhubaneswar, India. E-mail: kkhort12@gmail.com

H.M. Prathibhani C. Kumarihami, Department of Crop Science, Faculty of Agriculture, University of Peradeniya, Peradeniya, Sri Lanka. E-mail: prathibhani@agri.pdn.ac.lk

Van Tri Mai, Southern Horticultural Research Institute, Tien Giang, Vietnam. E-mail: maivantri2017@gmail.com

Sisir Mitra, Former Professor, Faculty of Horticulture, Bidhan Chandra Krishi Viswavidyalaya, Mohanpur, West Bengal, B-12/48 Kalyani, Nadia 741235, India. E-mail: sisirm55@gmail.com

Munirah Mohamad, Horticulture Research Centre, Malaysian Agricultural Research and Development Institute, Selangor, Malaysia. E-mail: muni@mardi.gov.my

Rosmah Murdad, Faculty of Sustainable Agriculture, Universiti Malaysia Sabah Sandakan Campus, Mile 10, 90000, Sandakan, Malaysia. E-mail: rosmahm@ums.edu.my

Asanga D. Nagalla, Department of Agricultural and Environmental Biology, Graduate School of Agricultural and Life Science, University of Tokyo, Tokyo, Japan. E-mail: adnagalla@gmail.com

Pradyot K. Pathak, KVK, 24 Pargana (N), West Bengal University of Animal and Fishery Science, West Bengal, India. E-mail: pathakpradyot@gmail.com

Dinh Thi Yen Phuong, School of Biotechnology, Tan Tao University, Long An, Vietnam. E-mail: Phuong.dinh@wsu.edu

D.K.N. Gamini Pushpakumara, Department of Crop Science, Faculty of Agriculture, University of Peradeniya, Peradeniya, Sri Lanka. E-mail: ngpkumara@agri.pdn.ac.lk

Selladurai Rajendiran, Central Horticultural Experiment Station (ICAR-IIHR), Chettalli-571 248, India, E-mail: Rajendiran.S@icar.gov.in

Prabhat Kumar Ray, Dr Rajendra Prasad Central Agricultural University, Samastipur, Bihar, India. E-mail: pkraypusa@gmail.com

Ankita Sahu, ICAR-Central Institute for Women in Agriculture, Bhubaneswar, India. E-mail: sahuankita29@gmail.com

Thandavarajan Sakthivel, Division of Fruit Crops, ICAR-Indian Institute of Horticultural Research, Bengaluru, Karnataka 560 089, India. E-mail: Sakthivel.T@icar.gov.in

Parlakoti M. Sampath, Department of Fruit Science, University of Agricultural Sciences, GKVK, Bengaluru, India. E-mail: sampathhort@gmail.com

Subbareddy Shyamalamma, Department of Biotechnology, University of Agricultural Sciences, GKVK, Bengaluru, India. E-mail: shyamala_reddyrs@yahoo.co.in

Hari S. Singh, ICAR-Central Institute for Subtropical Horticulture, Rehmankhera, Lucknow-226101, Uttar Pradesh, India. E-mail: Hari.Singh1@icar.gov.in

Nor Elliza Tajidin, Department of Crop Science, Faculty of Agriculture, Universiti Putra Malaysia, 43400 UPM, Serdang, Selangor, Malaysia. E-mail: elliza.tajidin@upm.edu.my

Josephine G. Tangonan, College of Human Ecology and Food Sciences, University of Southern Mindanao, Kabacan, Cotabato, The Philippines. E-mail: jgtangnan@yahoo.com

Anbazhagan Thirugnanavel, ICAR-Central Citrus Research Institute, Nagpur, Maharashtra 440033. India. E. mail: lotus.thiru@gmail.com

Prakash C. Tripathi, Division of Fruit Crops, ICAR-Indian Institute of Horticultural Research, Bengaluru, Karnataka 560 089, India. E-mail: prakash.tripathi@icar.gov.in

Marina Dsouza Valeeta, Department of Biotechnology, University of Agricultural Sciences, GKVK, Bengaluru, India. E-mail: valeeta.marina22@gmail.com

Huliyar V. Vidyashree, Department of Biotechnology, University of Agricultural Sciences, GKVK, Bengaluru, India. E-mail: vidyaramyashreehv@gmail.com

Preface

The jackfruit is indigenous to the rainforests of the Western Ghats of India and is cultivated throughout the tropical lowlands in South and South-east Asia. Major jackfruit-growing countries are India, Vietnam, Bangladesh, Malaysia, Sri Lanka, China, the Philippines, Indonesia and Nepal. Jackfruit is also found in Australia, the USA (Florida and Hawaii) and East Africa (e.g. Uganda, Tanzania and Mauritius), as well as in Brazil, Mexico and Caribbean nations such as Jamaica.

Jackfruit is a multi-purpose species providing food, timber, fuel, and medicinal and industrial products. It produces the largest fruit (up to 80 kg) in the world. The highly versatile fruit is used green as a cooked vegetable, eaten fresh when ripe and included in many regional cuisines. Certainly, the large size of the fruit, unusual appearance and strong aroma contribute to the jackfruit's slow acceptance. As global warming is expected to cause a significant effect on food production in the coming years, jackfruit could emerge as a nutritious staple crop as it is drought resistant and requires little maintenance. Despite the enormous nutritional and medicinal value, the jackfruit is still considered a minor or under-utilized or under-exploited fruit in many growing countries.

In the last 20 years, jackfruit researchers around the world have selected many good varieties for cultivation in their own country and have standardized the propagation, production and postharvest management technologies. The International Society for Horticultural Science (ISHS) organized the first symposium on jackfruit in Bangladesh in 2012 and as a result of the interest of the jackfruit researchers, ISHS developed a working group on Jackfruit and Other Moraceae. A follow-up symposium was held in Turkey in 2018 where it was felt that a compilation of research inputs was necessary with up-to-date information in the form of a book.

This publication presents the current state of knowledge concerning the history, physiology, culture, postharvest management and value addition of jackfruit throughout the world. Although it contains a wealth of documented information from literature and extensive personal experience of the authors within their respective disciplines in association with the crop, it can be concluded that further research into all aspects is required.

Sisir Mitra
Kalyani, India, 2022

Acknowledgements

I wish to express my appreciation and gratitude to all the authors who have given their time, without any financial reward, to contribute to this book. I am grateful to the scientists who have provided information and/or reviewed the chapters: Dr Fabio G. Enriquez, Department of Agriculture, CEBU, Philippines; Dr Borairin Buntong, Royal University of Agriculture, Cambodia; Dr F.M. Oloyde, Osun State University, Nigeria; Dr T. Chutinanthakun, Horticulture Research Institute, Thailand; Dr Y.M. Somasekhara, College of Agriculture, UAS, Bengaluru, India; Dr C. Dessanayake, Sabaragamuwa University of Sri Lanka, Sri Lanka; Dr C.J.R. de Moura, Chicago Botanic Garden, USA; Dr Jeff Wasielewski, UF/IFAS Extension Mimai-Dade County, USA; Mr Noor B.A. Aziz, MARDI, Malaysia; Dr P.C. Tripathi, Indian Institute of Horticultural Research, Bengaluru, India; Dr S. Jha, Bidhan Chandra Krishi Viswavidyalaya, India; Dr R. Cronje, ARC Institute of Horticultural Research, South Africa; Dr Jonathan Crane, University of Florida, USA; Mr Faming Zhang, Institute of Tropical and Subtropical Cash Crops, PR China; Mr Chelsea Moore, Department of Industry, Tourism and Trade, Northern Territory Government, Australia; Dr Allan H. Chambers, University of Florida, USA; Dr B. Ghosh, Bidhan Chandra Krishi Viswavidyalaya, India; Dr S. Debnath, Bidhan Chandra Krishi Viswavidyalaya, India; Dr P. Rajendran, Sai Agro-Innovations Research Foundation, Wayanad, Kerala, India; Dr Elahadi M. Yahia, Autonomous University of Querto, Mexico; Dr Leila Pio, University of Lavras, Brazil; Dr A.G. Moges, Ethiopian Institute of Agricultural Research, Ethiopia; Dr S. Karki, Ministry of Agriculture, Food Technology and Land Management, Nepal; Dr N.M. Chau, SOFRI, Vietnam; Mrs Peyanoot Naka, Horticulture Research Institute, Thailand; and Mr Yacob Ahmad, International Tropical Fruits Network, Malaysia.

I also wish to thank the following people for providing the photographs used in this book: Dr Helena de G. Bergallo, Universidade do Estado do Rio de Janeiro, Brazil; Dr N.J.C. Zerega, Northwestern University, Evanston, Illinois, USA; Dr C.K. Benergama, University of Paradeniya, Sri Lanka; Dr Mai Van Tri, Southern Fruit Research Institute, Tien Giang, Vietnam; Dr W.Z.W. Sembok, Universiti Malaysia Trengganu, Malaysia; Dr Lanxi Su, Spices and Beverage Institute, PR China; Dr Narayan Chawda, VNR Seeds, Raipur, India; Dr J.D.F. Basoy, University of Southern Mindanao, Philippines; Mr Shree Padre, Executive Editor, Adike Pathrike, India; Dr M.J. Gupta, ICAR-Central Coastal Agricultural Research Institute, Ela, Goa, India; Dr H. Lembisana Devi, ICAR, Regional Station, Tripura, India; Dr A. Sarkar, Nagaland University, India; Dr L. Wangchu, Central Agricultural University, Pasighat, India, Dr P.E. Rajsekharan, Indian Institute of Horticultural Research, Bengaluru, India; Dr I. Bandypadhyaya and Mr, P. Mukherjee, Santiniketan, India; Dr S. Shyamalamma, University of Agricultural Science, Bengaluru, India; Dr Nor Elliza Tajidin, Universiti Putra Malaysia, Malaysia;

Dr G. Karunakaran, Indian Institute of Horticultural Research, Bengaluru, India; Dr D. Maji, Bidhan Chandra Krishi Viswavidyalaya, India; Dr B. Das, West Bengal University of Animal and Fishery Science, India; and Dr D.K. Sarengi, Government of West Bengal, India.

My special thanks go to Dr P.K. Pathak, KVK, 24-Pargana (N), West Bengal University of Animal and Fishery Science, India, for his editorial assistance.

The continued support and encouragements received from various editors and assistants of CABI, particularly from Rebecca Stubbs and Emma McCann, are gratefully acknowledged.

On a personal note, I thank my wife, Ipsa, who has unselfishly allowed me to edit this book. Without her sacrifices, I could not possibly have finished.

Sisir Mitra

1 Botany, Origin, History and Production

Michel Chauvet[1]* and Sisir Mitra[2]
[1]15 Avenue du Quartier Bas, 34230 Vendémian, France; [2]Former Professor, Faculty of Horticulture, Bidhan Chandra Krishi Viswavidyalaya, West Bengal, India

1.1 The Family Moraceae

Moraceae is a family of 37 genera and approximately 1100 species, which extends worldwide, mainly in tropical areas, with some species in subtropical and temperate regions.

The family consists of trees, shrubs, climbers and herbs, usually with milky latex. The leaves are alternate, rarely opposite or subverticillate, most frequently simple and entire, sometimes dentate, palmately veined, pinnately or palmately incised, and rarely pinnate or palmate. Some species show both entire and lobed leaves on the same tree (*Broussonetia papyrifera*), or lobed leaves when juvenile (*Artocarpus heterophyllus*). Stipules are present and often conspicuous.

The inflorescences are extremely diverse. They may be bisexual or unisexual, and racemose or cymose to capitate or urceolate. Pistillate inflorescences are sometimes uniflorous. The flowers are unisexual, perianth simple with (0–)4(–10) persistent tepals. The flowers are staminate with (1–)4(–6) stamens. The flowers are pistillate with a unilocular, superior to inferior ovary, one or two styles and stigmas, and one ovule.

The fruits are rarely dry, more frequently drupaceous, usually enveloped by a fleshy perianth or immersed in a fleshy receptacle, often with the whole infructescence forming a syncarp. In their extreme form (e.g. *Artocarpus*, *Parartocarpus*), these syncarps are so highly integrated that their constituents (tepals, bracts and axial tissue) cannot be identified with certainty. The large syncarps of *Artocarpus* reach a weight of up to 50 kg.

Ficus pollination has attracted a lot of attention. The pollinators are agaonid wasps, and each *Ficus* sp. depends on a particular wasp. Little is known about the pollination of other Moraceae.

Individual drupes or small syncarps are usually brightly coloured and dispersed by birds. Larger syncarps are less coloured but odoriferous, and are dispersed by mammals. In some cases, sticky fruits suggest exozoochory.

A common feature of the family is the latex. Its composition is rather variable, with rubber, proteins, waxes, triterpenoid resins, polyphenols and cardenolides being the main components. In some species of *Ficus*, papain-like enzymes can be found.

1.2 Taxonomy

In the past, members of the order Urticales (Urticaceae, Moraceae, Cannabaceae, Ulmaceae) were kept separate from the Rosaceae, largely because of their very reduced and usually wind-pollinated flowers, but they have now been

*Corresponding author: Michel.chauvet34@orange.fr

 Jackfruit : Botany, Production and Uses (ed. S. Mitra)
DOI: 10.1079/9781800622319.0001

merged into the order Rosales (Stevens, 2001; Sytsma *et al.*, 2002).

The subdivision of the family Moraceae is not yet stabilized. According to Zerega *et al.* (2010), the tribes are Moreae, Artocarpeae, Castilleae, Dorstenieae and Ficeae. Ficeae has only one genus, *Ficus*. Within Moreae (ten genera), the genus *Maclura* is sometimes put into a separate tribe, Maclureae. Artocarpeae (five genera) includes *Parartocarpus* and *Hullettia*. Castilleae (11 genera), previously Olmedieae, now excludes the genus *Olmedia*, which is included in the genus *Trophis*. Dorstenieae contains nine genera, but the genus *Treculia* is sometimes included in Artocarpeae.

1.3 Distribution

The Moraceae are a pantropical family, with some members in subtropical and northern temperate regions. Most Moraceae are components of humid tropical lowland forests, but many species are also found in drier tropical or subtropical habitats. *Ficus* is the only pantropical genus, with a centre of diversity in Asia–Australasia. Dorsteniae are represented in tropical America and tropical Africa, while Castilleae are mainly neotropical. Moreae are the only tribe including truly temperate representatives (species of *Morus* and *Maclura*); the majority of its species, however, are tropical, with a centre of diversity in South-east Asia. Artocarpeae are neotropical and Asian, with the genus *Treculia* found in Africa.

Most species are evergreen, but *Broussonetia*, *Morus* and some species of *Artocarpus*, *Ficus* and *Maclura* are deciduous.

1.4 Economic Importance

The most important products of the Moraceae are fruits: breadfruit (*Artocarpus altilis*), jackfruit (*Artocarpus heterophyllus*), figs (*Ficus* spp., especially *Ficus carica* and *Ficus sycomorus*) and to a lesser extent mulberries (*Morus* spp.) and *Antiaris toxicaria*. *Brosimum alicastrum* gives an edible nut.

Artocarpus odoratissimus is cultivated in Borneo and the Philippines for the sweet pulp around the seed. *Artocarpus nitidus*, with the uniformly succulent kind of infructescence, is sometimes found in gardens in West Malesia, where it is also used for jam and conserves.

Many species yield valuable timbers: iroko (*Milicia excelsa*) in tropical Africa, letterwood or snakewood (*Brosimum guianense*) in the Guyanas, muirapiranga (*Brosimum rubescens*) in Brazil, Osage orange (*Maclura pomifera*) from North America, *Antiaris toxicaria*, *Streblus elongatus* and several species of *Artocarpus* in Asia.

A fine paper is prepared from the bark of the paper mulberry (*Broussonetia papyrifera*) in Asia. Tapas (bark tissues) are also obtained from this species, as well as from *Artocarpus elasticus*, *Artocarpus altilis* and *Ficus prolixa* in Oceania.

Morus alba is grown extensively for its leaves, which feed the larvae of the silkmoth (*Bombyx mori*). *Maclura pomifera* is sometimes used in India for that purpose.

The latex of many species is used locally, usually as an arrow poison (*Antiaris toxicaria* and *Parartocarpus venenosus*) or as a glue (*Artocarpus* spp.). Some species (*Castilla elastica*, *Ficus elastica*) have been grown as a source of rubber but are largely replaced nowadays by *Hevea brasiliensis* (Euphorbiaceae). Other species yield a yellow dye, such as *Maclura pomifera*, *Maclura cochinchinensis* and *Artocarpus heterophyllus* (Cardon, 2007). *Brosimum acutifolium* produces a psychotropic drug called takini (Grenand *et al.*, 2004), as does *Maquira sclerophylla*. *Dorstenia contrajerva* has long been renowned as a medicinal against snake bites.

Many species also have ornamental value (e.g. *Ficus benjamina*, *Ficus elastica*, *Ficus lyrata*, *Broussonetia papyrifera*, *Maclura pomifera*), or are used as shade trees. The banyan (*Ficus benghalensis*) and the peepal (*Ficus religiosa*) are sacred trees in India, while *Ficus sycomorus* has a high prestige in the Near East.

1.5 The Tribe Artocarpeae

Artocarpeae has long been a heterogeneous, ill-defined and ever-changing tribe within the Moraceae (Jarrett, 1959a; Corner, 1962; Rohwer, 1993; Berg, 2001; Datwyler and Weiblen, 2004; Berg *et al.*, 2006; Clement and Weiblen, 2009). Based on evidence from phylogenetic analyses of molecular data, and considering morphological characters, Zerega *et al.* (2010) circumscribed a monophyletic Artocarpeae comprising two neotropical (*Clarisia* and *Batocarpus*) and one paleotropical (*Artocarpus* including *Prainea*) genus. Three genera have been removed from recent circumscriptions of Artocarpeae: *Parartocarpus* and *Hullettia* form a monophyletic group that may warrant tribal status, and *Treculia* has been transferred to Dorstenieae.

Artocarpeae as defined by Zerega *et al.* (2010) are supported by a reduction in stamen number within the family. Members of the genera *Artocarpus*, *Prainea*, *Clarisia* and *Batocarpus* all have one (or less frequently two in *Batocarpus* and *Artocarpus annulatus*, and one to three in *Clarisia*) stamen per flower compared with the more typical two to five in other Moraceae. The typical number of stamens per flower in the Moraceae is four, but exceptions exist in all of the tribes.

Artocarpeae bear unisexual flowers, as do all Moraceae, and typically have unisexual inflorescences. Pistillate inflorescences are condensed capitate heads (or rarely uniflorous as in some *Clarisia* spp.), which develop into syncarps (Jarrett, 1959c). Both pistillate and staminate inflorescences typically comprise numerous small, tightly packed flowers, which sit on a fleshy receptacle, and have a perianth (Jarrett, 1959a).

Staminate inflorescences are spicate or less frequently globose. They have numerous flowers with two to four connate tepals, and typically one stamen (occasionally two to three may be present) that is straight in bud.

Pistillate inflorescences have one to more typically numerous flowers with two to four connate tepals, and adjacent perianths may be free or partially to completely fused to one another. This allows the entire inflorescence to develop into a highly specialized syncarp formed by the enlargement of the entire female head, even if only a portion of the flowers develop seeds (Jarrett, 1977). However, in *Prainea*, *Batocarpus* and *Clarisia* spp., adjacent pistillate perianths remain free so that only fertilized ovules enlarge.

Members of the Artocarpeae can be latex-producing trees, shrubs or, rarely, climbers, and are either monoecious or dioecious. The leaves are simple (rarely compound), spirally alternate or distichous, and have either amplexicaul or non-amplexicaul stipules. The inflorescences are typically axillary but are cauliflorous in a few species. Interfloral bracts may be present or absent.

1.5.1 Key to the genera of Artocarpeae

1. Staminate inflorescences spicate with a distinct abaxial sterile strip; staminate flowers crowded in longitudinal rows; leaves alternate and distichous with lateral stipules; neotropical in distribution............2

1. Staminate inflorescences spicate to globose to obovoid or clavate, lacking a distinct abaxial sterile strip; staminate flowers crowded but not in longitudinal rows; leaves alternate and distichous with lateral stipules or spirally alternate with amplexicaul stipules; paleotropical in distribution............*Artocarpus*

2. Inflorescences axillary; pistillate inflorescences multiflorous and globose-capitate; fruiting perianth green............*Batocarpus*

2. Inflorescences often cauliflorous, or, if axillary, then bark at base of the trunk reddish; pistillate inflorescence suniflorous or multiflorous and discoid-capitate; fruiting perianth red, orange, pale yellow or greenish-yellow ....*Clarisia*

1.6 The Genus *Artocarpus*

Artocarpus J.R. Forster & G. Forster (1776), nom. cons.

Synonyms:

- *Sitodium* Banks & Sol. ex Parkinson (1773)
- *Radermachia* Thunb. (1776) ('Rademachia')
- *Prainea* King ex Hook. f. (1888)
- *Saccus* O. Kuntze (1891)

Artocarpus is a *nomen conservandum*, against the earlier name *Sitodium*. The name is formed from the Greek ἄρτος (artos) for 'bread' and καρπός (karpos) for 'fruit'. Forster, who coined the name, writes: 'The fruit is locally used as bread.' Breadfruit, *Artocarpus altilis* (Parkinson)

Table 1.1. Major species of *Artocarpus* and their vernacular/common names. From Ghosh (2000).

Species	Vernacular names
Artocarpus altilis (Parkinson) Fosberg (syn: *A. communis* J.R. & G. Forster	Breadfruit (English), sukun (Indonesia), sukun (Malaysia), kamansi/rimas (Philippines), sule (Vietnam, Thailand)
Artocarpus camansi Blanco Kulur (Malaysia)	kamansi (Philippines)
Artocarpus anisophyllus Miq.	Bintawa (Malaysia)
Artocarpus chaplasha Roxb.	Toung-peing-nai (Myanmar)
Artocarpus dadah Miq.	Chempedak air, tampang (Malaysia)
Artocarpus elasticus Reinw. ex Blume	Terkalong (Malaysia), gumihan (Philippines)
Artocarpus fulvicortex F.M. Jarrett	Tampang jingga (Malaysia)
Artocarpus heterophyllus Lam. (syn: *A. philippensis* Lam., *A. brasiliensis* Gomez, *A. maxima* Blanco)	Jackfruit (English), nangka (Indonesia, Malaysia), kathal (Bangladesh, India), langka (Philippines), peignai (Myanmar), khnaor (Cambodia), mit (Vietnam)
Artocarpus integer (Thunb.) Merr. (syn: *A. integrifolia* L.f., *A. polyphema* Persoon, *A. champeden* (Lour) Stokes)	Chempedak (English, Indonesia, Malaysia), sonekadat (Myanmar), champada (Thailand)
Artocarpus camansi Blanco	Kulur (Malaysia), kamansi (Philippines)
Artocarpus kemado Miq.	Padau (Malaysia)
Artocarpus lacucha Buch.-Ham	Tampanmgberok (Malaysia)
Artocarpus nitidus Trécul	Kubi (Philippines), selangking (Malaysia), sampor (Cambodia), mahat (Thailand), cay chay (Vietnam)
Artocarpus odoratissimus Blanco	Marang/terap (Malaysia), marang (Philippines)
Artocarpus rigidus Blume	Monkey jack (English), pussar (Java-Indonesia), mandalika (Philippines), tempunik (Malaysia)
Artocarpus sarawakensis F.M. Jarrett	Pingan (Malaysia)
Artocarpus sericicarpus F.M. Jarrett	Pedalai/terapbulu (Malaysia)

Fosberg, is indeed an important starchy vegetable in Oceania.

The genus *Artocarpus* J.R. Forst. & G. Forst. (Moraceae) contains approximately 70 species of monoecious trees. Its range extends from India to the Solomon Islands, with a centre of diversity in Borneo (Williams *et al.*, 2017). Notable species include the widely cultivated *A. altilis* (Parkinson) Fosberg (breadfruit) and *Artocarpus heterophyllus* Lam. (jackfruit), in addition to other species of more regional importance, such as *Artocarpus integer* (Thunb.) Merr. (chempedak) and *Artocarpus odoratissimus* Blanco (terap) (Table 1.1). Sixteen species of *Artocarpus* occur in Singapore, of which 14 are indigenous to the island (Table 1.2) (Gardner and Zerega, 2021).

Artocarpus was revised by Jarrett (1959a, b) with subsequent taxonomic work by Kochummen (1978) and Berg *et al.*, 2006Berg *et al.*, 2011. The subgenera were revised following a phylogenetic study by Zerega *et al.* (2010), leading to four subgenera being distinguished. Additional phylogenetic work has provided a molecular framework for taxonomic revisions (Williams *et al.*, 2017; Gardner *et al.*, 2021). Gardner and Zerega (2020) updated the *Artocarpus* subgenus *Pseudojaca* Trécul, with a particular focus on the taxa in Singapore. *Artocarpus griffithii* (King) Merr. has been reinstated as distinct from *Artocarpus lamellosus* Blanco (which is called *Artocarpus nitidus* Trécul in the earlier literature), also requiring the reinstatement of the following taxa not found in Singapore: *Artocarpus borneensis* (Merr.) F.M. Jarett, *Artocarpus humilis* Becc., *Artocarpus vrieseanus* Miq. var. *subsessilis* F.M. Jarrett and *Artocarpus xanthocarpus* Merr. *Artocarpus dadah* Miq. has been reinstated as distinct from *Artocarpus lacucha* Roxb. ex Buch.-Ham., thereby

Table 1.2. *Artocarpus* species occurring in Singapore, with relevant taxonomic changes noted. Adapted from Gardner and Zerega (2021).

Species	Differences from Flora Malesiana
Artocarpus altilis (Parkinson) Fosberg*	In the strict sense following Zerega *et al*. (2005)
Artocarpus anisophyllus Miq.	
Artocarpus camansi Blanco*	Included by Berg *et al*. (2006) in *A. altilis* (Parkinson) Fosberg
Artocarpus dadah Miq.	Made consistent with the other entries, in *A. lacucha* Roxb. ex Buch.-Ham.
Artocarpus elasticus Rainw. ex Blume	In the strict sense not including *A. corneri* Kochumnen and *A. scortechinii* King
Artocarpus fulvicortex F.M. Jarrett	
Artocarpus gomezianus Wall. ex Trécul	
Artocarpus griffithii (King) Merr.	Included by Berg *et al*. (2006) in *A. nitidus* Trécul
Artocarpus heterophyllus Lam.*	
Artocarpus hispidus F.M. Jarrett	
Artocarpus integer (Thunb.) Merr.	
Artocarpus kemando Miq.	In the strict sense, not including *A. maingayi* King
Artocarpus lanceifolius Roxb.	In the strict sense, not including *A. cleventis* Merr.
Artocarpus lowii King	
Artocarpus rigidus Blume	
Artocarpus scortechinii King	Included by Berg *et al*. (2006) in *A elasticus* Reinw. ex Blume

*species found primarily in cultivation. Taxa in bold belong to *Artocarpus* subgenus *Artocarpus*.

necessitating the reinstatement of the following taxa not found in Singapore: *Artocarpus fretessii* Teijsm. & Binn. ex Hassk., *Artocarpus ovatus* Blanco and *A. vrieseanus* var. *refractus* (Becc.) F.M. Jarrett. *Artocarpus gomezianus* Wall. ex Trécul is restricted to the type subspecies, and *A. zeylanicus* (F.M. Jarrett) E.M Gardner & Zerega, formerly a subspecies, has been elevated to species level.

Gardner and Zerega (2021) recently made taxonomic updates of *Artocarpus* subgenus *Artocarpus* based on 517 nuclear genes. They proposed the following: *Artocarpus* subgenus *Cauliflori* (F.M. Jarrett) Zerega is reduced to a section within the *Artocarpus* subgroup *Artocarpus*, and *Artocarpus* section *Glandulifolium* F.M. Jarrett is raised to subgenus status. The new monotypic subgenus *Artocarpus* subgenus *Aenigma* E.M. Gardner & Zerega is proposed for *Artocarpus sepicanus* Diels, whose phylogenetic position remains uncertain and may be of ancient hybrid origin. *Artocarpus elasticus* Reinw. ex Blume, *Artocarpus scortechinii* King and *Artocarpus corneri* Kochummen are recognized as distinct species. *Artocarpus clementis* Merr. is reinstated as a distinct form of *Artocarpus lanceifolius* Roxb. *Artocarpus calophyllus* Kurz and *Artocarpus melinoxylus* Gagnep. are reinstated as distinct from both *Artocarpus chama* Buch-Ham. and *Artocarpus rigidus* Blume. Elmer is reinstated as distinct from *Artocarpus treculianus* Elmer.

Artocarpus species are monoecious trees, except in the subgenus *Prainea*, which is dioecious and includes a climber (*Artocarpus scandens*). The leaves are simple (rarely compound: *Artocarpus anisophyllus*), alternate and spiral (subgenera *Artocarpus* and *Cauliflori*) or alternate and distichous (subgenera *Pseudojaca* and *Prainea*), with large fully amplexicaul stipules (subgenera *Artocarpus* and *Cauliflori*) or small lateral stipules (subgenera *Pseudojaca* and *Prainea*).

The inflorescences are capitate, the head being cylindric to clavate, obovoid, ellipsoid or globose, and occasionally rami- or cauliflorous, with numerous flowers. Staminate flowers are free, usually mixed with peltate to clavate interfloral bracts, and the perianth is tubular, entire or two to four lobed, with one stamen and no pistillode. Pistillate flowers are at least distally connate with each other or with interfloral bracts, the perianth is tubular, thin-walled below and thick-walled above, and the apices either completely fused, forming an areolate surface, or free, forming variously shaped processes on the surface. The ovary is free, and the stigmas are one or two, equal or unequal.

The fruit is a syncarp, with at least the lower part of the perianth (the middle layer of the syncarp) becoming fleshy, and sometimes the entire head.

Pollination syndromes in Artocarpeae are still largely unknown, and empirical studies will be necessary for further elucidation. Phylogenetic classification will be a useful tool to understand and interpret pollination in an evolutionary context.

1.6.1 Key to the subgenera of *Artocarpus*

1. Perianths of adjacent pistillate flowers entirely free from one another; only fertilized flowers enlarging at maturity............subgenus *Prainea*

1. Perianths of adjacent pistillate flowers partially to completely fused to form a syncarp; entire pistillate head enlarging at maturity............2

2. Leaves distichous; stipules lateral and non-amplexicaul, less than 1 cm, stipule scar not annulate; pistillate inflorescences with adjacent perianths fused apically, giving the inflorescence and syncarp a smooth/uniform surface............subgenus *Pseudojaca*

2. Leaves spirally alternate; stipules fully amplexicaul, 1 cm or longer, leaving an annulate stipule scar; pistillate inflorescences with adjacent perianths fused only medially, leaving perianth apices free and giving the inflorescence and syncarp a spiky or bumpy surface............3

3. Inflorescences always axillary ....subgenus *Artocarpus*

3. Inflorescences develop directly from the trunk or branches of the previous year's growth (cauli- or ramiflorous). Axillary inflorescences may also be present and male inflorescences may have ring-like constrictions (in *Artocarpus annulatus*) .... subgenus *Cauliflori*.

1.6.2 Subgenus *Artocarpus* (synonym: subgenus *Jaca* Trécul (1847))

Breadfruit, *A. altilis* (Parkinson) Fosberg (1941), is by far the most important economic species of the subgenus *Artocarpus*. Its starchy fruits are a staple in Oceania. It probably originates from New Guinea and spread early on over all Oceania up to Hawaii and later to all tropical countries. Two types are known, one with seeds (breadnut) and one without seeds, which is reproduced vegetatively. It belongs to a complex of species that Jarrett (1959c) treated as one highly variable pantropical species (Ferrer-Gallego and Boisset, 2018). Zerega *et al.* (2005) distinguish two more species from *A. altilis*, *Artocarpus camansi* Blanco (1837) from New Guinea, and *Artocarpus mariannensis* Trécul (1847) from Palau and the Northern Mariana Islands, and their hybrids. Berg *et al.* (2006) included *Artocarpus horridus* F.M. Jarrett (1959) (the Moluccas), *Artocarpus blancoi* (Elmer) Merr. (1923), *Artocarpus multifidus* F.M. Jarrett (1959) and *Artocarpus pinnatisectus* Merr. (1921) (the last three from the Philippines), considering them all to represent a range of variation from wild to cultivated forms of breadfruit.

Recently, Gardner *et al.* (2021) identified a new species in the breadfruit clade from North Maluku, Indonesia, *Artocarpus bergii* E.M. Gardner, Arifiant and Zerega. This is a monoecious tree, of height at least 22 m; the leaves are spirally arranged, with the lamina subcoriaceous, elliptical to subobovate, and 11–29 × 5–14.5 cm; the apex is short-acuminate, acute to nearly rounded, base acute to obtuse. The staminate inflorescence (at anthesis) is in the leaf axil and solitary; the peduncle is 38–54 × 2 mm and pale green. The pistillate inflorescence (postreceptive) is in the leaf axil and solitary; the peduncle is 66–70 × 2.5 mm and pale green. The syncarp is cylindrical, 45–70 × 18–25 mm, and the surface is yellow when ripe.

Breadfruit is an evergreen tree (12–15 m up to 21 m). *A. altilis* tends to have a denser, more spreading canopy than *A. camansi*. The leaves (15–60 cm or longer) are alternate and broadly obovate to ovate, almost entire to deeply pinnately dissected with one to six pairs of lobes. The fruits (10–30 cm long × 9–20 cm wide) are variable in shape, size and surface texture. They are usually round, oval or oblong, weighing 0.25–6 kg. Skin texture varies from smooth rough to spiny. The colour is light green, yellowish-green or yellow when mature, although an unused variety (Afara, from French Polynesia) has pinkish or orange-brown skin (Ragone, 2006a). The flesh is creamy white to pale yellow. The fruits are typically mature and ready to cook as a starchy staple in 15–19 weeks (Worrell *et al.*, 1998). Ripe fruits have a yellow or yellow-brown skin and a soft, sweet, creamy flesh that can be eaten raw or cooked. The fruits contain none to many seeds depending on the cultivar. The seeds are obovoid or rounded, irregularly compressed, 1–3 cm long and pale to dark brown in colour (Ragone, 2007; Zerega and Ragone, 2016).

Breadnut (*A. camansi* Blanco) is one of the wild progenitors of breadfruit (Ragone, 2006b). Breadnut is scattered throughout the lowlands of New Guinea where it occurs naturally, and is also found cultivated in home gardens. It only occurs in cultivation in the Philippines where it is typically grown as a backyard tree. It is infrequently grown in the Pacific outside its native range. A few trees are found in New Caledonia, Pohnpei, Marquesas, Tahiti, Palau and Hawaii, introduced by immigrants from the Philippines (Ragone, 2007). The tree typically forms buttresses at the base of the trunk and has a more open branching structure (Fig. 1.1A) than *A. altilis* or *A. mariannensis*. The leaves are alternate, large (40–60 cm long) and moderately dissected with four to six pairs of lobes and sinuses cut halfway to the midrib. The fruits are oval (10–15 cm × 7–12 cm) weighing approximately 800 g (Ragone, 2006a). The spiny skin, with pointed, flexible 5–12 cm elongated sections, is dull green to greenish-brown when ripe (Fig. 1.1B). The fruit contains numerous seeds comprising 30–50% or more of the fruit weight (Bennett and Nozzolillo, 1987). The seeds are typically obovoid or flattened by compression, 2.5 cm long with a thin, light brown seedcoat patterned with darker veins (Ragone, 2007).

Chebiei or Dugdug (*A. mariannensis* Trécul) is cultivated through Micronesia for its edible fruits and seeds (Ragone, 1997, 2001) and has naturally hybridized and introgressed with *A.*

(A) (B)

Fig. 1.1. The tree (A) and fruit (B) of *Artocarpus camansi* tree. Photographs courtesy of Dr N.J.C. Zerega.

Fig. 1.2. The tree (**A**) and fruit (**B**) of *Artocarpus odoratissimus*. Photographs courtesy of Dr N.J.C. Zerega.

altilis (Fosberg, 1960; Ragone, 1997, 2001; Zerega *et al.*, 2004, 2005). Trees of *A. mariannensis* and hybrid cultivars tend to be more massive than *A. altilis* and *A. camansi*, with large trunks, extensive buttresses and full, rounded canopies. The leaves are alternate, 15–30 cm long, broadly obovate to broadly elliptic, and typically entire or shallowly one to three lobed on the upper third of the blade. The fruit is small (8–15 cm long), weighing 0.25–0.5 kg. The skin is dark green, even when ripe, with a sweet aroma and taste. The fruit is not as solid or dense as breadfruit and contains up to 15 rounded or obovoid, dark brown, shiny seeds (1.5 cm long) (Ragone, 2007).

Marang or terap (*A. odoratissimus* Blanco (1837)) trees are less cold tolerant than breadfruit and grow well in coastal areas where the temperature never goes below 7°C (Fig. 1.2A). It resembles both jackfruit and breadfruit in appearance. The fruits are round to oblong (Fig. 1.2B) and are regarded to superior to jackfruit and chempadak. The ripe fruit is light brown and succulent, weighs 1–3 kg and contains 60–200 seeds (bin Osman, 1993). The fruit is said to have a finer and more delicate flavour than jackfruit. The seeds are eaten roasted or boiled. Young fruits are also cooked.

A. elasticus Reinw. ex Blume (1825) is native to Burma, Thailand, Malaysia, Brunei, Indonesia and the Philippines. The yellowish-brown syncarp is oblong, with long flexible spines. The fleshy perianths are whitish in colour and sweet, and have a strong aroma. The seeds are edible on roasting. It is mostly used to provide barkcloth and timber.

A. chama Buch.-Ham. (1826) (synonym: *A. chaplasha* Roxb. (1832)) is native to northeastern India, lower Burma, and the Andaman and Nicobar Islands. It is an important timber in India; its leaves are also used for elephant fodder.

Artocarpus hirsutus Lam. (1789), from India, is a tall evergreen tree reaching up to 50 m in height with a girth of 5 m or more, with a straight clean stem and dense foliage. The fruits are ovoid (Fig. 1.3), developing an orange hue when mature. The tree is valued for its timber and its wood is also used for furniture. The leaves are used as fodder for elephants, goats, etc.

Artocarpus lakoocha Roxb is known as monkey jack, lokhat and tampang. The tree grows 6–9 m tall with deciduous, large, leathery leaves. It is a native of the sub-Himalayan regions of India up to 1200 m and also Malaysia and Sri Lanka. The fruits are round or irregular (Fig. 1.4), 5–12 cm diameter, velvety, and dull yellow tinged with pink. The sweet–sour pulp is eaten raw or made into curries and chutney. The wood is hard and is used in the preparation of boats and furniture.

Fig. 1.3. *Artocarpus hirsutus* fruit. Photograph courtesy of Dr G. Karunakaran.

A. rigidus Blume is distributed in the lowland forests of India, Myanmar and Malaysia, and in Sumatra, Kalimantan and Java in Indonesia where it is cultivated in homestead gardens. It is a tall tree of 30–35 m in height with a dense, dark green, spreading crown (Fig. 1.5A). The fruits are roundish, 7.5–12.5 cm diameter, and greenish-yellow becoming dull yellow (Fig. 1.5B). The pulp is yellow and has a sweet, pleasant flavour. The seeds are covered by the waxy pulp and are roasted and eaten in many parts of tropical Asia. The wood is yellow to orange-red, fairly heavy, durable and resistant to termites, and is used in making furniture, native boats and houses.

Many other species are used for their timber, latex or fruit. They include:

- *Artocarpus anisophyllus* Miq. (1861) (Malaysia, Sumatra, Borneo);
- *Artocarpus kemando* Miq. (1861) (Malaysia, Sumatra, Borneo);
- *Artocarpus lanceifolius* Roxb. (1832) (Thailand, Malaysia, Sumatra, Bangka, the Riau and Lingga Archipelago, and Borneo);
- *Artocarpus lowii* King (1888) (Malaysia, Sumatra);
- *Artocarpus maingayi* King (1888) (Malaysia, Sumatra);
- *Artocarpus sarawakensis* F.M. Jarrett (1975) (Malaysia);
- *Artocarpus scortechinii* King (1888) (Malaysia, Sumatra and the Lingga Archipelago);
- *Artocarpus sepicanus* Diels (1935) (New Guinea);
- *Artocarpus sericicarpus* F.M. Jarrett (1959) (Philippines, Borneo, Sulawesi and the Moluccas);
- *Artocarpus tamarin* Becc. (1902) (Borneo);
- *Artocarpus teysmannii* Miq. (1861) (Nicobar Islands, Malaysia, Sumatra, Borneo, Sulawesi, the Moluccas and New Guinea); and
- *Artocarpus treculianus* Elmer (1909) (Philippines).

Two kinds of timber are distinguished: terap, a light-weight wood, and keledang, a medium-weight wood.

Fig. 1.4. *Artocarpus lacucha* fruit. Photograph courtesy of Dr G. Karunakaran.

(A) **(B)**

Fig. 1.5. The tree (**A**) and fruit (**B**) of *Artocarpus rigidus*. Photographs courtesy of Dr N.J.C. Zerega.

1.6.3 Subgenus *Prainea* (King) Zerega, Supardi & Motley (2010)

The hard and heavy wood of members of the subgenus *Prainea* is used for posts and beams in house and bridge building. The fleshy perianth surrounding the fruit is edible. Three species are known:

- *Artocarpus frutescens* (Becc.) Renner (1907) (Sarawak, Kalimantan);

- *Artocarpus limpato* Miq. (1861) (Malaysia, Sumatra and Borneo); and
- *Artocarpus papuanus* (Becc.) Renner (1907) (Moluccas and New Guinea).

1.6.4 Subgenus *Pseudojaca* Trécul (1847)

Some species are mostly known for their fruits:

- *Artocarpus nitidus* Trécul (1847) (extends to southern China, Indo-China, Thailand, Malaysia, Borneo, Sumatra and Philippines). It includes four subspecies, with subspecies *Lingnanensis* being cultivated in Vietnam.
- *Artocarpus lacucha (lakoocha)* Buch.-Ham. (1826) (south of Himalayas);
- *Artocarpus tomentosulus* F.M. Jarrett (1960) (Malaysia, Indonesia);
- *Artocarpus tonkinensis* A. Chev. (1918) (China, Vietnam, Cambodia); and
- *Artocarpus vrieseanus* Miq. (1867) (Philippines, Sulawesi, Moluccas, New Guinea, Bismarck Archipelago, Japan, Salawati, Aru Island, Louisiade Archipelago, Solomon Islands).

Other species are used for timber:

- *Artocarpus altissimus* (Miq.) J.J. Sm. (1907) (Sumatra, West Kalimantan);
- *Artocarpus dadah* Miq. (1861) (Burma, Thailand, Malaysia, Sumatra and Borneo);
- *Artocarpus fretessii* Teijsm. & Binn. ex Hassk. (1866) (Philippines, Sulawesi, Moluccas, Irian Jaya);
- *Artocarpus fulvicortex* F.M. Jarrett (1960) (Indonesia, Malaysia);
- *Artocarpus glaucus* Blume (1825) (Malaysia, Sumatra, Borneo, Java, Lesser Sunda Islands);
- *Artocarpus gomezianus* Wall. ex Trécul (1847) (Burma, Thailand, Indo-China, Malaysia, Sumatra, Java, Philippines);
- *Artocarpus hypargyreus* Hance ex Benth. (1861) (East Asia, South China);
- *Artocarpus ovatus* Blanco (1837) (Philippines);
- *Artocarpus rubrovenius* Warb. (1904) (Philippines); and
- *Artocarpus subrotundifolius* Elmer (1908) (Philippines).

1.6.5 Subgenus *Cauliflori* (F.M. Jarrett) Zerega (2010)

This subgenus contains only two related species, *A. heterophyllus* (see section 1.7) and *Artocarpus integer*.

Artocarpus integer (Thunb.) Merr. (1917) has the following synonyms:

- *Radermachia integra* Thunb. (1776);
- *Polyphema champeden* Lour. (1790);
- *Artocarpus integrifolia* L.f. (1781) non Trimen (1898); and
- *Artocarpus champeden* (Lour.) Stokes (1812).

It is commonly known as chempedak (cempedak in Indonesian) and is widely distributed in Burma (Tenasserim), Peninsular Thailand, Peninsular Malaysia, the Lingga Archipelago, Sumatra, Borneo, Sulawesi, the Moluccas and Irian Jaya. It is also cultivated in these areas as well as in western Java. It is an important crop in Malaysia and is also popularly cultivated in southern Thailand and parts of Indonesia, but is underutilized elsewhere (Wang *et al.*, 2018).

Chempedak is similar to jackfruit but has smaller and narrower fruits (Fig. 1.6B) with a thinner rind and more juicy flesh, and is darker yellow or orange when ripe. It also has a characteristic aroma when ripe, described as both harsh and penetrating like durian (*Durio zibethinus* L.) and musky sweet like overripe jackfruit (Corner, 1938). The starchy seeds are also edible when cooked. Young plants bear scattered, reddish, wiry hairs on the leaves, and twigs and the mature plants usually are smaller than the jackfruit trees (Fig. 1.6A).

Chempedak's putative wild relative was first described by Corner (1938) as *Artocarpus integer* (Thunb). Merr. var. *silvestris* Corner, commonly known as bangkong. Bangkong is most commonly found from primary and secondary lowlands to montane forests of Peninsular Malaysia, although it has been recorded in Sumatra, Lingga Archipelago, and south/south-east Borneo (Jarrett, 1959a; Kochummen, 1978; Turner, 1995). Wang *et al.* (2018) stated

(A) (B)

Fig. 1.6. The tree (**A**) and chempedak fruit (**B**) of *Artocarpus integer*. Photographs courtesy of Dr G. Karunakaran.

that chempedak and bangkong are sister taxa and distinct genetically and morphologically, but the direction of domestication origin is unclear. Genetic diversity is generally higher in bangkong than in chempedak.

Some hybrids between chempedak and jackfruit are known, called 'nangka-chempedak' in Malaysia. One of these has been cloned and named 'CH/NA'.

1.7 The Jackfruit, *Artocarpus heterophyllus*

1.7.1 Nomenclature

Artocarpus heterophyllus Lam. (1789) ('*heterophylla*') has the following synonyms:

- *Sitodium cauliflorum* Gaertn. (1788)
- *Artocarpus philippensis* Lam. (1789)
- *Artocarpus nanca* Noronha (1790), nom. inval.
- *Polyphema jaca* Lour. (1790)
- *Artocarpus brasiliensis* Gomes (1803)
- *Arctocarpus maxima* Blanco (1837)
- *Saccus heterophyllus* (Lam.) Kuntze (1891)

According to Art. 62.2 of the International Code of Nomenclature for algae, fungi and plants (Shenzhen Code, 2017), all names ending in *-carpus* are treated as masculine. Hence, it is *A. heterophyllus* and not *A. heterophylla*.

Artocarpus integrifolius L.f. (1782) is often but erroneously used as the correct name; it is a synonym for *Artocarpus integer* (Thunb.) Merr. (1917).

The etymology of *heterophyllus* comes from Greek ἕτερος (heteros), meaning 'other, different', and φύλλον (phullon) meaning 'leaf'. Thus, the meaning is 'which exhibits different leaves'.

1.7.2 History

The first Westerner to observe jackfruit was Ibn Battuta, who travelled in India in 1333. He said that '*shaki* and *barki* are the best fruits of India. This name is given to long-lasting trees; their leaves are like those of walnuts and their fruits are born on the trunk of the tree itself. Those near the soil form *barki*; their sweetness is greater and their taste more aggregable than *shaki*. Those to be found above are called *shaki*, the fruit of which is similar to big bottle gourds and the bark like an ox skin. When it has become yellow, in autumn, it is collected, cut up, and one to two hundred seeds are to be found in each fruit, like gherkins (*Cucumis sativus*). When this kernel is roasted or boiled, it tastes like broad beans' (Samuel Lee/Ibn Battuta, 2005). In fact, the names cited by Ibn Battuta represent two cultivar groups distinguished in southern India, *chakka*, which has a soft and sweet pulp, and *barka* or *varaka*, which has a firm and tasty pulp.

Later on, one of the first Europeans who described jackfruit was the traveller Jean de Marignolli (Yule, 1350):

> There is again another wonderful tree called *Chake-baruhe*, as big as an oak. Its fruit is produced from the trunk and not from the branches, and is something marvellous to see, being as big as a great lamb, or a child of three years old. It has a hard rind like that of our pine-cones, so that you have to cut it open with an axe; inside it has a pulp of surpassing flavour, with the sweetness of honey and of the best Italian melon; and this also contains some five hundred chestnuts of like flavour, which are capital eating when roasted.

The Portuguese were the first Europeans to become established in southern India, and spread the knowledge of jackfruit to Europe. The name *jaca* appears in Portuguese in 1535, and Garcia da Orta describes the fruit in his *Coloquios dos simples* (1563) (Dalgado, 1919–1921). The name was further borrowed by the French (*jaca* in 1553, and *jaque* in 1611) (Arveiller, 1963) and English (*jaca* in 1613, later *jacq*, and then *jack fruit* in 1830) as described in the *Oxford English Dictionary*.

The first illustrations of jackfruit were in 1656 by Michał Boym in his book *Flora Sinensis* and Jan Nieuhof (1665) in his *Description Générale de la Chine*. A better description and illustration can be found in the *Hortus Malabaricus* of Rheede tot Drakenstein (1678–1703), who gives the Malayalam name *tsjaka-maram*, where *tsjaca* is *chakka* in the Dutch spelling, and *maram* means 'tree'.

The first part of the Malayalam name is supposed to come from Proto-Dravidian *kā(y)*, meaning 'fruit', but there are many other names in India, such as *panasa* or *phanas*, and *kanthal*. In the Malay area, it is called *nangka*. All these different names show that the use and cultivation of jackfruit is very old. It is thought that jackfruit originated in India, but it must have spread very early to South-east Asia.

Jackfruit spread to China in the 16th century, and to South America in the 18th, but has never gained popularity outside southern Asia.

1.7.3 Description

Jackfruit is an evergreen, monoecious tree up to 20–30 m tall and 80–200 cm in diameter. All living parts exude a viscid, white latex when injured. The bark is rough to somewhat scaly, and dark grey to greyish-brown. The crown is dense, conical in young and shaded trees, becoming rounded or spreading in the older tree (Fig. 1.7A). New shoots, twigs and leaves are usually glabrous but occasionally short-haired and scabrid. The stipules are ovate–acute, 1.5–8 cm × 0.5–3 cm, deciduous and leaving annular scars on the twigs.

(A) (B)

Fig. 1.7. The tree (**A**) and fruit (**B**) of *Artocarpus heterophyllus*. Photographs courtesy of Dr G. Karunakaran.

The leaves are thin, leathery, obovate–elliptic to elliptic, 5–25 cm × 3.5–12 cm, broadest at or above the middle, the base cuneate, with the margin entire or in young plants often with one to two pairs of lobes; the apex is rounded or blunt with a short, pointed tip. The leaves are dark green and shiny above, dull pale green underneath, and the petiole is 1.5–4 cm long, shallowly grooved on the adaxial side, and sparsely hairy. The leaves are inserted alternately on horizontal branches but tend to be spiral on ascending branches with a 2/5 phyllotaxy. The stipules are present, ovate, about (1.5–)4(–8) cm long and (0.5–)1(–3) cm wide, deciduous noticeable scars on the stem.

The inflorescences are solitary and borne axillary on special lateral, short leafy shoots arising from older branches and the main trunk; the male flower heads are barrel-shaped or ellipsoid, 3–8 cm long and 1–3 cm across, composed of sterile and fertile flowers closely embedded on a central core (receptacle), dark green, with a stalk 1.5–3.5 cm long and 0.5–1 cm wide, bearing an annular ring near the distal end. Sterile male flowers have a solid perianth, while fertile male flowers have a tubular, bilobed, 1–1.5 mm long perianth, with stamens 1–2 mm long. Female heads are borne singly or in pairs distal to the position of the male heads, and are cylindrical or oblong, dark green, 5–15 cm long and 3–4.5 cm wide, with a distinct annulus at the top end of the stout stalk, subtended by a spathaceous, deciduous bract, 5–8 cm long; female flowers have tubular perianths, which are fused at both ends and projecting as three to seven angled, blunt or pointed, minute pyramidal protuberances topped by spathulate or ligulate styles and stigmas.

Female flowers have a tabular perianth of free segments when young but fusing in the midpoint; the perianth encloses a basal compressed ovary with an obliquely inserted terminal style with a club-shaped stigma. The stigmatic points are persistently sharp (becoming blunt in mature fruits). The stigmas protrude after 4–6 days of opening of the spathes. The female spikes become woolly with numerous stigmas within another 4–6 days. The receptivity of the stigma remains for 28–36 days (Sambamurthy and Ramalingam, 1954; Azad, 1989).

Moncur (1985) suggested a pollination mechanism that may involve the interplay of both wind and insect vectors in which the activity of insects on the male inflorescences may promote pollen release, with subsequent transportation by wind or gravity. However, Pushpakumara (2006) suggested that wind is primarily responsible for the pollination of *A. heterophyllus* and this is supported by the observations of the deposition of pollen grains on pollen traps and pollen grains and/or grains with pollen tubes counted on open-pollinated stigmas. Gardner *et al.* (2018) stated that jackfruit is thought to be pollinated by a gall midge, *Clinodiplosis ultracrepidata* (Diptera: Cecidomyiidae).

The fruit is a syncarp, barrel- or pear-shaped, 30–100 × 25–50 cm, with short pyramidal protuberances or warts; the stalk is 5–10 cm long, 1–1.5 cm wide; the rind is approximately 1 cm thick, together with the central core (receptacle) inseparable from the waxy, firm or soft, golden-yellow, fleshy perianths surrounding the seeds.

The seeds are numerous (100–500 per fruit), oblong–ellipsoid, 2–4 × 1.5–2.5 cm in size and 2.5–14 g in weight, enclosed by a horny endocarp and subgelatinous exocarp. The testa are thin and leathery. The embryo has a ventral radicle and fleshy cotyledons that are unequal (the smaller one being one-third to one-half of the larger). The endosperm is very small or absent (Wealth of India, 1985; Morton, 1987; Soepadmo, 1991; Campbell and Ledesma, 2003; Simon *et al.*, 2007; Prakash *et al.*, 2009; Witherup *et al.*, 2019).

1.7.4 Distribution around the world

Jackfruit originated in the rainforest of the Western Ghats of India. Archaeological findings in India have revealed that jackfruit was cultivated in India 3000–6000 years ago. Findings also indicate that Indian Emperor Ashoka the Great (274–237 BC) encouraged arbori-horticulture of various fruits including jackfruit. It then spread its cultivation to neighbouring South and South-east Asian countries and to southern China (Haq, 2006).

Jackfruit is believed to have been introduced to the Philippines during the 12th century AD, when the coastal Filipinos started to trade with India, Malaysia and China (Pelzer, 1948). Jackfruit is now widely grown in many

Asian countries especially India, Bangladesh, Myanmar, Nepal, Sri Lanka, Thailand, Malaysia, Indonesia and the Philippines. It is also grown in southern China and in the Indo-Chinese region in Laos, Cambodia and Vietnam. From Asia, it spread to tropical Africa, especially the eastern part: Zanzibar, Kenya, Uganda and Madagascar (Haq, 2006). Jackfruit spread further to tropical and subtropical America (Brazil, Surinam, Caribbean and USA) and Australia between the mid-17th and late 19th century.

India is the largest producer of jackfruit in the world producing 183,000 t of fruit in 2017–2018. The major producing states are Assam, Tripura, West Bengal, Kerala, Arunachal Pradesh, Tamil Nadu, Karnataka and Jharkhand. It is widely grown in the homesteads in most of the states, with very few commercial orchards.

Jackfruit is considered the national fruit of Bangladesh and is one of the most widely distributed and cultivated crops in this country. Major jackfruit-growing areas are the former districts of Dhaka, Mymensingh, Sylhet, Rajshahi, Dinajpur and Chittagong. In most rural areas, the jackfruit plant is grown in the homestead.

In China, jackfruit is grown over about 12,000 ha of which more than 60% is in Hainan province. Oher relatively small growing areas are the Guangdong, Guangxi, Yunnan, Sichuan and Fujian Provinces.

In Thailand, jackfruit is grown mostly as a backyard plant and in mixed fruit orchards, mostly in the east and south of the country. The total area under jackfruit cultivation increased from 7140 ha in 2005 to 8951 ha in 2020. Jackfruit is locally known as 'Khanun' in Thailand.

Jackfruit is widespread in Vietnam but is grown mainly in the south. The area under jackfruit cultivation has increased rapidly during the past 5 years, with an average growth of 20% year^{-1} from 2016. The total area planted to jackfruit is about 59,705 ha, excluding jackfruit trees grown as a backyard tree, with an annual production of 550,000 t in 2020. The Mekong Delta is known as the top jackfruit-producing area in Vietnam. The total cultivation is 30,045 ha and contributes about 50% of its share to the country. After the Mekong Delta, the south-east region and central highland rank as the second and third largest producers with an estimated acreage of 12,597 ha and 5312 ha, respectively, in 2020. The top five jackfruit-producing provinces in Vietnam are Tien Giang (13,141 ha), Dong Nai (7645 ha), Hau Giang (6966 ha), Dong Thap (2692 ha) and Dak Lak (2081 ha) (General Statistics Office, 2021). The area under cultivation of jackfruit reached an estimated 72,200 ha in 2021.

Most jackfruit produced in Vietnam is exported, and only a small percentage is consumed locally. The rate of processed jackfruit is estimated to be not more than 10% of the production output, and a small percentage is frozen pulp. The main processing product is jackfruit dried slices/chips. The export value of jackfruit has increased continuously in recent years, from more than US$61 million in 2016 to more than US$99.8 million in 2019 and not less than UD$138.1 million in 2020. China was the largest export market for Vietnamese jackfruit, accounting for more than 94% in 2018, 86% in 2019 and nearly 96% in 2020 (Anon, 2021b). Jackfruit is also exported to many other markets such as the Philippines, Papua New Guinea, Canada and Australia.

European colonists, perhaps the Portuguese and later the British, introduced jackfruit into Malaysia from southern India. It is locally known as 'nangka', 'tsjaka' and 'jaka'. The main jackfruit-growing areas in Malaysia are Johor, Sarawak, Sabah, Kelantan and Kedah. The area planted with jackfruit in Malaysia in 2020 was 4675 ha with production of 35,624 t of fruit. Malaysia exports jackfruit to the Middle East, Europe, China, Australia, Hong Kong and Singapore.

Jackfruit is thought to have been introduced in the Philippines in 12th century AD and is now widely grown. The major growing regions are Cagayan Valley, Calabarzon, Bicol, Western Visayas, Central Visayas, Eastern Visayas, Zamboanga Peninsula, Northern Mindanao and Davao Region. Jackfruit is known locally by many names such as 'langka', 'nangka', 'ananka', 'yanka', ubiyen', 'badak' and 'sagakat'. However, the area and production of cultivated jackfruit in the Philippines has been decreasing since 2012. The total planted area in 2012 was 14.42 thousand ha, which decreased to 13.81 thousand ha in 2021.

Indonesia introduced jackfruit from India and it is now grown in Java, Sumatra, Bali, Nusa Tenggara, Kalimantan, Sulawesi, Maliku and Papua. It is mostly grown as backyard trees and in mixed fruit orchards. There are several local names of jackfruit in Indonesian regions such as 'panch', 'nongko', 'langgae', 'anane', 'sibodak', 'naka', 'baduh', 'binaso', 'malasa' and 'kuloh'. The combined production of jackfruit and chempadak increased from 537,186 t in 2002 to 824,068 t in 2020.

Tropical exotic fruits including jackfruit were introduced into northern Australia as early as the late 1800s. In northern Australia, about 9300 jackfruit trees were recorded in 2012.

Jackfruit was reported in Hawaii prior to 1888. It is not widely cultivated in the USA but may be found in south Florida and Hawaii.

In Israel, jackfruit was introduced in the 1930s. The trees bore fruit for several seasons but eventually died. The fresh fruit is not marketed in Israel (Avner Galili, Juran Technologies, Israel, personal communication).

There is very little commercial production of jackfruit in Taiwan. It is grown mostly in backyards or private orchards for personal consumption. Most local jackfruit is used for fried fruit, while dry processed fruits are imported from Vietnam, Malaysia and Thailand (Dr Chung-Ruey Yen, National Pingtung University of Science and Technology, Taiwan, personal communication).

A few jackfruit trees were planted at the Melkassa and Jimma Agricultural Research Centres in Ethiopia for evaluation and adaptation. It is not grown elsewhere in Ethiopia (Dr A.D. Moges, Ethiopian Institute of Agricultural Research, Ethiopia, personal communication).

Jackfruit was introduced into Mauritius from India in 1750 by Governor David (Rouillard and Guého, 1999) and is locally known as 'zak' or 'jacques'. It is grown mostly as a backyard tree throughout the island.

Jackfruit has become an important fruit and a production option in tropical areas of Nayarit and Jalisco, Mexico. In Mexico, its cultivation is recent – it was introduced about 30 years ago and has had an important increase in growth since (Fuentes *et al.*, 2021). In 2020, Mexico produced 23,994 t of fruit with a productivity of 16.82 t ha^{-1}.

The oldest known record of jackfruit in Brazil dates back to 1683, in the colonization period, when 11 jackfruit trees of unknown precedence were introduced in the state of Bahia (Ferrão, 1993). The favourable climatic conditions (rainfall and mean temperature) and altitudinal range in the coastal Atlantic rainforest in Brazil have facilitated the spread of jackfruit throughout Brazil (Bergallo *et al.*, 2016).

The areas and production of jackfruit in the major growing countries are presented in Table 1.3.

Table 1.3. Areas and production of jackfruit in the major growing countries.

Country	Year(s)	Area (ha)	Production (t)	Productivity (t ha^{-1})	Reference
Australia	2011–2012	(9,271 trees)	742	(80 kg per tree)	Diczbalis (2012)
Bangladesh	2017–2018	10,120	1,076,000	9.8	Anon (2019a)
Cambodia	2020	4,526	83,090	18.4	Anon (2021a)
China	2019	27,000	150,000	12.5	Lanxi *et al.* (2019)
India	2017–2018	185,000	1,830,000	9.9	Anon (2018)
Indonesia (*A. heterophyllus* + *A. integer*)	2020	–	824,068	–	Anon (2020a)
Malaysia	2020	4,675	35,624	7.7	Anon (2020b)
Mexico	2020	1,684	23,994	16.82	Fuentes *et al.* (2021)
Nepal	2019–2020	3,265	35,244	1079	Anon (2020c)
Philippines	2019	14,000	40,000	2.86	Anon (2020d)
Sri Lanka	2018	44,730	(256 million fruits)	(5700 fruits ha^{-1})	Anon (2019b)
Thailand	2020	8,951	33,703	13.45	Anon (2020e)
Vietnam	2020	59,705	550,000	9.21	Anon (2021b)

References

Anon (2018) *Horticultural Statistics at a Glance 2018*. Horticulture Statistics Division, Department of Agriculture and Farmers Welfare, Government of India.

Anon (2019a) *Yearbook of Agricultural Statistics – 2019*, 31st Series. Bangladesh Bureau of Statistics, Ministry of Planning, Bangladesh.

Anon (2019b) Department of Agriculture, Ministry of Agriculture, Sri Lanka. Available at: www.agrimin.gov.lk/web/ (accessed 28 June 2022).

Anon (2020a) Department of Agriculture, Government of Indonesia. Available at: www.pertanian.go.id/index_en.php (accessed 28 June 2022).

Anon (2020b) Department of Agriculture, Government of Malaysia, Malaysia. Available at: www.doa.gov.my/ (accessed 28 June 2022).

Anon (2020c) Ministry of Agriculture, Food Technology and Land Management, Government of Nepal, Nepal. Available at: https://nepal.gov.np:8443/NationalPortal/view-page?id=199 (accessed 28 June 2022).

Anon (2020d) *Supply utilization accounts of selected agricultural commodities (2018 to 2020)*. Philippine Statistics Authority, Republic of the Philippines. Available at: https://psa.gov.ph/sites/default/files/SUA-2018-2020-ao13Nov_ONS-signed.pdf (accessed 28 June 2022).

Anon (2020e) Office of Agricultural Economics, Department of Agriculture. Bangkok, Thailand. Available at: www.oae.go.th (accessed 3 October 2022).

Anon (2021a) *Annual Results Report 2020*. Cambodia. United Nations, Cambodia.

Anon (2021b) General Statistics Office, Government of Vietnam. Hanoi, Vietnam. Available at: www.gso.gov.vn/ (accessed 3 October 2022).

Arveiller, R. (1963) *Contribution à l'Étude des termes de voyage en Français (1505–1722)*, D'Artrey, Paris.

Azad, A.K. (1989) Studies on floral biology, pattern of fruit set and quality of fruit borne in different sections of jackfruit plants. MSc thesis, Faculty of Agriculture, BAU, Mymemsing, Bangladesh.

Battuta, I. (2005) *The Travels of Ibn Battuta in the Near East, Asia and Africa 1325–1354*. Translated by Samuel lee. Dover Publications, Mineola, NY.

Bennett, F.D. and Nozzolillo, C. (1987) How many seeds in a seeded breadfruit, *Artocarpus altilis* (Moraceae)? *Economic Botany* 41(3), 370–374. DOI: 10.1007/BF02859053.

Berg, C.C. (2001) *Moreae, Artocarpeae, and* Dorstenia *(Moraceae): With Introductions to the Family and* Ficus *and with Additions and Corrections to Flora Neotropica Monograph 7*. New York Botanical Garden, New York.

Berg, C.C., Corner, E.J.H. and Jarrett, F.M. (2006) Moraceae genera other than *Ficus*. *Flora Malesiana* Series 1. In: *Spermatophyta 17*. pp. 1–146.

Berg, C.C., Pattharahirantricin, N., Chantarasuwan, B., Santisuk, T. and Larsen, K. (2011) *Cecropiaceae and Moraceae Flora of Thailand Part. 4*, Vol. 10. Forest Herbarium, Royal Forest Department, Bangkok.

Bergallo, H.G., Bergallo, A.C., Rocha, H.B. and Rocha, C.F.D. (2016) Invasion by *Artocarpus heterophyllus* (Moraceae) in an island in the Atlantic Forest Biome, Brazil: distribution at the landscape level, density and need for control. *Journal of Coastal Conservation* 20(3), 191–198. DOI: 10.1007/s11852-016-0429-9.

bin Osman, M. (1993) Genetic resources and utilization of tropical fruits. In: Anthony, K., de Groot, P. and Haq, N. (eds) *Underutilized Fruits and Nuts in Asia*. International Centre for Underutilized Crops, Southampton, UK, pp. 31–48.

Campbell, R.J. and Ledesma, N. (2003) *The Exotic Jackfruit*. Fairchild tropical botanic garden: Coral Gables, FL.

Cardon, D. (2007) *Natural Dyes: Sources, Tradition, Technology and Science*. Archetype Books, London (English translation by Caroline Higgitt of Le Monde des Teintures Naturelles, Paris/Berlin, 2003).

Clement, W.L. and Weiblen, G.D. (2009) Morphological evolution in the mulberry family (Moraceae). *Systematic Botany* 34(3), 530–552. DOI: 10.1600/036364409789271155.

Corner, E.J.H. (1938) Notes on the systematy and distribution of Malayan phanerograms, II: the jack and chempedak. *Gardens' Bulletin* 10, 56–81.

Corner, E.J.H. (1962) The classification of Moraceae. *Gardens' Bulletin* 19, 187–252.

Dalgado, S.R. (1919–1921) *Glossário Luso-Asiático*, Vol. 1 and 2. Reprinted 1988, Asian Educational Services, New Delhi.

Datwyler, S.L. and Weiblen, G.D. (2004) On the origin of the fig: phylogenetic relationships of Moraceae from ndhF sequences. *American Journal of Botany* 91(5), 767–777. DOI: 10.3732/ajb.91.5.767.

Diczbalis, Y. (2012) Tropical Exotic Fruit Industry – Strategic Direction Setting (2012–2015). RIRDC Publication No. 12/050. Australian Government, Canberra.

Ferrão, J.E.M. (1993) *A Aventura Das Plantas e Os Descobrimentos Portugueses*. Instituto de Investigação Científica Tropical, Lisbon.

Ferrer-Gallego, P.P. and Boisset, F. (2018) The naming and typification of the breadfruit, *Artocarpus altilis*, and breadnut A. *camansi* (Moraceae). *Willdenowia* 48, 125–135.

Fosberg, F.R. (1960) Introgression in *Artocarpus* (Moraceae) in Micronesia. *Brittonia* 12(2), 101. DOI: 10.2307/2805210.

Fuentes, L.M.H., Monreal, J.J.V. and Jaimes, R.G. (2021) *Plagas y Enfermedades Del Cultivo de Yaca Artocarpus heterophyllus: Recomendaciones de Control*. Instituto National de Investigaciones, Santa Catarina, Mexico.

Gardner, E.M. and Zerega, N.J.C. (2020) Taxonomic updates to *Artocarpus* subgenus *Pseudojaca* (Moraceae), with a particular focus on the taxa in Singapore. *Gardens' Bulletin Singapore* 72(2), 173–213. DOI: 10.26492/gbs72(2).2020-06.

Gardner, E.M. and Zerega, N.J.C. (2021) Taxonomic updates to *Artocarpus* subgenus *Artocarpus* (Moraceae) and allied taxa with a particular focus on the species native to Singapore. *Gardens' Bulletin Singapore* 73(2), 309–374. DOI: 10.26492/gbs73(2).2021-08.

Gardner, E.M., Arifiani, D. and Zerega, N.J.C. (2021) *Artocarpus bergii* (Moraceae), a new species in the breadfruit clade from the Moluccas. *Systematic Botany* 46(1), 91–95. DOI: 10.1600/036364421X16128061189387.

Gardner, E.M., Gagné, R.J., Kendra, P.E., Montgomery, W.S., Raguso, R.A. *et al*. (2018) A flower in fruit's clothing: pollination of jackfruit (*Artocarpus heterophyllus, Moraceae*) by a new species of gall midge, *Clinodiplosis ultracrepidata* sp. nov. (diptera: cecidomyiidae). *International Journal of Plant Sciences* 179(5), 350–367. DOI: 10.1086/697115.

Ghosh, S.P. (2000) *Status Report of Genetic Resources of Jackfruit (Artocarpus heterophyllus Lam.) in India and South East Asia*. IPGRI, Malaysia.

Grenand, P., Moretti, C., Jacquemin, H. and Prévost, M.-F. (2004) *Pharmacopées Traditionnelles En Guyane, Créoles, Wayãpi, Palikur*, 2nd edn. IRD, Paris.

Haq, N. (2006) *Jackfruit (Artocarpus heterophyllus)*. Southampton Centre for Underutilised Crops, University of Southampton, Southampton, UK.

Jarrett, F.M. (1959a) Studies in *Artocarpus* and allied genera, I. General Considerations. *Journal of the Arnold Arboretum* 40(1), 1–29. DOI: 10.5962/p.36682.

Jarrett, F.M. (1959b) Studies in *Artocarpus* and allied genera. II. a revision of *Prainea*. *Journal of the Arnold Arboretum* 40(1), 30–37. DOI: 10.5962/p.324658.

Jarrett, F.M. (1959c) Studies in *Artocarpus* and allied genera, III. a revision of *Artocarpus* subgenus *Artocarpus*. *Journal of the Arnold Arboretum* 40(2), 113–155. DOI: 10.5962/p.186026.

Jarrett, F.M. (1977) The syncarp of *Artocarpus* – a unique biological phenomenon. *Gardens' Bulletin* 29, 35–39.

Kochummen, K.M. (1978) Moraceae. In: Ng, F.S.P. (ed.) *Tree Flora of Malaya*, Vol. 3. Springer, Kuala Lumpur, pp. 119–168.

Lanxi, S.U., Tingyu, B., Gang, W.U., Huan, Y.U. and Lehe, T. (2019) Research status and development trend of jackfruit cultivation. *Chinese Journal of Tropical Agriculture* 39, 10–15.

Moncur, M.W. (1985) Floral ontogeny of the Jackfruit, *Artocarpus heterophyllus* Lam. (Moraceae). *Australian Journal of Botany* 33(5), 585. DOI: 10.1071/BT9850585.

Morton, J.F. (1987) Jackfruit. In: *Fruits of Warm Climates*. J.F. Morton, Miami, Florida, pp. 58–64.

Pelzer, K.J. (1948) *Pioneer Settlement in the Asiatic Tropics*. American Geographic Society, New York, NY.

Prakash, O., Kumar, R., Mishra, A. and Gupta, R. (2009) *Artocarpus heterophyllus* (jackfruit): an overview. *Pharmacognosy Reviews* 3, 353–358.

Pushpakumara, D.K.N.G. (2006) Floral and fruit morphology and phenology in *Artocarpus heterophyllus* Lam (Moraceae). *Sri Lankan Journal of Agricultural Science* 43, 82–107.

Ragone, D. (1997) *Breadfruit.* Artocarpus altilis *(Parkinson) Fosberg*. International Plant Genetic Research Institute, Rome.

Ragone, D. (2001) Chromosome numbers and pollen stainability of three species of Pacific Island breadfruit (*Artocarpus*, Moraceae). *American Journal of Botany* 88(4), 693–696.

Ragone, D. (2006a) *Artocarpus altilis* (breadfruit). In: Elevitch, C.R. (ed.) *Traditional Trees of Pacific Islands*. Premanet Agricultural Resources, Holualoa, Hawaii, pp. 85–100.

Ragone, D. (2006b) *Artocarpus camansi* (breadnut). In: Elevitch, C.R. (ed.) *Traditional Trees of Pacific Islands*. Permanent Agricultural Research, Holualoa, Hawaii, pp. 101–110.

Ragone, D. (2007) Breadfruit: diversity conservation and potential. *Acta Horticulturae* 757, 19–30. DOI: 10.17660/ActaHortic.2007.757.1.

Rohwer, J.G. (1993) Moraceae. In: Kubitzki, K. (ed.) *The Families and Genera of Vascular Plants*, Vol. II. *Flowering Plants · Dicotyledons: Magnoliid, Hamamelid and Caryophyllid Families*, Springer, Berlin, Heidelberg, pp. 438–453. DOI: 10.1007/978-3-662-02899-5.

Rouillard, G. and Guého, J. (1999) *Les Plantes et Leur Histoire à l'Ile Maurice*. Mauritius.

Sambamurthy, K. and Ramalingam, V. (1954) Preliminary studies in blossom biology of the jack (*Artocarpus heterophyllus* Lam) and pollination effects. *Indian Journal of Horticulture* 11, 24–29.

Simon, L., Shyamalamma, S. and Narayanaswamy, P. (2007) Morphological and molecular analysis of genetic diversity in jackfruit. *The Journal of Horticultural Science and Biotechnology* 82(5), 764–768. DOI: 10.1080/14620316.2007.11512302.

Soepadmo, E. (1991) *A. heterophyllus* Lamk. In: Verheij, E.W.M. and Coronel, R.E. (eds) *Plant Resources of South-East Asia (PROSEA), Vol. 2. Edible Fruits and Nuts*. PROSEA Foundation, Bogor, Indonesia. Available at: www.prota4u.org/prosea/view.aspx?id=1476 (accessed 19 July 2022).

Stevens, P.F. (2001) Angiosperm phylogeny website. Version 14, July 2017 [and more or less continuously updated since]. Available at: www.mobot.org/MOBOT/research/APweb (accessed 28 June 2022).

Sytsma, K.J., Morawetz, J., Pires, J.C., Nepokroeff, M., Conti, E. *et al.* (2002) Urticalean rosids: circumscription, rosid ancestry, and phylogenetics based on *rbcL*, *trnL-F*, and *ndhF* sequences. *American Journal of Botany* 89(9), 1531–1546. DOI: 10.3732/ajb.89.9.1531.

Turner, I.M. (1995) A catalogue of the vascular plants of Malaya. *Gardens' Bulletin* 47, 1–757.

Wang, M.M.H., Gardner, E.M., Chung, R.C.K., Chew, M.Y., Milan, A.R. *et al.* (2018) Origin and diversity of an underutilized fruit tree crop, cempedak (*Artocarpus integer*, Moraceae). *American Journal of Botany* 105(5), 898–914. DOI: 10.1002/ajb2.1094.

Wealth of India (1985) *A Dictionary of Indian Raw Materials and Industrial Products*, Vol. 1. Revised edition, Council of Scientific and Industrial Research, New Delhi.

Williams, E.W., Gardner, E.M., Harris, R., Chaveerach, A., Pereira, J.T. *et al.* (2017) Out of Borneo: biogeography, phylogeny and divergence date estimates of *Artocarpus* (Moraceae). *Annals of Botany* 119(4), 611–627. DOI: 10.1093/aob/mcw249.

Witherup, C., Zuberi, M.I., Hossain, S. and Zerega, N.J.C. (2019) Genetic diversity of Bangladeshi Jackfruit (*Artocarpus heterophyllus*) over time and across seedling sources. *Economic Botany* 73(2), 233–248. DOI: 10.1007/s12231-019-09452-5.

Worrell, D.B., Sean Carrington, C.M. and Huber, D.J. (1998) Growth, maturation and ripening of breadfruit, *Artocarpus altilis* (Park.) Fosb. *Scientia Horticulturae* 76(1–2), 17–28. DOI: 10.1016/S0304-4238(98)00134-4.

Yule, H. (1350) The travels of John de Marignolli (1339–1353). In: *Cathay and the Way Thither: Being a Collection of Medieval Notices of China*, Vol. II. Hakluyt Society (printed 1913–1916), London, pp. 209–269. Available at: https://depts.washington.edu/silkroad/texts/marignolli.html

Zerega, N.J.C. and Ragone, D. (2016) Toward a global view of breadfruit genetic diversity. *Tropical Agriculture* 93, 77–91.

Zerega, N.J.C., Ragone, D. and Motley, T.J. (2004) Complex origins of breadfruit (*Artocarpus altilis*, Moraceae): implications for human migrations in Oceania. *American Journal of Botany* 91(5), 760–766. DOI: 10.3732/ajb.91.5.760.

Zerega, N.J.C., Ragone, D. and Motley, T.J. (2005) Systematics and species limits of breadfruit (*Artocarpus*, Moraceae). *Systematic Botany* 30, 603–615.

Zerega, N.J.C., Supardi, M.N. and Motley, T.J. (2010) Phylogeny and recircumscription of *Artocarpeae* (Moraceae) with a focus on *Artocarpus*. *Systematic Botany* 35(4), 766–782. DOI: 10.1600/036364410X539853.

2 Composition and Uses

Anup K. Bhattacherjee*
ICAR-Central Institute for Subtropical Horticulture, Rehmankhera, Lucknow-226101, India

2.1 Introduction

Jackfruit (*Artocarpus heterophyllus*) is the largest edible tropical fruit. It is commonly known as 'poor man's food' because poor people in jackfruit-growing areas are used to consuming this fruit (raw and ripe) as one of their daily meals instead of rice or wheat (Goswami and Chacrabati, 2016). Jackfruit is a composite or multiple fruit (syncarp) consisting of edible (pulp/flake and seed) and non-edible (rind and rachis) portions. Jackfruit has a relatively high productivity of around 25.71 t ha^{-1}. A mature jack tree can yield from ten to 200 fruits throughout the year (Samaddar, 1985). The tender green fruit is consumed as a vegetable, while the juicy pulp of the ripe fruit is used as fresh fruit. The seeds can be eaten after cooking, roasting or frying. In the Indian subcontinent, jackfruit is popularly known as 'kathal' or 'kanthal'. There are mainly two types of jackfruit available in the world – khaja (with a firm perianth at the edible ripe stage) and ghila (with a soft and juicy perianth when ripe). Besides these, another type of jackfruit called dorosha is also found in some Asian countries with characteristics of both the khaja and ghila types (Rahman *et al.*, 1999).

India is the world leader in jackfruit production with 1.4 million t, followed by Bangladesh (0.926 million t), Thailand (0.392 million t) and Indonesia (0.34 million t) (Anon, 2017). In India eastern, north-eastern and southern states like Tripura, Odisha, Assam, West Bengal, Karnataka, Kerala, Chhattisgarh and Jharkhand are the major jackfruit-growing states. As a commercial crop, jackfruit is less popular than others, even in India and Bangladesh, because of the wide variation in fruit size and quality, the long gestation period of seed-grown plants and the widespread belief that excessive consumption of flakes can lead to digestive problems (Samaddar, 1985). Jams, beverages, candies, preserves and dehydrated forms are some of the value-added products of jackfruit. It is also known to be beneficial against asthma, ulcers, indigestion, hypertension, nervousness and constipation (Goswami and Chacrabati, 2016). It has the ability to slow down ageing and cell degeneration. Jackfruit is a good source of macronutrients such as carbohydrates, proteins, fats and organic acids, micronutrients (minerals and vitamins), and some non-nutrients such as phenolic compounds including flavonoids, carotenoids and volatile compounds. The wood of the jack tree, which ages to an orange and/or red-brown colour, is highly durable, resisting termites and decay (Elevitch and Manner, 2006). A yellow dye is sometimes extracted from the wood and used for dyeing cloth, especially in India and the Far East (Seddon and Lennox, 1980). This chapter deals with the various phytochemical constituents of jackfruit along with

*dadabhatu@gmail.com

DOI: 10.1079/9781800622319.0002

traditional, pharmacological and culinary uses of the plant parts.

2.2 Nutritional Composition

The nutritional composition of jackfruit varies with its developmental stage, flake colour, variety and agroclimatic conditions. There is also a wide variation observed in fruit weight, which varies from 3.5 to 30 kg making it one of the largest fruits of the world. The ripe fruit consists of pulp (29%), seeds (12%) and rind including peel (59%). The edible pulp of jackfruit contains water (74%), carbohydrates mainly sugars (23%), protein (2%), fat (1%) and dietary fibre. It is a rich source of vitamin B complex and a moderate source of vitamin C, as well as possessing significant amounts of macro-minerals such as potassium, magnesium, calcium and phosphorus. The nutritional, vitamin and mineral compositions of young, mature and ripe jackfruit along with those of the seeds are presented in Table 2.1. It is a low-calorie fruit, containing only 95 kcal (400 kJ) 100 g^{-1} fruit. When compared with other tropical fruits such as mango, banana, orange, pineapple, papaya and ber, jackfruit pulp and seeds possess more protein, calcium, iron and thiamine (Kumar *et al.*, 1988). The fruit also contains many bioactive phytonutrients such as fatty acids, amino acids, phenolic compounds, flavonoids, stilbenoids, morin, dihydromorin, cynomacurin, artocarpin, isoartocarpin, cycloartocarpin, cycloheterophyllin, artonins A, artonins B, artocarpesin, oxydihydroartocarpesin, artocarpetin, artocarpetin A, artocarpanone, norartocarpetin and cycloartinone, which

Table 2.1. Proximate and nutritional composition of immature fruit, mature fruit, ripe fruit and seed of jackfruit (100 g fresh weight of edible portion). Data from Wealth of India (1985); Gunasena *et al.* (1996); Azad (2000).

Proximate/nutritional composition	Immature fruit	Mature fruit	Ripe fruit	Seed
Energy (kJ)	50–210	98	88–410	133–139
Water (g)	76.2–85.2	72	72.0–94.0	51.0–64.5
Protein (g)	2.0–2.6	1.9	1.2–1.9	6.6–7.04
Fat (g)	0.1–0.6	0.6	0.1–0.4	0.40–0.43
Carbohydrate (g)	9.4–11.5	8.5	16.0–25.4	25.8–38.4
Dietary fibre (g)	2.6–3.6	1.0	1.0–1.5	1.0–1.5
Total sugars (g)	–	19.08	20.6	–
Vitamins				
Vitamin A (IU)	30	40	175–540	10–17
Vitamin B_1 (mg)	0.05–0.15	0.105	0.03–0.09	0.25
Vitamin B_2 (mg)	0.05–0.20	0.055	0.05–0.40	0.11–0.30
Vitamin B_3 (mg)	0.5	0.92	0.9	0.5
Vitamin C (mg)	12–14	13.8	7–10	11
Minerals				
Total minerals (g)	0.9	0.7	0.87–0.9	0.9–1.2
Calcium (mg)	30.0–73.2	22.0	20.0–37.0	50.0
Magnesium (mg)	–	29.0	27.0	54.0
Potassium (mg)	287–323	407	191–407	246
Sodium (mg)	3.0–35.0	2.0	2.0–41.0	63.2
Phosphorus (mg)	20.0–57.2	38.0	38.0–41.0	38.0–97.0
Iron (mg)	0.4–1.9	0.6	0.5–1.1	1.5

makes it a potent antioxidant, antibacterial, antifungal, anti-inflammatory, anti-cancer, anti-neoplastic, cicatrizant and hypoglycaemic fruit (Pavanasasivam *et al.*, 1973; Rama Rao and Venkataraman, 1973; Ko *et al.*, 1998; Prakash *et al.*, 2009).

2.2.1 Minerals

Among the various nutrient compositions, minerals are one of the important components of our daily diet, although only 4–6% of our body weight is composed of minerals. There are two types of minerals – macro (required at more than 100 mg day^{-1}), i.e. calcium, magnesium, phosphorus, potassium, sodium, sulfur and chlorine, and micro (required at less than 100 mg day^{-1}), i.e. zinc, iron, copper, manganese, molybdenum, selenium and iodine. The major minerals found in jackfruit are potassium, calcium, magnesium and phosphorus, although it also contains iron, sodium, zinc and manganese in appreciable quantities (Table 2.1). Jackfruit is particularly rich in potassium, which helps to maintain sodium–potassium balance in the human body, thereby maintaining blood pressure along with strengthening of bones and teeth. It also facilitates the contraction and relaxation of muscles in the body. Potassium together with other electrolytes such as magnesium, calcium and sodium also conduct electricity in the body, which is important for proper functioning of the heart and other body organs (Soetan *et al.*, 2010). Calcium also occurs in high levels and helps in strengthening of bones and teeth. Inadequate intake of calcium has been associated with illnesses such as osteoporosis (Soetan *et al.*, 2010). Magnesium levels were also found to be moderately high. This mineral plays an important role in the regulation of blood sugar levels, acting as a cofactor in many enzymes such as insulin in glucose metabolism (Takaya *et al.*, 2004). The fruit contains moderate levels of iron, which is important in the synthesis of haemoglobin. Similarly, the concentration of red blood cells in the body is strongly determined by the presence of iron (Soetan *et al.*, 2010). Zinc, in contrast, is required by the body only in small quantities but also plays a major role in a number of metabolic processes, being utilized as a cofactor by a number of enzymes. Copper is also found in low levels in the seeds and in the pulp. This mineral is required in the body in small quantities and considered to be crucial in the formation of connective tissues. It is also used, alongside iron, in the formation of red blood cells (Soetan *et al.*, 2010).

The presence of 88 mg potassium, 90 mg calcium, 41 mg phosphorus and 0.5 mg iron 100 g^{-1} fruit was reported by Bhatia *et al.* (1955). Jackfruit pulp contains 1.2 g of total mineral matter, which includes 246 mg potassium, 28 mg phosphorus, 21 mg calcium and 0.8 mg iron 100 g^{-1} mature edible portion (Faruque and Omar, 1985). In another study, it was reported that a 100 g edible portion of immature and ripe jackfruit contains 0.7 and 0.5–1.2 g minerals including 323 and 88–407 mg potassium, 53 and 19–23 mg calcium, 20 and 18–68 mg phosphorus, 0.4 and 0.7–1.1 mg iron, and 3 and 2 mg, sodium, respectively (Wealth of India, 1985). Every 100 g of ripe flake contains 287–323 mg potassium, 30.0–73.2 mg calcium, 20.0–57.2 mg phosphorus and 3.0–35.0 mg sodium (Samaddar, 1985). The mineral content in fruit varies considerably depending on pre- and postharvest factors such as variety, maturity period/stage of utilization, and location or environmental conditions. Young or immature fruit does not contain magnesium, while ripe fruit contains 27 mg 100 g^{-1} fruit. Ripe fruit pulp has slightly higher amounts of phosphorus and potassium compared with immature fruit pulp (Gunasena *et al.*, 1996). The calcium, copper and lead contents increase slightly with advancement of the maturity period, while the magnesium and iron contents decrease marginally, and no change is observed in manganese content (Tiwari and Vidyarthi, 2015).

Jackfruit seed also has substantial amounts of important minerals. Calcium, phosphorus and iron content in fresh seeds were reported to be 0.05–0.55%, 0.13–0.23% and 0.002–1.2%, respectively, while those in roasted seeds were 0.13%, 0.54% and 0.05%, respectively. The amounts of these three minerals in fresh and oven-dried seeds were recorded as 0.55%, 0.23% and 0.002%, and 0.13%, 0.54% and 0.005%, respectively. Jackfruit seeds also contained 54 mg magnesium, 63 mg sodium, 246 mg phosphorus, 0.2 mg copper, 256 mg sulfur and 14 mg chlorine in 100 g of fruit (Brown, 1951).

Gupta *et al.* (2011) detected nine minerals (concentrations are on dry weight (DW) basis): potassium (786.6 p.p.m.), calcium (29.47 p.p.m.), sodium (28.39 p.p.m.), barium (0.275 p.p.m.), zinc (2.28 p.p.m.), argon (0.047 p.p.m.), tin (0.031 p.p.m.), chromium (0.018 p.p.m.) and cadmium (0.01 p.p.m.) in jackfruit seeds. In Bangladesh, researchers could not find much difference in the mineral content of raw and roasted seed flour of three varieties of jackfruit, collected from Valuka and Modhupur, two jackfruit-growing regions (Goswami and Chacrabati, 2016). The authors detected the presence of calcium, magnesium, sulfur and phosphorus in khaja, ghila and dorosha types of jackfruit. Potassium (705.71 mg 100 g^{-1}), calcium (115.85 mg 100 g^{-1}) and magnesium (96.75 mg 100 g^{-1}) were found to be at higher levels compared with other minerals in jackfruit seed flour (Miah *et al.*, 2017; Sultana *et al.*, 2017). Amadi *et al.* (2018) estimated the contents of potassium (0.22 g 100 g^{-1}), calcium (0.04 g 100 g^{-1}), iron (18.25 mg 100 g^{-1}), manganese (9.50 mg 100 g^{-1}) and zinc (9.28 mg 100 g^{-1}) in fresh seeds. Both the seeds and pulps were found to be rich in potassium (2297.50 and 2836.66 mg 100 g^{-1} DW), sodium (193.75 and 185.33 mg 100 g^{-1} DW), calcium (176.73 and 141.67 mg 100 g^{-1} DW), magnesium (84.25 and 76.28 mg 100 g^{-1} DW), iron (23.09 and 14.05 mg 100 g^{-1} DW), zinc (7.12 and 5.31 mg 100 g^{-1} DW) and copper (2.84 and 2.21 mg 100 g^{-1} DW) (Ojwang *et al.*, 2018). The seeds contained higher amounts of these minerals than pulp, except in case of potassium, which was to be found higher in pulp. In one study in Malaysia, phosphorus, potassium, aluminium, calcium, magnesium, manganese, iron, copper, zinc, boron and sodium were identified as minerals in jackfruit seed powder, while manganese, copper, boron and iron were found in minute quantities (Sy Mohamad *et al.*, 2019). The mineral composition of jackfruit leaves was reported to be 0.52 g calcium, 0.21 g potassium, 59.50 mg iron, 12.75 mg manganese and 5.73 mg zinc in 100 g of fruit (Amadi *et al.*, 2018).

2.2.2 Carbohydrates and sugars

Usually, two types of carbohydrates are found in fruits – simple (monosaccharides and disaccharides, e.g. glucose, fructose, galactose, sucrose) and complex (polysaccharides, e.g. starch, cellulose, pectin). The starch content resembled a close correlation with maturity, variety, and different climatic and agronomic conditions. The presence of a high starch percentage in the fruit perianth and seed has been reported where the starch and total free sugar content of the flesh increases with maturity (Rahman *et al.*, 1995). Raw fruit does not contain any sugar, but ripe fruit has 20.6 g 100 g^{-1} total sugars. Hydrolysis of starch to free sugars and other soluble substrates is common in fruits during ripening, and jackfruit is no exception. After isolating starch from the perianth of raw and ripe fruit, researchers observed that raw fruit had much higher starch content of 31.6% and 29.8% for soft and firm varieties, respectively, compared with 2.2% and 9.0% for ripe fruit (Rahman *et al.*, 1995). The carbohydrate content of raw and ripe fruit varies from 9.4 to 11.5 g and from 16.0 to 25.4 g in 100 g of the edible portion (Azad, 2000). The total carbohydrate content in flakes was reported to be maximum (31.2%) in the Mondan variety, followed by 28.4% in the Venkanni and Nettadi varikka varieties, and minimum (23.8%) in the Valayan variety. The total soluble sugar in flakes ranged between 15.3% (Venkanni) and 21.3% (Valayan) (Chrips *et al.*, 2008). The starch content in the pulp of three types of jackfruit grown in two different areas of Bangladesh (Valuka and Modhupur) ranged from 6.11% (Valuka ghila) to 8.34% (Modhupur khaja) (Goswami and Chacrabati, 2016). Jackfruit contains 23.25 g carbohydrates 100 g^{-1} fruit, of which 19.08 g (82%) is sugars (USDA, 2016).

Several studies have also reported a high carbohydrate content in seeds, which varied from 66.2% (Bobbio *et al.*, 1978) to 76.1% (Kumar *et al.*, 1988) to 38.4% (Morton, 1987). In the case of off-season and on-season jackfruit seeds of two germplasms in Bangladesh, the starch content varied from 38.79% to 44.65% in Ah-15 germplasm and from 41.93% to 49.18% in Ah-24 germplasm (Hasan, 2002). Chrips *et al.* (2008) stated that the total carbohydrate concentration in seeds of different neglected varieties of jackfruit in Kerala, India, varied from 37.4% (Chemparathy) to 42.8% (Mondan). Gupta *et al.* (2011) found 26.20% carbohydrate in jackfruit seeds. In the case of raw and roasted seed flour, the maximum starch content (73.34%) was in raw seeds of the Modhupur ghila type and the minimum (39.76%) in raw seeds of Valuka khaja type (Goswami and Chacrabati, 2016). The authors noticed a trend of increased starch content in roasted seeds compared with raw seeds, which was reversed in seeds of ghila-type fruits. Both jackfruit seeds and pulp have very high carbohydrate levels (31.41–34.95% and 21.65–24.91% for seeds and pulp, respectively), which means that both seeds and pulp are good sources of energy (Ojwang *et al.*, 2018). In jackfruit seed and its flour, the carbohydrate content was similar (42.49%) (Miah *et al.*, 2017; Sultana *et al.*, 2017). A recent study described 69.39 g carbohydrate 100 g^{-1} seed powder (Sy Mohamad *et al.*, 2019), while another estimation reported 83.68 g carbohydrate 100 g^{-1} seed powder in Malaysian jackfruit (Zuwariah *et al.*, 2018). Proximate composition analysis of fruit pulp, seeds and leaves have resulted in the carbohydrate content of 7.74, 7.89 and 5.33%, respectively (Amadi *et al.*, 2018). In Sri Lanka, the slowly available glucose percentage of a jackfruit-based meal was calculated as 30%, which was double that of the standard. The boiled jackfruit flesh contained disintegrated starch granules, while the seeds contained intact swollen and disintegrated granules. Jackfruit seeds, which contain 22% starch, can be considered a good source of starch (Hettiaratchi *et al.*, 2011).

Among 24 jackfruit selections, collected from the Western Ghats of Karnataka, India, starch, total sugar and reducing sugar contents varied from 5.13% (SMG-3) to 0.63% (SMG-5), 32.1% (SMG-5) to 19.1% (SMG-27), and 14.6% (SMG-15) to 8.63% (SMG-21), respectively (Jagadeesh *et al.*, 2007). The starch content in all jackfruit selections was lower than the sugar content at the edible ripe stage. Similarly, researchers in Bangladesh reported that total sugar, reducing sugar and non-reducing sugar content varied from 11.29%, 4.90% and 6.39% (Valuka dorosha) to 17.89%, 8.19% and 9.70% (Modhupur ghila), respectively (Goswami and Chacrabati, 2016). Among 40 genotypes from Tamil Nadu, India, the contents of total, reducing and non-reducing sugars ranged from 9.24% (AH-5) to 25.71% (AH-13), 2.94% (AH-31) to 7.26% (AH-13), and 5.40% (AH-5) to 18.45% (AH-13), respectively (Aseef *et al.*, 2017). Significant variations in the amounts of total sugars (11.4–50.0%), reducing sugars (2.4–9.5%) and non-reducing sugars (7.6–42.3%) were recorded in 20 jackfruit genotypes collected from the alluvial zone of West Bengal, India, where the maximum amounts of total and non-reducing sugars were observed in G-20, while the maximum amount of reducing sugar was in G-16 (Pebbuli and Bauri, 2018).

Sengupta and Rao (1963) identified some sugars such as rhamnose, xylose, arabinose and galactose in fruit pulp. Higher levels of glucose, fructose and sucrose were recorded in ripe fruits. Ripe fruits possess a high percentage of glucose, fructose and sucrose, which impart sweetness to the fruit (Chowdhury *et al.*, 1997). Trehalose and psicose, two sugars that are rarely found in nature and are good for human health, were detected in the fruit and peel of jackfruit, respectively, in Thailand (Muangthai and Katinted, 2014). Trehalose is one of the most stable disaccharides comprising two glucose molecules bound with an α,α-(1→1) linkage and is one of the many energy sources in living organisms, having 45% sweetness compared with sucrose. Similarly, D-psicose is another rare sugar, which is a C-3 epimer of D-fructose synthesized by the enzymatic reaction of D-tagatose 3-epimerase (Kim *et al.*, 2006). It has 70% sweetness compared with sucrose but provides zero energy and can be used as a sweetening agent to help weight loss. Li *et al.* (2014) in China reported that 80% methanol was the most effective solvent for simultaneous extraction of glucose, fructose and sucrose, and the fruit pulp contained 14.01 g sugars 100 g^{-1} on a fresh weight (FW) basis, with sucrose being the major component. Li *et al.* (2017) further noticed a rapid increase in the content of total soluble sugar and sucrose during ripening, while the content of glucose and fructose increased slowly. During the ripening process, glucose is released from starch via hydrolysis followed by sucrose hydrolysis, resulting in an increase in glucose and fructose content. The amylose content (25%) in jackfruit seed starch was reported to be similar to that of potato starch (27%) but better than that of rice starch (17%) (Wong *et al.*, 2021). The abundance of different types of sugars and fatty acids in jackfruit flakes indicates an active mechanism of sugar to fatty acid conversion.

2.2.3 Dietary fibre and pectin

Dietary fibre helps in improving glucose tolerance by delaying the movement of carbohydrate to the small intestine, which reduces the risk of heart disease and facilitates smooth bowel movement and eases constipation. Consumption of high-fibre food can change the water content, viscosity and microbial mass inside the intestine, which in turn changes the rate and ease of passage through the intestine (Anderson *et al.*, 1994). Dietary fibre is of two types, soluble and insoluble, where fruits contain mostly soluble-type dietary fibre. Pectin is a complex polysaccharide found mostly in plant cell walls, with applications in food, pharmaceuticals and a number of other industries as a gelling and stabilizing agent (Mohnen, 2008). Pectin is used in the preparation of jam, jelly, marmalade, frozen foods and low-calorie foods as a fat/sugar substitute in the food industry (Thakur *et al.*, 1997). Blood cholesterol levels and gastrointestinal problems can be minimized with the help of pectin. The edible portion of jackfruit is a good source of soluble dietary fibre, while the non-edible portion is rich in pectin.

In the Philippines, 2.6%, 0.8% and 1.4% fibre content in immature fruits, ripe fruits and seed, respectively, has been recorded in jackfruit (Coronel, 1983). In India, one report gave the fibre content in immature fruits, mature fruits, ripe fruits and seeds as 2.6, 1.0, 0.8–1.8 and 1.4–1.8 g 100 g^{-1} edible portion, respectively (Wealth of India, 1985). Tender jackfruit pulp was found to contain a much higher amount of fibre (4.4 g 100 g^{-1}) compared with mature (1.3 g 100 g^{-1}) and ripe (1.1 g 100 g^{-1}) pulp (Faruque and Omar, 1985), which indicated that raw pulp is richer in fibre content than ripe pulp. The portion of insoluble dietary fibre has been recorded to be much higher than the portion of soluble dietary fibre (Nahar *et al.*, 1993). The fibre content in young fruit, ripe fruit and seeds was reported to be 2.6–3.6, 1.0–1.5 and 1.0–1.5 g 100 g^{-1} edible portion, respectively (Gunasena *et al.*, 1996). In another study, the total dietary fibre content in the perianth was found to be almost the same in firm and soft varieties (Rahman *et al.*, 1999). There were no significant variations in crude fibre content between different parts of the fruit or at different stages of ripening. The amount of crude fibre fluctuates within the range of 0.33–0.40% during ripening of fruit (Ong *et al.*, 2006). A slight variation in the fibre content in different varieties of jackfruit pulp was reported by Goswami and Chacrabati (2016) in Bangladesh where it varied from 0.51% (Modhupur ghila) to 0.90% (Modhupur khaja). Dietary fibre in jackfruit has been reported as 1.5 g 100 g^{-1}, which is lower than banana (2.6 g 100 g^{-1}) and fig (2.9 g 100 g^{-1}) but on a par with mango (1.6 g 100 g^{-1}) and pineapple (1.4 g 100 g^{-1}) (Srivastava and Singh, 2020). The mean fibre content of jackfruit seed has been reported to be 1.33% (Bhatia *et al.*, 1955). A study in Bangladesh has suggested that the fibre content of jackfruit flesh ranged from 0.57% to 0.86% depending on the germplasm and the season, while the fibre content in on-season and off-season jackfruit seeds of Ah-15 and Ah-24 germplasms was 1.38%, 1.45%, 1.50% and 1.58%, respectively (Hasan, 2002). A variation in fibre content among raw and roasted seeds flour was observed, where Modhupur ghila roasted seed flour had the maximum fibre content (1.71%) and Valuka ghila roasted seed flour had the minimum (1.26%) (Goswami and Chacrabati, 2016). In Sri Lanka, the total dietary fibre content of seeds has been reported as 11.1% on FW basis in a jackfruit meal comprising flesh and seeds (Hettiaratchi *et al.*, 2011). In jackfruit seed, the crude fibre content was found to be 3.56% (Miah *et al.*, 2017). The crude fibre content in jackfruit pulp, seed and leaves was estimated as 3.01%, 3.92% and 4.91%, respectively, in a study in Nigeria (Amadi *et al.*, 2018).

The pectin extracted from jackfruit has been purified and characterized for its yield, moisture content and methoxyl value (Berry and Kalra, 1988). The enzymes present in the fruit include amylase, invertase, pectinase and protease. Furthermore, both intra- and extracellular secretions of a strong pectolytic enzyme have been reported, which could account for the rapid cell degradation noticed after ripening. Pectin is extracted mainly from fruit waste, in this case jackfruit rind and perianth, the inedible parts of the fruit. In India, in a study on characterization of pectin from different fruit waste, it has been observed that pectin extracted from rind had 66% anhydrouronic acid (AUA) and 7.67% methoxyl content with 159% gel grade, which was of medium quality (Madhav and Pushpalatha, 2002). In another study, different parts of jackfruit were used in pectin extraction and it was found that the bulbs, perigones, rind, core and seed of firm-fleshed jackfruit contained 2.46%, 3.38%, 5.12%, 5.62% and 0.58% pectin as calcium pectate at harvest, respectively, which declined on ripening to 2.00%, 3.03%, 4.65%, 4.01% and 0.49%, respectively. Similarly, in the case of soft-fleshed jackfruit, the pectin content was estimated as 2.71%, 3.52%, 5.24%, 5.75% and 0.70%, respectively, which upon ripening reduced to 2.06%, 3.24%, 4.59%, 4.45% and 0.60%, respectively (Patil *et al.*, 2011). The yield of pectin from jackfruit waste was reported to be 14–17% in another study (Koh *et al.*, 2014), which might be due to the methods used for pectin extraction, the varieties tested and the variations in pectin content. Begum *et al.* (2014) have mentioned that the yield of pectin from fruit waste (rind and core) varied from 8.94% to 15.14% on a DW basis, depending on the extraction method, which was lower than the yield (22.5%) reported by Mohamed and Hasan (1995). A decrease in pectin content upon maturity and ripening has also been reported by Lauzon *et al.* (2017) where the pectin content in fruit pulp significantly declined from 3.63 g at 85 days of maturity to 2.62 g at 94 days of maturity. The authors suggested that the pectin content was indirectly proportional to the fruit maturity period and moisture content of the pulp. The significant reduction in the amount of pectin upon maturity was due to the enzymatic action on pectin, which resulted in the production of soluble pectic substances.

The levels of pectin as calcium pectate vary according to variety as well as different parts of the fruit of the same variety. The yield of pectin from peel and core was reported to be 0.5% and 1.2% on a FW basis and 6.0% and 14.0% on a DW basis, respectively, with 3.43% and 2.98% methoxyl content and 56.78% and 53.62% AUA content, respectively (Ahmmed *et al.*, 2017). Yield, galacturonic acid content and the degree of esterification (DE) for freeze-dried, spray-dried, vacuum oven-dried and oven-dried pectin samples were observed to be in the ranges of 7–11%, 62–68% and 59–66%, respectively, from fruit waste (Begum *et al.*, 2017a). The authors categorized this pectin as high methoxyl pectin because DE was noticed to be more than 50% for pectin obtained through the various drying techniques. Begum *et al.* (2017b) observed that ammonium oxalate as an extractant was highly suitable for pectin recovery from fruit waste on the basis of yield and functional and chemical properties, and that it contained high levels of AUA (57.34%), DE (62.77%) and viscosity. Sundarraj and Ranganathan (2017a) reported a pectin content of 7.52% extracted from the peel of *Artocarpus integer* (Thunb.). Response surface methodology was used to determine the optimum temperature and time for the extraction of pectin from fruit waste using oxalic acid, which were reported as 90°C and 60 min, respectively, with a yield of 38.42%, and classified it as low-methoxyl pectin based on low methoxyl content and DE (Sundarraj *et al.*, 2018). In the Philippines, brown-coloured pectin (yield 5.74%) was obtained from ripe peel, which contained 4.05% methoxyl content and 1.77% AUA along with 26.34% DE (Arollado *et al.*, 2018). In a study in China, pectin from jackfruit peel was extracted using ultrasonic microwave-assisted extraction (UMAE) using different organic acids and mineral acids. The optimum UMAE conditions for extraction of pectin with citric acid were 86°C extraction temperature, 29 min extraction time, and a solid:liquid ratio of 1:48 (w/v) with a yield of 21.5% pectin. This pectin was found to be rich in galactose, rhamnose, arabinose, glucose and galacturonic acid with a high degree of methyl esterification (Xu *et al.*, 2018). In Uganda, yellow-, orange- and white-coloured jackfruit flakes yielded 917, 868 and 662 g pectin mol^{-1} (yield 15–30%), respectively (Balamaze *et al.*, 2019).

2.2.4 Protein and fats

Usually, fruits are low in protein and fat; however, jackfruit seeds contain a considerable amount of protein and a very low amount of fat, which enables its use in the preparation of protein-enriched, low-fat products. The fruit contains about 1.9% protein and 0.1% fat on a FW basis (Wealth of India, 1985). The protein and fat contents in tender, mature and ripe pulp were 2.6%, 6.6% and 1.9%, and 0.3%, 0.4% and 0.1%, respectively, indicating that the levels of these two nutrients decrease during ripening (Faruque and Omar, 1985). The protein content in immature fruit, ripe fruit and seeds is 2.0–2.6, 1.2–1.9 and 6.6–7.04 g 100 g^{-1} edible portion, respectively, while the fat content in young fruit, ripe fruit and seeds was 0.1–0.6, 0.1–0.4 and 0.40–0.43 g 100 g^{-1} edible part, respectively (Azad, 2000). Total protein content in the flakes of five jackfruit varieties from South India varied from 1.4% (Velayan) to 2.3% (Nettadivarikka) and that in seed from 5.3% (Chemparathy) to 6.8% (Nettadivarikka) and the seeds also had more protein than the flakes (Chrips *et al.*, 2008). The amount of protein in three types of jackfruit pulp, collected from two different locations of Bangladesh, ranged from 0.575 (Valuka khaja) to 0.97% (Modhupur ghila), while the protein in seeds varied between 6.07% (Modhupur khaja raw) and 7.32% (Valuka dorosha roasted) (Goswami and Chacrabati, 2016). In a comparative study of five different fruits, it was observed that jackfruit contains a significantly higher amount of protein (1.72 g 100 g^{-1}) and a slightly higher amount of fat (0.64 g 100 g^{-1}) compared with mango, banana, pineapple and fig (Table 2.2) (Srivastava and Singh, 2020).

In seeds from on-season and off-season germplasms Ah-15 and Ah-24, the protein content was recorded as 10.55% and 9.86%, and 12.40% and 11.55%, respectively. Similarly, the fat content in seeds of these two germplasms was found to be 0.43% and 0.93%, and 0.88%

Table 2.2. Comparison of the nutritional composition of various subtropical fruits (100 g^{-1} of edible part). Data from USDA (2016) and Srivastava and Singh (2020).

Nutritional composition	Jackfruit	Mango	Banana	Pineapple	Fig
Energy (kcal)	95	60	89	50	74
Water (g)	73.50	81.81	74.91	86.00	79.06
Carbohydrates (g)	23.25	15	22.84	13	19.18
Total sugars (g)	19.08	13.70	12.23	9.85	16.26
Dietary fibre (g)	1.5	1.6	2.6	1.4	2.9
Protein (g)	1.72	0.82	1.09	0.54	0.75
Fat (g)	0.64	0.38	0.33	0.12	0.30
Thiamine (mg)	0.105	0.028	0.031	0.079	0.060
Riboflavin (mg)	0.055	0.038	0.073	0.032	0.050
Niacin (mg)	0.92	0.669	0.665	0.50	0.40
Pantothenic acid (mg)	0.235	0.197	0.334	0.213	0.300
Pyridoxine (mg)	0.329	0.119	0.400	0.112	0.113
Folate (µg)	24	43	20	18	6
Ascorbic acid (mg)	13.8	36.4	8.7	47.8	2.0
Calcium (mg)	24	11	–	13	35
Magnesium (mg)	29	10	27	12	17
Potassium (mg)	448	168	358	109	242
Sodium (mg)	2	1	1	1	1
Manganese (mg)	0.043	0.063	0.270	0.927	0.128
Phosphorus (mg)	21	14	22	8	14
Iron (mg)	0.23	0.16	0.26	0.29	0.37
Zinc (mg)	0.13	0.09	0.15	0.12	0.15

and 0.35%, respectively (Hasan, 2002). In a jackfruit-based breakfast meal containing flesh (80%) and seeds (20%) in Sri Lanka, 4.7% protein (FW basis) was reported in seeds (Hettiaratchi *et al.*, 2011). Similarly in Bangladesh, a jackfruit seed meal was reported to have 5.69% protein and 0.86% fat (Awal and Gheyasuddin, 1991). Jackfruit seed contains more protein (6.6%) but less fat (0.4%) compared with edible pulp (0.6% protein and 0.6% fat) (Singh *et al.*, 2015). Goswami and Chacrabati (2016) noted that Valuka ghila raw and Valuka dorosha raw seeds contained the maximum amount of fat (0.80%), while Modhupur ghila roasted seeds had the minimum fat content (0.39%). Sultana *et al.* (2017) in Bangladesh recorded that the crude protein and crude fat contents of jackfruit seed flour were 16.01% and 0.98%, respectively. In 20 different genotypes, collected from a new alluvial zone of West Bengal, the soluble protein content in dry seed powder was found to be at maximum levels (144.75 mg g^{-1}) in G-3 and at minimum levels (33.86 mg g^{-1}) in G-5, and a wide variation was observed (Pebbuli and Bauri, 2018). Overall, jackfruit has 10.06% protein and 1.49% fat in the pulp, 10.09% protein and 4.29% fat in the seeds, and 1.19% protein and 0.72% fat in the leaves, as reported from Nigeria (Amadi *et al.*, 2018).

The occurrence of a lectin, jacalin, in the seeds of jackfruit was first reported by Chatterjee *et al.* (1979). Jacalin is the major protein isolated from jackfruit seeds and is a tetrameric two-chain lectin (molecular mass 65 kDa) binding a heavy chain of 133 amino acid residues and a light β-chain of 20–21 amino acid residues. It is extremely specific for the *O*-glycoside of the disaccharide Thomsen–Friedenreich antigen (Galβ1→3 GalNAc), even in its sialylated form (Kabir, 1998; Prakash *et al.*, 2009; Swami *et al.*, 2012). The suitability of jacalin in studying various *O*-linked glycoproteins, especially human immunoglobulin (IgA1), is because of this specificity (Suresh Kumar *et al.*, 1982). Chatterjee *et al.* (1988) later isolated and characterized another lectin, artocarpin (molecular weight 70 kDa), from *Artocarpus lakoocha* seeds, purified it using rivanol (6,9-diamino-2-ethoxyacridine lactate) and showed it to be homogeneous by polyacrylamide disc gel electrophoresis and immunoelectrophoresis. Artocarpin (also known as KM+) is a metalloprotein with two sugar exothermic combining sites per lectin molecule of four subunits, and is inhibited by α-glycosides of D-Gal/D-GalNAc with the α-glycoside of β-D-Gal-(1→3)-D-GalNAc being the most potent (Chatterjee *et al.*, 1988). It is specific to mannose and D-glucose but not D-galactose, and agglutinated human and several normal and enzyme-treated animal erythrocytes, and bound to rat lymphocytes and mouse ascites cells. It is made up of a single polypeptide chain of 159 amino acids sharing 52% identity with the jacalin sequences (Rosa *et al.*, 1999).

In terms of amino acids, the highest levels observed in jackfruit were of phenylalanine followed by aspartic acid and glutamine. A slightly higher amino acid content was noticed in immature fruit compared with ripe fruit (Azad, 2000). Singh *et al.* (2015) reported that arginine, cysteine, histidine, leucine, isoleucine, lysine, methionine, phenylalanine, threonine, tryptophan and valine were present as essential amino acids in tender fruit, while the seeds did not have arginine and histidine, and the amounts of amino acids in tender fruits were comparatively higher than in the seeds. Some essential amino acids such as lysine (10.30%), isoleucine (8.61%), leucine (6.73%), methionine (4.82%), threonine (3.90%), histidine (1.92%) and valine (1.73%) along with some non-essential amino acids such as tyrosine (6.67%), glycine (4.94%), serine (4.46%), aspartic acid (4.40%), glutamic acid (4.34%) and arginine (2.44%) were isolated from jackfruit seed flour protein where the maximum protein content (89.94%) was extracted at pH 10 (Miah *et al.*, 2017). Total fatty acid (free and bound) contents in whole fruit ranged from 22.2 to 306.0 mg 100 g^{-1} in different parts of the fruit where capric, myristic, lauric, palmitic, oleic, stearic, linoleic and arachidic acids were identified as the major fatty acids (Chowdhury *et al.*, 1997). Fernandes *et al.* (2017) determined eight essential and 12 non-essential amino acids along with 19 fatty acids in the jackfruit seed kernel and seed coating membrane. All the amino acids were found to be present in the seed kernel and seed coating membrane, but dodecanoic acid and tridecanoic acid among the fatty acids were not detected in the seed kernel and *cis*-10-pentadecenoic acid was not found in the seed coating membrane (Tables 2.3 and 2.4).

Table 2.3. The content of various amino acids in the jackfruit seed kernel and the seed coating membrane. From Fernandes *et al.* (2017).

Amino acids	Seed kernel (mg kg^{-1} DM)	Seed coating membrane (mg kg^{-1} DM)
Essential amino acids		
Threonine	387.00 ± 1.40	48.20 ± 0.11
Valine	290.54 ± 3.62	37.43 ± 0.44
Isoleucine	157.01 ± 0.79	11.82 ± 0.29
Leucine	396.43 ± 7.61	71.39 ± 0.60
Tryptophan	94.69 ± 0.91	8.30 ± 0.14
Phenylalanine	210.92 ± 2.56	29.51 ± 0.10
Lysine	242.59 ± 0.63	29.04 ± 0.24
Histidine	104.90 ± 0.04	32.08 ± 0.25
Total	1884.08 ± 17.56	267.76 ± 2.18
Non-essential amino acids		
Aspartic acid	247.64 ± 20.35	25.83 ± 0.02
Glutamic acid	703.31 ± 30.18	20.39 ± 6.03
Asparagine	759.79 ± 14.07	158.01 ± 0.16
Glutamine	2670.13 ± 4.00	338.52 ± 0.73
Serine	309.86 ± 2.83	26.60 ± 0.13
Glycine	154.17 ± 4.81	18.46 ± 0.11
Alanine	241.32 ± 2.23	32.08 ± 0.07
Proline	2585.93 ± 10.58	194.38 ± 0.52
Arginine	1250.50 ± 3.64	63.36 ± 1.26
Cysteine	395.73 ± 0.64	147.36 ± 0.28
Ornithine	27.74 ± 0.67	9.18 ± 0.02
Tyrosine	504.10 ± 2.07	42.70 ± 0.14
Total	9850.21 ± 96.07	1076.87 ± 9.47

Values are mean ± standard deviation. DM, dry matter.

2.2.5 Total soluble solids and acidity

Various researchers have proposed a number of theories regarding the increase in total soluble solids (TSS) during ripening, which include: (i) decomposition of the cell wall resulting in the release of water-soluble compounds (Rees *et al.*, 1981); (ii) conversion of starch to sugars (Sharaf and El-Saadany, 1987); and (iii) the enzymatic activity of polygalacturonase in degrading water-insoluble pectic substances causing an increase in water-soluble galacturonic acid (Burns, 1991). Selvaraj and Pal (1989) reported titratable acidity of 0.21–0.13% during ripening. In 24 selections collected from the Western Ghats of India, TSS and acidity contents varied from 19.87°Bx (SMG-3) to 35.00°Bx (SMG-5) and from 0.19% (SMG-22) to 0.483% (SMG-6), respectively, while the TSS:acid ratio ranged from 54.2 (SMG-25) to 160 (SMG-22) (Jagadeesh *et al.*, 2007). Usually, the top portion of fruit had the highest TSS and acidity, followed by the middle and lower portions (Saxena *et al.*, 2011). In three types of Bangladeshi jackfruit, the TSS content ranged from 19.3% (Valuka ghila) to 27.0% (Valuka dorosha), while the acidity ranged from 0.46% (Modhupur khaja) to 0.91% (Valuka ghila) (Goswami and Chacrabati, 2016). Similarly, among 40 local genotypes from Tamil Nadu, India, significant variation in TSS and acidity was reported, with the maximum (32.40°Bx) and minimum (13.72°Bx) TSS

Table 2.4. Fatty acid content of the jackfruit seed kernel and seed coating membrane. From Fernandes *et al*. (2017).

Fatty acid	Seed kernel (mg kg^{-1} DM)	Seed coating membrane (mg kg^{-1} DM)
Dodecanoic acid	–	12.69 ± 0.27
Tridecanoic acid	–	1.70 ± 0.06
Tetradecanoic acid	20.17 ± 0.22	53.38 ± 0.48
cis-10-Pentadecenoic acid	3.43 ± 0.10	–
Pentadecanoic acid	24.55 ± 0.26	31.93 ± 0.94
cis-9-Hexadecenoic acid	21.57 ± 0.41	40.41 ± 0.85
Hexadecanoic acid	836.70 ± 11.94	864.79 ± 15.42
cis-10-Heptadecenoic acid	9.29 ± 0.08	11.17 ± 0.27
Heptadecanoic acid	23.42 ± 0.28	24.14 ± 0.33
cis-9,12-Octadecadienoic acid	801.19 ± 10.44	147.59 ± 1.02
cis-9-Octadecenoic acid	109.81 ± 2.48	189.42 ± 1.55
trans-9-Octadecenoic acid	17.62 ± 1.92	23.31 ± 0.51
Octadecanoic acid	181.62 ± 4.52	254.75 ± 1.80
cis-9,12,15-Octadecatrienoic acid	2.17 ± 0.33	3.91 ± 0.11
Eicosanoic acid	74.71 ± 3.79	78.52 ± 1.03
Heneicosanoic acid	24.13 ± 0.68	20.79 ± 0.60
Docosanoic acid	112.41 ± 4.96	91.12 ± 1.58
Tricosanoic acid	20.65 ± 1.06	25.60 ± 0.07
Tetracosanoic acid	64.62 ± 4.30	85.98 ± 0.63
Total	2348.06 ± 47.77	1961.20 ± 27.52

Values are mean ± standard deviation. DM, dry matter.

recorded in the AH-13 and AH-5 genotypes, respectively, and the maximum (0.37%) and minimum (0.07%) acidity observed in the AH-40 and AH-35 genotypes, respectively (Aseef *et al.*, 2017). The TSS and acidity were also found to change with fruit maturity, where 16.93°Bx for TSS was recorded in 85-day-old mature pulp from jackfruit var. Evirac, which increased to 33.88°Bx after 88 days but decreased to 27.35°Bx at 94 days of maturity (Lauzon *et al.*, 2017). The authors observed that acidity in the pulp increased from 0.0596% at 85 days of maturity to 0.2334% after 94 days of maturity and suggested that, upon maturity, sugars in the pulp were converted to acid, thereby increasing the acidity. Among the 20 genotypes collected from West Bengal, India, maximum values for TSS (26.9°Bx), acidity (0.55%) and TSS:acid ratio (181) were found in G-17, G-18 and G-5, respectively, whereas these values were at minimum levels (TSS 16°Bx, acidity 0.13% and TSS:acid ratio 45) in G-2, G-5 and G-18, respectively (Pebbuli and Bauri, 2018). Balamaze *et al.* (2019) reported that the TSS and acidity varied significantly with varieties and flake colour, with the white-flaked variety having a lower TSS, acidity and TSS:acid ratio than the orange- and yellow-flaked varieties.

2.2.6 Vitamins

Vitamins are a group of organic compounds that are essential for normal physiological functioning, although they are not synthesized endogenously by the body and therefore have to be sequestered in small quantities from the diet. Usually, there are two types of vitamins available in fruits: (i) fat soluble (vitamins A, D, E and K); and (ii) water soluble (vitamins from the B complex group and vitamin C). Generally, vitamins are essential to boost the body's immune system. Jackfruit is a moderate to good source of many vitamins including vitamins C and A and is one of the rare fruits that is rich in vitamin B complex containing significant amounts of pyridoxine (vitamin B_6), niacin (vitamin B_3), riboflavin (vitamin B_2) and folic acid (vitamin B_9). Vitamin B_3 is essential for energy metabolism, nerve function and synthesis of some hormones, while vitamin B_6 is required for reducing homocysteine levels in blood, thereby lowering the risk of heart disease, and vitamin B_9 is helpful for pregnant and lactating women. Berry and Kalra (1988) reported that ascorbic acid and carotene content varied markedly between different cultivars and was influenced by the maturity stage of the fruit and the agroclimatic conditions. According to Morton (1987), 100 g of edible, ripe, fresh pulp possesses 540 IU vitamin A, 0.03 mg thiamine, 4 mg niacin and 8–10 mg ascorbic acid. Young fruit, ripe fruit and seed contain 30, 175–540 and 10–17 IU vitamin A, 0.05–0.15, 0.03–0.09 and 0.25 mg thiamine, 0.05–0.2, 0.05–0.4 and 0.11–0.3 mg riboflavin, and 12–14, 7–10 and 11 mg ascorbic acid, respectively, per 100 g edible portion (Table 2.1) (Gunasena *et al.*, 1996). Jackfruit contains 61 μg β-carotene (provitamin A), 157 μg lutein and zeaxanthin (carotenoids), 0.105 mg thiamine (vitamin B_1), 0.055 mg riboflavin, 0.92 mg niacin, 0.235 mg pantothenic acid (vitamin B_5), 0.329 mg pyridoxine, 24 μg folate, 13.8 mg ascorbic acid and 0.34 mg tocopherol (vitamin E) in 100 g of fruit. Compared with mango, pineapple, banana and fig, jackfruit possesses higher levels of vitamin B_1, vitamin B_3 and vitamin B_6 (Table 2.2) (USDA, 2016; Srivastava and Singh, 2020).

The change in vitamin content (vitamins B_1, B_2 and C) at different maturity periods of fruit was studied by Tiwari and Vidyarthi (2015). They noted that the amount of all three vitamins varied significantly during the maturity period where vitamin contents increased up to 75 days and decreased thereafter. In Bangladesh, maximum ascorbic acid content (8.18 mg 100 g^{-1}) was found in Valuka dorosha pulp, while khaja types contained lower ascorbic acid content (5.20 and 4.57 mg 100 g^{-1} in Valuka khaja and Modhupur khaja, respectively) compared with ghila-type fruits (7.26 and 7.13 mg 100 g^{-1} in Valuka ghila and Modhupur ghila, respectively) (Goswami and Chacrabati, 2016). They also observed a significant variation in carotene content where Valuka khaja, Valuka ghila, Valuka dorosha, Modhupur khaja and Modhupur ghila types possessed 334.06, 470.91, 380.45, 346.03 and 520.46 μg carotene 100 g^{-1}, respectively. Here also, ghila types contained higher levels of carotene than khaja types. A large variation in ascorbic acid and carotene content among 40 genotypes in Tamil Nadu, India, were found where ascorbic acid content varied from 12.30% (AH-20) to 2.90% (AH-32) with a mean value of 7.22%, and carotene from 0.897 (AH-35) to 0.128 (AH-38) mg 100 g^{-1} with a mean value of 0.365 mg 100 g^{-1} (Aseef *et al.*, 2017). Among the 20 genotypes from West Bengal, ascorbic acid content differed from 2 (G-8) to 8 (G-1) mg 100 g^{-1} with a mean value of 4.8 mg 100 g^{-1} (Pebbuli and Bauri, 2018). Ascorbic acid content in fruit pulp, seed and leaves was reported to be 2.10, 0.11 and 0.99 mg 100 g^{-1}, respectively, in Nigeria (Amadi *et al.*, 2018). Ascorbic acid content in yellow-, white- and orange-coloured flakes of Ugandan jackfruit varied from 1.6 (yellow) to 4.3 (orange) mg 100 g^{-1}, while the carotenoid content ranged from 0.3 (white) to 2.4 (orange) mg 100 g^{-1} (Balamaze *et al.*, 2019). The orange flakes had higher levels of carotenoids compared with the yellow and white flakes.

Jagadeesh *et al.* (2007) also reported great diversity in the carotenoid content of 24 jackfruit selections from Karnataka, India, where the maximum carotenoid content (0.879 mg 100 g^{-1}) was recorded in SMG-2 and the minimum (0.363 mg 100 g^{-1}) in SMG-13, with a mean value of 0.592 mg 100 g^{-1}. The intensity of the flakes' colour was found to be directly correlated with the carotenoid content in each selection where deep saffron-coloured flakes contained higher carotenoid levels than cream-coloured flakes. Similarly, there was wide variation in β-carotene content among 20 genotypes of West Bengal, India, where the maximum amount of β-carotene was found in G-15 (105.74 mg g^{-1} fresh bulb) and the minimum in G-13 (12.84 mg g^{-1} fresh bulb) with a mean value of 64.69 mg g^{-1} fresh bulb (Pebbuli and Bauri, 2018). Eighteen carotenoid compounds were identified in jackfruit pulp using high-performance liquid chromatography photodiode array and mass spectrometry (HPLC-PDA-MS/MS) (de Faria *et al.*, 2009). Here, lutein and β-carotene were the dominant carotenoids, and total carotenoid content was determined as 107.98 μg 100 g^{-1} FW. The fruit kernel also contained six carotenoids: β-carotene, α-carotene, β-zeacarotene, α-zeacarotene, β-carotene-5,6-α-epoxide (a dicarboxylic carotenoid) and crocetin, which theoretically corresponded to 141.6 retinol equivalents 100 g^{-1} (Chandrika *et al.*, 2004). Lycopene was also detected in fruit pulp, despite its soft colour (Setiawan *et al.*, 2001).

2.2.7 Phenolic components

Phenolic components or polyphenols are widely distributed secondary plant metabolites found in various plant parts. Natural polyphenols can be of two types: (i) simple molecules such as phenolic acids, phenylpropanoids and flavonoids; and (ii) complex or highly polymerized molecules such as lignins, tannins and melanins. Flavonoids and phenolic acids are the most common and widely distributed subgroups of phenolic compounds (Soobrattee *et al.*, 2005). Fruits and vegetables are natural sources of phenolic components, which are bioactive phytochemicals useful for human health. Although they are not considered vital nutrients for human health, today's researchers are interested in them because of their antioxidant properties through scavenging of free radicals, quenching of lipid peroxidation, preventing DNA oxidative damage and inhibition of cell communication, thereby reducing the risk of many degenerative diseases including cancer and heart problems (Duthie *et al.*, 2003). The distribution of phenolic compounds in fruits varies according to cultivar/variety, maturity period, soil composition, and storage and environmental conditions.

Jackfruit is a rich source of phenolic components, mainly flavonoids and phenolic acids. The total phenolic content in jackfruit was reported to be 0.36 mg gallic acid equivalents (GAE) 100 g^{-1} on a DW basis (Wongsa and Zamaluddien, 2005). The amounts of total phenolics and total flavonoids were found to be 45 and 23 mg 100 g^{-1}, respectively, in modified packaged jackfruit bulb during the initial storage period under modified atmospheric conditions, which further decreased with the increases in the storage period (Saxena *et al.*, 2009). Addition of chemical additives to stabilize the bulbs and the synergistic effect of modified atmospheric storage conditions could reduce the loss of phenolics and flavonoids during storage. High levels of phenolic components were detected in immature fruits, giving them an astringent characteristic, which diminished with fruit growth and maturity (Saxena *et al.*, 2011). The amount of total phenolics in fruit pulp after extraction with water, methanol, ethanol and acetone were recorded to be 0.25, 0.21, 0.46 and 0.18 mg GAE g^{-1}, respectively, and the amount of total flavonoids with the same solvents were 1.20, 0.24, 0.23 and 0.19 mg rutin equivalents (RE) g^{-1}, respectively (Jagtap *et al.*, 2010). Jackfruit contains lignans, isoflavones and saponins, which have widespread health benefits such as anti-cancer, anti-hypertensive, anti-ageing and anti-ulcer properties (Swami *et al.*, 2012). Significant increases in the total phenolic content from 41.91 (unripe) to 109.85 μg g^{-1} (ripe) FW in fruits during ripening has been reported in a study from Jamaica (Benkeblia, 2019). Zhang *et al.* (2017) observed that fruit peel extract had 4.65, 4.12 and 4.95 times higher total phenolic content than pulp, flake and seed extracts, respectively. The amounts of total phenolics in peel, pulp, flake and seed extracts were 48.04, 10.34, 11.57 and 9.71 mg GAE g^{-1} dry matter (DM), respectively. Similarly, the amounts of total flavonoids were 2.79, 2.27, 2.29 and 1.62 mg quercetin equivalent (QE) g^{-1} DM in peel, pulp, flake and seed, respectively. Identification of 53 bioactive compounds was possible using HPLC quadrupole time-of-flight MS/MS where prenylflavonoids, hydroxycinnamic acids and glycosides were the major compounds. The maximum phenol content among the 20 genotypes from West Bengal, India, was 34.52 mg GAE g^{-1} DW in G-20 and the minimum was 8.50 mg GAE g^{-1} DW in G-5 (Pebbuli and Bauri, 2018). Similarly, fruit pulp and seeds, collected from selected regions in Kenya and Uganda, contained 12.10–14.55 and 17.37–18.69 mg total phenolics g^{-1} and 0.18–0.29 and 0.50–0.89 mg total flavonoids g^{-1}, respectively (Ojwang *et al.*, 2018). Estimation of polyphenols using Folin–Ciocalteu reagent found 316, 355 and 382 mg tannic acid equivalents (TAE) polyphenols g^{-1} in spine, skin and rind of jackfruit peel, respectively

(Meera *et al.*, 2018). The ethyl acetate fraction of fruit peel contains high levels of phenolics (49.67 g GAE 100 g^{-1}) and flavonoids (70.20 g QE 100 g^{-1}), as reported recently from Indonesia (Yamin *et al.*, 2021).

Extraction with different solvents resulted in different amounts of total phenolics from seeds, for example 125 μg ml^{-1} from an ethyl acetate extract (Burci *et al.*, 2015), 1.45 and 2.12 μg GAE mg^{-1} from acetone and dichloromethane:methanol (1:1) extracts, respectively (Gupta *et al.*, 2011), 0.4 μg GAE g^{-1} DW from a methanol extract (Bhat *et al.*, 2017), 4.16, 2.30, 2.77 and 1.18 mg GAE g^{-1} at 100 mg g^{-1} seed extract from ethanol, acetone, ethyl acetate and water, respectively (Shanmugapriya *et al.*, 2011), and 437 mg GAE g^{-1} DW from a methanol extract (Shirajum, 2014). The total flavonoid content from various seed extracts was reported to be 290.6 μg RE mg^{-1} extract in acetone and 457.1 μg RE mg^{-1} extract in dichloromethane:methanol (1:1) (Gupta *et al.*, 2011), 10 mg QE g^{-1} DW in methanol (Bhat *et al.*, 2017), 4.05 mg g^{-1} in ethanol, 2.21 mg g^{-1} in acetone, 2.67 mg g^{-1} in ethyl acetate and 0.86 mg g^{-1} in water of QE at 100 mg g^{-1} seed extract (Shanmugapriya *et al.*, 2011). Loizzo *et al.* (2010) reported that total aqueous extract of leaves possessed the highest phenolic content (523.2 mg g^{-1}), followed by an ethyl acetate (361.2 mg g^{-1}) and the aqueous fraction (294.5 mg g^{-1}). Ajiboye *et al.* (2016) reported that a stem bark extract had a high level of total phenols (12.31 mg TAE g^{-1}), total flavonoids (10.49 mg QE g^{-1}) and total proanthocyanidins (8.34 mg g^{-1} catechin equivalents) at 5 mg ml^{-1} ethanol, and with an increase in ethanol concentration, the levels of these compounds also increased. The authors identified six phenolic compounds in the extract using gas chromatography/mass spectrometry (GC-MS). The highest (59.00–67.37 mg g^{-1}) and lowest (28.34–38.14 mg g^{-1}) phenolic levels were recorded in roots and bark, respectively, with mean phenolic contents of 35.18, 65.56 and 23.59 mg g^{-1} in leaves, roots and barks, respectively, with an overall mean from all three plant parts of 41.44 mg g^{-1} (Ojwang *et al.*, 2017). Similarly, the mean flavonoid contents in leaves, roots and bark were estimated as 5.74, 8.73 and 2.24 mg g^{-1}, respectively, where the range of flavonoid content in roots was 7.31–10.74 mg g^{-1} and that in bark was 1.49–3.09 mg g^{-1}. The tannin and flavonoid contents in fruit pulp, seed and leaves were estimated as 0.03, 0.06 and 0.07, and 3.91, 2.03, 3.91 mg 100 g^{-1}, respectively (Amadi *et al.*, 2018).

The tetracyclic triterpenoids, 9,19-cyclolanost-3-one-24,25-diol and 9,1-cyclolanost-3-one-24,25-diol, along with cycloartenone and cycloartenol were isolated by Barik *et al.* (1994). Resveratrol (*trans*-3,5,4-trihydroxystilbene) is one of the polyphenols naturally present in jackfruit (Lagouge *et al.*, 2006) and is well known for its health-promoting antioxidant, cardioprotective and anti-inflammatory activities (Shen *et al.*, 2009). Caffeic acid was found as the main free phenolic acid in jackfruit (Lee *et al.*, 2013). HPLC analysis of various phenolic acids in different parts of jackfruit revealed that raw fruit skin contained the maximum amounts of gallic acid (22.73 μg g^{-1} FW) and tannic acid (6.70 μg g^{-1} FW), and ripe fruit skin had the maximum amount of ferulic acid (13.41 μg g^{-1} FW), while caffeic acid (2.84 μg g^{-1} FW) was only detected in raw fruit seed (Singh *et al.*, 2015). Yuan *et al.* (2017) isolated flavonoids such as artoheteroids A–D, morin, artocarpin A, albanin A, euchrenone A, norartocarpanone and steppogenin from jackfruit. A high concentration of ellagic acid was reported in jackfruit by Swami and Kalse (2019). Liu *et al.* (2020) isolated seven prenylated chromones and five prenylated flavonoids, including two new prenylated chromones, artoheterophines A and B. Fu *et al.* (2020) purified artocarheterones A–E from the ripe fruits of *A. heterophyllus*. HPLC diode-array detection was used to detect two phenolic components, gallic acid and protocatechuic acid, in jackfruit wine (Jagtap *et al.*, 2011).

The heartwood of *A. heterophyllus* is reported to contain brosimone, cudraflavone B, prenylapigenin, norartocarpin, kuwanon C, artocarpin, cudraflavone C, apigenin, albanin A, morin, 2′,4′-dihydroxyflavone, 3-methylbut-2-en-1-ol, 2-methylpent-2-ene, artocarpesin, norartocarpetin and oxyresveratrol (Venkataraman, 2001; Arung *et al.*, 2007). Zheng *et al.* (2008) isolated furanoflavone,7-(2,4-dihydroxyphenyl)-4-hydroxy-2-(2--hydroxypropan-2-yl)-2,3-dihydrofuro(3,2-g)chromen-5-one(artocarpfuranol), together with 14 known compounds, dihydromorin, steppogenin, norartocarpetin, artocarpanone, artocarpesin, artocarpin, cycloartocarpin, cycloartocarpesin, artocarpetin, brosimone I, cudraflavone B, carpachromene, isoartocarpesin and cyanomaclurin from the wood of *A. heterophyllus*. Later, Zheng *et al.* (2009) isolated phenolic compounds, including one isoprenylated 2-arylbenzofuran derivative, artoheterophyllin A and three isoprenylated flavonoids, artoheterophyllin B, C and D, from an ethanol extract of the twigs of the jackfruit tree. A new prenylated flavonoid, 3-prenyl luteolin, was isolated from *A. heterophyllus* wood extract (Arung *et al.*, 2010a). Zheng *et al.* (2014) isolated new phenolic compounds, artoheterophyllins E–J, 4-geranyl-2′,3,4′,5-tetrahydroxy-*cis*-stilbene and 5-methoxymorican M, along with two new natural compounds, 2,3-dihydro-5,7-dihydro xy-2-(2-hydroxy-4-methoxyphenyl)-4H-benz opyran-4-one and 6-[(1S,2S)-1,2-dihydroxy-3-methylbutyl]-2-(2,4-dihydroxyphenyl)-5--hydroxy-7-methoxy-3-(3-methyl-2-buten-1-yl)-4H-1-benzopyran-4-one from an ethanol extract of jackfruit wood. Similarly dihydromorin and norartocarpetin were isolated from the heartwood by Septama *et al.* (2020).

Two novel 2,4,6-trioxygenated flavanones, heteroflavanones A and B, were isolated from the root bark of jackfruit. Their structures were elucidated as 5-hydroxy-7,2,4,6-tetramethoxyflavanone and 8-(γ,γ-dimethylallyl)-5-hydroxy-7,2,4,6-tetramethoxyflavanone (Lu and Lin, 1993; Lin *et al.*, 1995). The bark from the stem also contains betulinic acid and two new flavone pigments, cycloheterophyllin and triterpenic compounds such as cycloartenyl acetate, cycloartenone, heterophylol and tannin (Lin and Lu, 1993; Barik *et al.*, 1994; Prakash *et al.*, 2009). The leaves and stem are also reported to contain sapogenins, cycloartenone, cycloartenol, β-sitosterol and tannins, which show oestrogenic activity (Prakash *et al.*, 2009). In another research, a 2-arylbenzofuran derivative, artocarstilbene B, and a benzaldehyde derivative, (*E*)-3,5-dihydroxy-4-(3-methylbut-1-enyl) benzaldehyde, were obtained from a leaf extract (Wang *et al.*, 2017). The roots are reported to contain β-sitosterol, ursolic acid, betulinic acid, cycloartenone, 2′,4′,6′-trioxygenated flavanones, and heteroflavanones A and B (Dayal and Seshadri, 1974; Prakash *et al.*, 2009). The fruit latex also has some phenolic compounds such as α-artostenone (Balakrishna and Seshadri, 1947), and *cis*- and *trans*-1,4-polyisoprene (Bhadra *et al.*, 2019), as well as proteins and mineral materials.

2.2.8 Organic acids

Organic acids are not only responsible for fruit sourness/acidity but also affect the sensory characteristics such as the taste, flavour and colour of fruits and their processed products (Kader, 2008). The metabolism of phenolic compounds is affected by organic acids as they alter pH and act as precursors of phenolics and flavour components (Galdón *et al.*, 2008). Therefore, they can be used as an authenticity index in fruit products in the processing industry (Kelebek, 2010).

Not much work has been done on the estimation of organic acids in jackfruit. Citric and malic acids as the major non-volatile organic acids were identified by Selvaraj and Pal (1989), where the total organic acid concentration increased up to 75% ripening and then declined in fully ripe fruit. The reduction in organic acid content during ripening of fruits might be due to the use of organic acids as respiratory substrates and as a carbon source for the production of new compounds (Kays, 1991). Ong *et al.* (2006) reported that the dominant organic acids in jackfruit were malic and citric acids, although oxalic and succinic acids were also detected. Malic acid content was maximum in unripe fruit, while citric acid content was maximum in ripe fruit. The total amount of organic acids decreased during the ripening period. Fernandes *et al.* (2017) identified nine organic acids, namely oxalic, aconitic, citric, pyruvic, malic, quinic, shikimic, acetic and fumaric acid in jackfruit seed kernel and seed coating membrane. Pyruvic, shikimic and acetic acids were not detected in the seed kernel, although it contained much higher levels of total organic acids than the seed coating membrane (Table 2.5). Susana and Gunawan (2018) estimated the levels of seven organic acids: oxalic acid (0.081 mg 100 g^{-1}), succinic acid (6.758 mg 100 g^{-1}), D-malic acid (3.932 mg 100 g^{-1}), L-malic acid (16.602 mg 100 g^{-1}), L-ascorbic acid (8.394 mg 100 g^{-1}), citric acid (1.653 mg 100 g^{-1}) and D-galacturonic acid (0.094 mg 100 g^{-1}) in

Table 2.5. Organic acid composition of the jackfruit seed kernel and seed coating membrane. From Fernandes *et al.* (2017).

Organic acid	Seed kernel (mg kg^{-1} DM)	Seed coating membrane (mg kg^{-1} DM)
Oxalic acid	649.45 ± 26.38	122.38 ± 12.36
Aconitic acid	650.51 ± 6.93	96.52 ± 6.53
Citric acid	8,086.95 ± 807.60	1,745.72 ± 120.31
Pyruvic acid	–	59.32 ± 6.25
Malic acid	3,539.64 ± 335.53	877.97 ± 137.97
Quinic acid	460.84 ± 10.50	230.38 ± 44.69
Shikimic acid	–	12.95 ± 0.30
Acetic acid	–	84.28 ± 5.82
Fumaric acid	535.88 ± 1.37	26.33 ± 1.47
Total	13,923.26 ± 1188.30	3,255.85 ± 335.70

Values are mean ± standard deviation. DM, dry matter.

minimally processed jackfruit. They also noted that the malic acid content decreased and the citric acid content increased during storage. They concluded that organic acids can protect jackfruit from microorganism contamination, thereby helping to extend its shelf life. The amounts of phytic acid and oxalic acid in pulp, seed and leaves of jackfruit were estimated as 6.14, 8.11, 7.51 mg 100 g^{-1}, and 3.69, 5.53, 0.42 mg 100 g^{-1}, respectively, showing that the seeds are richer in organic acids than the pulp and leaves (Amadi *et al.*, 2018). Phytic acid interferes with the absorption of minerals such as calcium, magnesium, iron and zinc by forming insoluble complexes, thereby reducing their bioavailability.

2.2.9 Volatile components

Volatile components are secondary plant metabolites that are volatile in nature, such as essential oils, and impart flavour in fruits. Flavour is one of the important sensory attributes responsible for overall acceptability. Flavour/aroma can be used to differentiate various kinds of fruits, or fruits of the same type but different varieties (Kays, 1991). Fruit flavour is particularly sensitive to compositional alterations. Aroma volatiles generated by a harvested fruit can be changed or affected by some pre- and postharvest conditions such as genetics, preharvest environment, harvest maturity, and postharvest handling and storage. In the case of fruits and vegetables, the volatiles responsible for their characteristic flavour are usually esters, aldehydes, alcohols and terpenes or their derivatives. Swords *et al.* (1978) isolated 20 aroma volatiles from an extract of distillate from jackfruit using gas GC-MS, among which 16 were esters and four aliphatic alcohols; however, no terpenoid was identified, which is rare in the plant kingdom and a unique feature of jackfruit. The major isolated chemicals were butyl isovalerate, butyl butyrate, isoamyl isovalerate, butyl acetate, isobutyl isovalerate and propyl isovalerate, which are all esters. The authors suggested that the mild floral aroma of jackfruit was due to the presence of esters and not aliphatic alcohols. Using two different GC-MS columns, identification of 21 volatile compounds in jackfruit was possible (Rasmussen, 1983). In Malaysian jackfruit, 45 volatile components were detected using capillary GC and GC-MS; among these, esters represented the highest proportion (31.9%) and are responsible for the flavour of the fruit (Wong *et al.*, 1992). Olias *et al.* (1995) reported that ester concentrations remain low at the initial stage of physiological maturity when the green fruit odour prevails and slowly increase with the advancement of maturity. They are formed possibly because of the reactions of alcohols and organic acids catalysed by the enzyme alcohol acyl transferase, which is highly active during fruit maturation. The major aroma compounds are produced via β-oxidation of fatty acids, and secondary metabolites are generated mainly as fatty acids oxidized through the lipoxygenase pathway (Beaulieu and Baldwin, 2002) A total of 39 volatile components were identified in two Amazonian jackfruit varieties (hard and soft) using GC-MS (Maia *et al.*, 2004). The main flavouring compounds in the hard variety were isopentyl isovalerate (28.4%), butyl isovalerate (25.6%), palmitic acid (8.3%) and ethyl isovalerate (6.2%), whereas in the soft variety they were isopentyl isovalerate (18.3%), butyl acetate (16.5%), ethyl isovalerate (14.4%), butyl isovalerate (12.9%), 2-methylbutyl acetate (12.0%) and isopentan-1-ol (5.0%). Fraga (2005) identified 59 aroma components in headspace jackfruit extracts, among which the majority were esters (68%), followed by alcohols (16%), acids (8%) and aldehydes (8%). The ethylic and butyric esters with C4 to C6 linear chains were the dominating compounds in fruit pulp. The major odour-active compounds found in fresh pulp were ethyl 3-methylbutanoate, ethyl butanoate, butyl butanoate and ethyl 2-methylbutanoate. A total of 23 aroma volatiles were tentatively identified by GC and GC-MS in jackfruit cultivar J3, and isovaleric acid 3-phenylpropyl ester, isopropyl myristate, 3-phenylpropanol, octadecane and methyl dodecanoate were earmarked as the main ester compounds (Ong *et al.*, 2006). The authors noticed that the number of volatile components increased with fruit maturity. The major volatile compounds were ethyl isovalerate, 3-methylbutyl acetate, 1-butanol, propyl isovalerate, isobutyl isovalerate, 2-methyl butanol and butyl isovalerate among the 37 compounds isolated from five jackfruit cultivars in Malaysia (Ong *et al.*, 2008). The authors observed that these chemicals were consistently present in all five cultivars, which indicated that these volatiles were responsible for the sweet and fruity flavour of the jackfruit.

Headspace solid-phase microextraction (SPME) and GC-MS was also effectively used as a simple, rapid and reliable tool to analyse the volatile fraction in jackfruit peel and pulp (Peng *et al.*, 2013). 2,3-Butanediol, butyrolactone, 1-methoxy-2-propyl acetate, butanoic acid, 3-methyl-butyl ester, *n*-amyl isovalerate, *n*-hexadecanoic acid were found in both peel and pulp, while compounds such as geranyl geraniol, lanosterol, pentanoic acid, 3-phenylpropyl ester, phthalic acid, butyl 3-methylbut-2-en-1-yl ester, hexadecanoic acid and methyl ester were only present in the peel. However, in the pulp, compounds such as dibutyl phthalate phenol and 2,4-bis(1,1-dimethylethyl)-phenols were confirmed, which were not found in the peel. Srinivasan and Kumaravel (2016) profiled 27 volatile compounds from jackfruit powder using GC-MS and the major compounds were 9,12-octadecadienoic acid (Z,Z)- (7.22%), lup-20(29)-en-3-one (11.06%), 9,19-cyclo-9β-lanostane-3β, 25-diol (9.59%), 5-hydroxymethyl furfural (9.91%), lanosterol (8.58%) and *n*-hexadecanoic acid (11.22%). They concluded that the majority of the compounds belonged to fatty acids, steroids and terpenoids. Fruit peel also contains many volatile aromatic compounds and 24 volatile components were identified from ethanol extract of jackfruit peel. Among these, the predominant components were hexadecanoic acid (21.90%), squalene (16.34%), ethyl linoleate (6.40%) and methyl-8,11,14-heptadecatrienoate (5.40%). Similarly, calophyllolide (38.64%), thialisopyine (10.28%), AD-glycopyranose, 4-*O*-α-D-galactopyranosyl (6.82%), (±)inophyllum D (6.40%) and 9-octadecannamide (5.04%) were the dominating volatile compounds in a methanol extract of peel (Sundarraj and Ranganathan, 2017b). GC-MS analysis showed the presence of 11 aroma phytochemicals, namely 1,3-oxazolidine carboxylic acid (8.754%), 3,3′-dithiobis-methyl-N-acetyl alanine (7.391%), bis(2-ethylhexyl)phthalate (4.025%), octadecyl ester of hexadecanoic acid (1.846%), 2-methoxy-4-vinyl phenol (1.784%), *N*,*N*-dimethyl-6-benzyloxyhexylamine (1.74%), 2-methoxy-1,3-benzodioxole (1.25%), 1-methyl-3,5-diisopropoxy benzene (1.022%), (Z),-2-hydroxy-1,9-octadecenoic acid (0.793%), *o*-amino benzohydroxamic acid (0.556%) and 7-propylquinoline (0.49%) in a methanol extract of fruit (Kurian *et al.*, 2018). Susana and Gunawan (2018) identified 16 volatile chemicals in minimally processed jackfruit grown in Indonesia using GC-MS, which included isopentyl isovalerate, butyl isovalerate, ethyl isovalerate and palmitic acid as the main chemicals. A slight decrease in volatile components during storage of minimally processed jackfruit was also observed. Fifty-three aroma chemicals were identified in jackfruit pulp and among these the chemical structures of 44 odorants were assigned. The major odorants with high odour activity values were ethyl 3-methylbutanoate (74,000), ethyl butanoate (1800), 3-methylbutanal (1500) and 2-methylpropanal (1400) (Grimm and Steinhaus, 2019).

SPME-GC-MS analysis revealed the presence of 66 odour-active compounds in four Malaysian jackfruit accessions (Malaysia No. 1, XYS17, XYS18 and Liulianmi) and these four accessions shared nine aromatic substances, with 3-methyl-1-butanol, dodecanal, 2-methylbutyl isovalerate, isoamyl acetate and propyl isovalerate being predominant (He *et al.*, 2019). Malaysia No. 1 was characterized by a high content of *n*-amyl isovalerate and esters (benzyl acetate), butyl acetate, isoamyl acetate, isobutyl isovalerate and 2-methylbutyl isovalerate; XYS17 had a high content of isobutyl isovalerate, 2-methylbutyl isovalerate, *n*-amyl isovalerate, octyl acetate and decanal; XYS18 had a high content of ethyl isovalerate, butyl isovalerate, ethyl butyrate, methyl isovalerate and octanal; and Liulianmi had 3-methyl-1-butanol, 2-methyl-1-butanol, 2-methylbutyl isovalerate, ethyl isovalerate and 3-methylbutyl 3-methylbutanoate. GC-MS analysis was used for the detection of five and seven bioactive compounds in petroleum ether and chloroform extracts, respectively, from wild jackfruit (*Artocarpus hirsutus* Lam.). Squalene (62.36%), medowlactone (3.64%), dodecanal (2.95%), 4H-pyran-3-carboxylic, 6-amino-5-cyano-2-methyl-4-phenylbenzyl ester (0.84%) and methyl-9-*cis*,11-*trans*-octadecadienolate (0.35%) were detected in the petroleum ether extract, while sitosterol (23.21%), 4H-pyran-4-one, 2,3-dihydro-3,5-dihydroxy-6-methyl (3.54%), heptadecenal (1.50%), heneicosene (1.46%), cetene (1.36%), heptacos-1-ene (1.10%) and hexadecanol (0.89%) were detected in the chloroform extract (Neha and Abdussalam, 2019). In China, Gao *et al.* (2021) identified 59 and 57 types of aroma volatiles in the juices of six jackfruit cultivars before and after pasteurization using headspace SPME and GC-MS. Pasteurization resulted in the disappearance of ethyl acetate, *n*-propyl acetate, 3-methyl-1-butanol, 3-hexen-1-ol acetate (Z)-, *n*-heptyl acetate, decyl acetate, 2,4-dodecadienal, 1,9-dodecadiene and (*E*)-9-octadecenoic acid ethyl ester. However, propyl butyrate, ethyl hexanoate, hexyl isovalerate, methylbutyl hexanoate, bicyclohept-2-ene, 2,6-dimethyl-6-(4-methyl-3-penten-1-yl)-benzoic acid, 2-methylbutyl ester, ethyl-9-hexadecenoate and ethyl-9,12,15-octadecatrienoate appeared in juice samples even after pasteurization. Eighty-six volatile components were characterized using a headspace SPME and GC-MS combined technique in five jackfruit cultivars, grown in Nayarit, Mexico, which included 51 esters, 13 aldehydes, seven alcohols, seven ketones, seven terpenes and one carboxylic acid (Barros-Castillo *et al.*, 2021). The predominant volatile components were ethyl 3-methylbutanoate, propyl 3-methylbutanoate, butyl 3-methylbutanoate, pentyl 3-methylbutanoate, 3-phenylpropanal, β-cyclocitral, 1-butanol, butanal, 3-methyl-1-butanol, 3-methylbutyl 3-methylbutanoate, ethylbutanoate and 2-methylpropyl 3-methylbutanoate, although their concentrations varied according to the cultivars. The ethanol extract of jackfruit stem bark contained 17 volatile components, among which eight were fatty acids and their ethyl esters, five were amino compounds, two were sterols, one each was an aldehyde and a ketone (Ajiboye *et al.*, 2016). Jackfruit leaves possessed significantly higher levels of alkaloids (12.43 mg 100 g^{-1}) compared with pulp (7.88 mg 100 g^{-1}) and seed (8.85 mg 100 g^{-1}) (Amadi *et al.*, 2018).

2.3 Uses

2.3.1 Traditional uses

Traditionally, every part of the jackfruit tree is used for various purposes in many countries, as summarized below. It is the national fruit of Bangladesh and the state fruit of the Indian states of Kerala and Tamil Nadu.

Tree

Jackfruit trees can be utilized to avoid soil erosion and protect other crops as wind breaks, along with use as an intercrop with mango, coconut and citrus fruits (Elevitch and Manner, 2006). Generally, jackfruit trees are used in landscaping of gardens and to provide shade for crops such as coffee, pepper, betel nut and cardamom (Gupta *et al.*, 2020).

Wood

The golden-yellow timber with a good grain is useful for preparing valuable items as well as furniture (Fig. 2.1), boats (Fig. 2.2) and construction of houses, as it is considered to be termite proof and superior to teak. In Indonesia, the hardwood is carved out to make the barrels of drums used in the *gamelan* (an indigenous orchestra). The softwood is used to make the bodies of *kutiyapi*, a type of boat lute in the Philippines (Elevitch and Manner, 2006). In Sri Lanka, the wood is exported to Europe where it is used in the manufacturing of doors, windows, roofs, fish sauce barrels, masts and oars (Morton, 1987). Jackfruit wood is considered best for making Buddhist statues in the temples of Vietnam. The chips of heartwood are used for the extraction of an orange-red dye, which is used to colour the robes of Buddhist monks (Seddon and Lennox, 1980). The ornate wooden plank known as *avani palaka* made of jack wood is used as the priest's seat during Hindu ceremonies in Kerala, India. The wood has a sedative property and its pith is believed to help in abortion. Splinters of the wood are inserted in bamboos used to collect coconut toddy, as the jackfruit wood imparts a yellow colour to palm sugar (Morton, 1987). The wood is also used in preparing musical instruments. In India, it is used to make the body of the string instrument *veena* and *mridangam*, *thimila* and *kanjira* drums (Arora and Parle, 2016). A crude methanol extract of stem has been shown to have antibacterial activity. The wood is nervine and antidiabetic, and is useful in convulsions.

Leaves and bark

The burnt ash of jackfruit leaves together with maize or coconut shells is used to heal ulcers either alone or in combination with coconut oil (Morton, 1987). Leaf extract is used in diabetic patients as a control measure. Heated leaves are reported to cure wounds, abscesses and ear problems, and to relieve pain. The leaves and latex extracts can cure asthma, prevent ring-worm infestation and heal cracked feet (Tyagi *et al.*, 2015). The leaves are useful in fever, boils and skin diseases. A tea made with dried and powdered leaves is consumed to relieve asthma. The leaves are believed to have vermifuge and anti-syphilitic activities and to induce lactation in women and domesticated mammals (Khan *et al.*, 2003). They are also used to feed goats (Fig. 2.3). An infusion of mature leaves and bark is used to treat gallstones (Tyagi *et al.*, 2015). The leaves can work as excellent adsorbents and they are used to remove methylene blue, an industrial dye, from aqueous solutions (Uddin *et al.*, 2009).

The leaves and bark have beneficial effects in the treatment of anaemia, dermatitis and coughs. The bark is made into a poultice and burnt bark ash can cure abscesses and ear

Fig. 2.1. Jackfruit wood is used in making furniture. Photograph courtesy of P. Mukherjee.

Fig. 2.2. Jackfruit wood being used in preparation of a boat. Photograph courtesy of Dr D. Sarangi.

Fig. 2.3. Jackfruit leaves are used to feed goats. Photograph courtesy of Dr B. Das.

problems (Morton, 1987). The bark of the mature tree is supposed to be helpful in treating dysentery and releasing the placenta after calving in cows (Tyagi *et al.*, 2015). Due to its fibrous nature, the bark is occasionally used in cloth manufacturing (Arora and Parle, 2016).

Fruit

The young fruits are acrid, astringent and carminative. The ripe fruits are sweet and soothing, have laxative and aphrodisiac properties, and are also used as a brain tonic. The fruit is believed to be good for indigestion and has anti-ulcer capabilities. In China, the pulp and seed are considered to be a cooling and nutritious tonic and are used as a remedy for overcoming the influence of alcohol in the body (Morton, 1987). The fruits may help to reduce and cure tension as well as nervousness. The fruit has high levels of potassium and is said to be helpful in controlling blood pressure (Gupta *et al.*, 2020). It is a moderate source of ascorbic acid and carotenoids, which are proven antioxidants. It is an excellent source of proteins, carbohydrates and B vitamins, which are necessary for the development of our body. The fruit is rich in dietary fibre and is used to treat leprosy, ulcers, constipation, heart disease and rheumatism in Unani and Ayurvedic medicines (Baliga *et al.*, 2011). The fruit possesses many phytonutrients with health benefits covering various claims from anti-cancer to anti-hypertensive properties. In terms of anti-ageing properties, the fruit may help to reduce damage to cells to make the skin look supple and younger. As it has very few calories and a low fat content, the fruits can be used as part of a low-calorie, low-fat diet for specific individuals (Swami *et al.*, 2012). It is also thought to help to fight wrinkles and its consumption is said to help achieve a glowing complexion and flawless skin (Arora and Parle, 2016). Fruit waste and leaves are used as fodder for livestock animals such as cattle, goats and pigs (Elevitch and Manner, 2006). The leaves and fruits are used for preparing jewellery (Fig. 2.4) and decorations (Fig. 2.5) in social ceremonies.

Seeds

The roasted seeds are considered to be an aphrodisiac, while the seed starch can be useful in relieving biliousness (Morton, 1987). A fresh seed extract can help in curing diarrhoea and dysentery (Baliga *et al.*, 2011). A extract from seeds and/or bark helps in digestion (Tyagi *et al.*, 2015). The seeds promote a healthy blood circulation and are also beneficial for the growth of hair (Arora and Parle, 2016).

Latex

The latex when mixed with vinegar is useful in healing abscesses, snake bites and glandular swellings. The dried latex yields artostenone, which can be converted to artosterone, a compound with significant androgenic action (Morton, 1987). The latex is also helpful in treating dysopia, ophthalmic disorders and pharyngitis, and is also used as an antibacterial agent (Swami *et al.*, 2012). The heated latex can be used as a household cement to mend cracks in chinaware and earthen pots, and to caulk boats and holes in buckets (Arora and Parle, 2016). The latex can be utilized as an irrigation solution (for washing a body cavity or wound with a stream of fluid) in dentistry, a denture cleaning solution and in the future as a potential dental filling material for cost-effectiveness (Rao *et al.*, 2014).

Root

The root of the jackfruit tree is useful as remedy for skin diseases and asthma (Morton, 1987). A root extract can help in healing fever and diarrhoea (Swami *et al.*, 2012). A crude methanol extract of root has shown promising antibacterial activity.

2.3.2 Pharmacological uses

Jackfruit has long been used in the traditional Indian Ayurvedic and Unani system because of the presence of many bioactive phytochemicals in the various plant parts. Jackfruit possesses many antioxidant chemicals such as phenolic compounds, flavonoids, ascorbic acid and carotenoids, and the fruit extract can be used in curing different types of cancer such as lung cancer, skin cancer, breast cancer, gastric cancer and prostate cancer. The antioxidant properties of jackfruit flesh/pulp extracts are mainly correlated with the presence of phenolics and flavonoids (Jagtap *et al.*, 2010). Jackfruit is one of the rare fruit sources of B vitamins. Vitamin B complex (B_1, B_2, B_3, B_6, B_9 and B_{12}) are essential for neuronal function, and severe deficiencies have been linked to increased risk of neurological disorders, psychiatric disease, dementia, anaemia, muscle weakness, beriberi and pellagra (Kennedy, 2016). Vitamin B_6 (pyridoxine) has also heart-friendly properties and helps in reducing homocysteine levels in the blood, thereby lowering the risk of heart disease (Fernando *et al.*, 1991). Ascorbic acid (vitamin C) is another well-known antioxidant, and jackfruit seed and pulp possess significant ascorbic acid

Fig. 2.4. Leaf along with other flowers are used to prepare ornaments. Photograph courtesy of Dr I. Bandyopadhyaya.

and gallic acid (phenolic compound) equivalent antioxidant effects and together contribute to about 70% of the total antioxidant activity (Soong and Barlow, 2004). Ascorbic acid also helps in protecting from skin damage that happens as a consequence of the normal ageing process and prolonged exposure to ultraviolet radiation from the sun by aiding in the production of collagen, resulting in improved firmness and strength to the skin (Pullar *et al.*, 2017). It is also essential to maintain oral health for the prevention of scurvy.

Jackfruit is a rich source of many minerals. The fruit has an abundance of magnesium, which is necessary for the absorption of calcium and helps in strengthening the bones and preventing osteoporosis. Iron in the fruit helps to prevent anaemia and aids in proper blood circulation, while copper plays an important role in thyroid gland metabolism (Singh *et al.*, 1991). Potassium in jackfruit aids in lowering blood pressure by maintaining the sodium–potassium balance in the blood system, thereby reducing the risk of heart attack, stroke and bone loss, and improving muscle and nerve function (Swami *et al.*, 2012). Sleeping disorders can be cured by consuming jackfruit because of its high levels of iron and magnesium, which assist in improving overall sleeping quality. Magnesium also helps in preventing anaemia, which is one of the main reasons for insomnia (Gupta *et al.*, 2020). Jackfruit is an

Fig. 2.5. Fruit is used for decoration in social ceremony. Photograph courtesy of Dr P. Rajendran.

excellent source of dietary fibre, which acts as a good laxative and reduces the exposure time between cancer-causing chemicals and the colon mucous membrane, thereby helping to protect against colon cancer. A high-fibre content also maintains smooth bowel movements and prevents constipation (Morton, 1987). Jackfruit is both gluten free and casein free and hence has anti-inflammatory benefits to the skin. The fruit has a low glycaemic load (glucose level) and is high in dietary fibre and therefore is widely accepted by diabetologists as an ideal food for diabetic patients. Research conducted by Sydney University in Australia on the glycaemic index of jackfruit had revealed that one cup of jackfruit had a glycaemic value of 17, which is nearly half the other commonly consumed foods like roti/chapatti (glycaemic value 27) and rice (glycaemic value 29) (Devi *et al.*, 2021). Similarly, an aqueous decoction of leaves was found to be effective in improving utilization of the external glucose load in a glucose tolerance test in rats (Fernando *et al.*, 1990), and when tried on both healthy human volunteers and non-insulin-dependent diabetic patients, the result was also satisfactory (Fernando *et al.*, 1991). The decoction, when administered at 10 ml kg^{-1} to overnight-fasted individuals 1 h before the oral glucose load, could lower the blood glucose levels at all evaluated time points up to 2.5 h.

Bioactive compounds in jackfruit have been proven to be clinically useful in, for example, fever, boils, wounds, skin diseases, convulsions, constipation, ophthalmic disorders and snake bites and as a diuretic (Rama Rao and Venkataraman, 1973; Prakash *et al.*, 2009). Lignans, saponins and isoflavones are found in jackfruit that contribute to its anti-cancer and anti-ageing activities as well as to prevention of the formation of cancer cells in the body and in the fight against stomach ulcers (Baliga *et al.*, 2011). Fruit pulp contains phytochemicals with chemoprotective properties that help to minimize the mutagenicity of aflatoxin B_1 and the proliferation of cancerous cells, whereas the flesh possesses compounds effective against lymphoma cancer (Ruiz-Montañez *et al.*, 2014). The anti-inflammatory action of jackfruit is due to the presence of bioflavonoids such as artocarpesin, norartocarpetin and oxyresveratrol, which inhibit the release of inflammatory mediators from mast cells, neutrophils and macrophages by suppressing the production of nitric oxide and prostaglandin E2 (Wei *et al.*, 2005; Fang *et al.*, 2008). The anti-inflammatory effects of these isolated compounds were evaluated by determining their inhibitory effects on the production of pro-inflammatory mediators in lipopolysaccharide-activated RAW 264.7 murine macrophage cells, and these three compounds exhibited potent anti-inflammatory activity (Jha *et al.*, 1997). Artocarpin-H, isolated from jackfruit stem latex, and the protease fraction have shown anti-inflammatory activity using carrageenan-induced rat paw oedema and the cotton pellet-induced granuloma method, which was dependent on concentration (Chanda *et al.*, 2009). Triterpenes and sterols, extracted from jackfruit seed with supercritical carbon dioxide, have shown potential in inhibiting the inflammatory action induced in RAW 264.7 cells at a concentration of 30 μg ml^{-1} extract (Tramontin *et al.*, 2021).

Broad-spectrum antibacterial activity was reported from methanol extracts of stem, root, bark, heartwood, leaves, fruits and seeds of jackfruit. Among all the fractions, the butanol fractions of root, bark and fruits were found to be the most effective (Khan *et al.*, 2003). Jackfruit shoot extracts have exhibited nematicidal activity against various nematodes including *Rotylenchulus reniformis*, *Tylenchorhynchus brassicae*, *Tylenchus filiformis* and *Meloidogyne incognita* (Sharma and Trivedi, 1995). Methanol extract of leaves and two phytochemicals, artocarpin and artocarpesin isolated from leaves, showed *in vitro* inhibitory effects on the six primary cariogenic bacteria and Actinomyces at doses ranged between 3.13 and 12.5 μg ml^{-1} (Sato *et al.*, 1996). An aqueous extract along with ethyl acetate and aqueous fractions of leaves were found to be effective against many food-borne pathogens such as *Escherichia coli*, *Listeria monocytogenes*, *Salmonella typhimurium*, *Salmonella enterica*, *Bacillus cereus*, *Enterococcus faecalis* and *Staphylococcus aureus* at varying concentrations (Loizzo *et al.*, 2010). The antibacterial activity of five extracts of jackfruit seed powder against multidrug-resistant methicillin-resistant *Staphylococcus aureus* revealed that ethanol and methanol extracts at 30 mg ml^{-1} showed better antimicrobial activity than methicillin at 5 μg ml^{-1} (Karthy *et al.*, 2009). Jacalin, a chitin-binding lectin isolated from jackfruit seed, has antifungal activities against *Fusarium moniliforme* and *Saccharomyces cerevisiae*, as well as haem agglutination properties against human and rabbit erythrocytes (Trindade *et al.*, 2006).

Jackfruit wood extract is also useful in inhibiting melanin biosynthesis. The phytochemical artocarpanone in wood extract has shown significant tyrosinase-inhibitory activity by inhibiting mushroom tyrosinase activity and melanin synthesis in B16 melanoma cells (Arung *et al.*, 2006a). Artocarpin, cudraflavone C, kuwanon C, norartocarpin, 6-prenyl apigenin and albanin A, some isoprenoid-substituted bioflavonoids, inhibited melanin biosynthesis/production of melanin in B16 melanoma cells but did not inhibit tyrosinase activity (Arung *et al.*, 2006b). Some prenyl-substituted flavone-based polyphenols, isolated from jackfruit wood, were found to inhibit *in vivo* melanin biosynthesis in B16 melanoma cells with little or no cytotoxicity (Arung *et al.*, 2007). Isoprenoid-substituted bioflavonoids, isolated from the methanol extract of jackfruit wood, such as norartocarpin, cudraflavone B and C, artocarpin, brosimone I, kuwanon C and 6-prenyl apigenin showed better cytotoxic activity and anti-neoplastic efficacy on B16 melanoma cells than the clinically used drugs vinblastine, carmustine and 5-fluorouracil (Arung *et al.*, 2010a). The isoprenylated 2-arylbenzofuran derivatives artoheterophyllin A, B, C and D, isolated from twigs, along with 3-prenyl luteolin (half-maximal inhibitory concentration (IC_{50}) 76.3 μM) were also found to be effective in inhibiting tyrosinase activity (Zheng *et al.*, 2009; Arung *et al.*, 2010b). The prenylflavones cycloheterophyllin and artonin A and B, isolated from jackfruit, can act as powerful antioxidants against lipid peroxidation when biomembranes are exposed to free oxygen radicals, DPPH (2,2-diphenyl-1-picrylhydrazyl) radicals, peroxyl radicals and hydroxyl radicals (Ko *et al.*, 1998). Artocarpin exhibited concentration-dependent potent cytotoxic activity on cultured human T47D breast cancer cells *in vitro* (Arung *et al.*, 2010b). These studies suggest that the jackfruit tree possesses bioactive chemicals that can be used in treating hyperpigmentation and as a skin-whitening agent after thorough investigation on skin safety issues in an animal model.

Subchronic treatment with a high dose of jackfruit seed (500 mg kg^{-1}) decreased sexual activity in male rats in terms of libido, sexual arousal, sexual vigour and sexual performance within 2 h of administration; these activities returned to normal levels after 6 h. These sexual behavioural actions were due to the noticeable sedative activity of seeds, which did not reduce male fertility or affect vital functions (Ratnasooriya and Jayakody, 2002). The observed effects were temporary and reverted on withdrawal of the seeds from the diet. This study concluded that jackfruit seeds do not possess aphrodisiac activity in rats. Jackfruit seeds contain jacalin, a tetrameric two-chain lectin protein, which has found diverse applications such as the isolation of human plasma glycoproteins (IgA1, C1-inhibitor, haemopexin and 2-HS glycoprotein), investigation of IgA nephropathy, analysis of *O*-linked glycoproteins and detection of tumours due to its immunomodulatory activity (Kabir, 1998). The uniqueness of jacalin in being strongly mitogenic for human $CD4^+$ T-lymphocytes can be used for the evaluation of the immune status of human patients affected by human immunodeficiency virus type 1 (Pereira-da-Silva *et al.*, 2006). Jacalin extract from jackfruit seed could be used to arrest the cell cycle in human breast cancer cells (MCF7) with an effective IC_{50} of 125 μg ml^{-1} (Zuraidah *et al.*, 2017). Xiao *et al.* (2015) reported that JACALIN-LECTIN LIKE1 (AtJAC1) regulates flowering time in *Arabidopsis thaliana*, where loss of function of AtJAC1 leads to precocious flowering, whereas overexpression of AtJAC1 causes delayed flowering. They concluded that AtJAC1 coordinates with GRP7 (GLYCINE-RICH RNA-BINDING PROTEIN7) in shaping plant development through the regulation of RNA processing in *Arabidopsis*. The abundance of source material (jackfruit seeds) for the production of jacalin, its ease of purification, and its high yield and stability have made it an attractive and cost-effective lectin. The ethanol extract of dried leaves and the various fractions such as petroleum ether, methanol, butanol and butanone were found to be markedly effective in healing wounds in albino rats, where the methanol fraction had the best effect in reducing the scar area (Patil *et al.*, 2005). Kotowaroo *et al.* (2006) observed that the aqueous leaf extract significantly decreased α-amylase activity in rat plasma *in vitro* by behaving as a competitive inhibitor of α-amylase and acting as a starch blocker, with the highest inhibitory activity seen at 1 mg ml^{-1}.

2.3.3 Culinary uses

Jackfruit is eaten and used in cooking in both its immature and ripe forms. Immature jackfruit is used in curry preparations, as well as to make pickles. The perianth of the fruit is usually cut into small pieces with a sharp knife and the pulp is known for its meaty texture. Curries made from the immature fruit are usually spiced with mustard powder during preparation. Similarly, large chunks of the immature fruit can be pickled in brine with mustard seed seasoning to obtain delicious pickles with an excellent texture (Saxena *et al.*, 2011). The young fruit is rich in starch, and in many tropical countries, it forms part of the staple diet. The ripe fruit with bulbs of a golden-yellow colour is consumed after pitting to remove the seeds. The unfertilized floral parts are removed from the fleshy bulbs to obtain delicious fruit segments with a firm texture, sweetish flavour and a typically strong aroma. The aroma of the fruit is intense and can be detected from a long distance away, making it easy to identify the presence of jackfruit in the near vicinity. A number of value-added products are prepared from jackfruit bulbs including leather, crisps, osmo-dehydrated products and various minimally processed and hurdle technology-based products (different jackfruit products are described in Chapter 3, this volume). The pulp is also used for flavouring in ice cream and beverages. The seeds can be consumed after boiling or roasting, and a delicious curry can also be prepared from the cut seeds. Dehydrated and salted snacks based on jackfruit seeds are also popular, and cooked and dried seeds can be milled to obtain a flour-like consistency and used as a starch extender in bread dough (Saxena *et al.*, 2011). The tender young leaves are cooked and consumed as a vegetable product, and young male flower spikes can be grated or mashed for use as vegetables, or pickled with salt and vinegar.

Jackfruit is commonly used in cuisines all over the world. In Bangladesh and Nepal, unripe fruit is used to prepare delicious curries and the seed is often dried and preserved to be used later in curry. Jackfruit is used as a canned sugary syrup in Thailand and Vietnam. In Vietnam, jackfruit is also used to make a sweet dessert soup called jackfruit chè. In Indonesia, the ripe fruit is usually sold separately and consumed on its own, or sliced and mixed with shaved ice as a sweet dessert (Arora and Parle, 2016). In the Philippines, the unripe fruit is usually cooked in coconut milk and is eaten as a viand together with rice, while the ripe fruit is used as an ingredient in local desserts. Jackfruit is used to make wine and pickles in Sri Lanka. In Africa, the seeds of jackfruit are ground into a flour-type meal and used mainly for bread and other baking purposes. Jackfruit is used to make patties, casseroles and salads in Australia, whereas in Malaysia, it is often added to rice dishes and salads, mixed with shrimp, or used in drinks such as lassi. In Florida, jackfruit is used in ice creams, jams and jellies, and to make sweet curry and fruit salad (Arora and Parle, 2016).

2.3.4 Other uses

Jack tree latex has the same composition as that of jackfruit gum. As well as lipids, rubbers, resins, sugars and proteins, including proteolytic enzymes, both contain a high percentage of waxy matter, which has α-artostenone occurring in the form of enol wax esters in the plant, which are converted to artostenone upon saponification (Balakrishna and Seshadri, 1947). The possibility of using the latex as an alternative source for natural rubber has been investigated and it has been found from Fourier-transform infrared spectroscopy, ^{1}H nuclear magnetic resonance and GC-MS studies that the latex contains both *cis-* and *trans*-1,4 polyisoprene along with a few other chemical constituents as impurities. The latex cannot be used to replace natural rubber because of its significantly lower molecular weight and chemical impurities, but it can be used as a low-molecular-weight natural resin, especially to improve dispersion of carbon black in rubber compounds and to enhance wet skid resistance for tyres (Bhadra *et al.*, 2019).When used as a coupling agent with a composite (calcium carbonate-filled natural rubber:low-density polyethylene:waste polyethylene at a ratio of 30:60:10), the latex at 1.5 p.p.h. of polymer loading showed tremendous improvements in physico-mechanical and chemical properties in terms of tensile strength, elongation at break, hardness, percentage swelling, gel content and water-absorption capacity (Sampath *et al.*, 2020).

2.4 Conclusion

Jackfruit is a rich source of many nutraceuticals including vitamins (particularly vitamin B complex), minerals, dietary fibre and bioactive phytochemicals. Almost all parts of the plant including the wood and latex have medicinal properties and have traditionally been used for curing many diseases from ancient times. Despite these facts, jackfruit remains an underutilized fruit on a commercial scale, even in the countries where it is grown in abundance because of reasons such as the larger size of fruits, the high percentage of inedible portions leading to generation of more waste material, the difficulty in peeling and separation of edible flakes from the rind, a lack of knowledge on the proper postharvest practices and inadequate processing facilities. Many validated clinical trials have proved the pharmacological efficacy of many phytonutrients isolated from various parts of the tree. Therefore, it has tremendous potential to be utilized as 'all-purpose food' at a commercial level. Problems in peeling the seed (i.e. removal of the inedible seedcoat) restrict its use in the food industry, despite it containing high levels of protein, starch and some minerals. The seeds contain very low levels of fat and can easily be used in the food industry as part of a low-fat diet, especially for people with diabetes. Jackfruit is grown only in the home or in kitchen gardens or as a boundary crop, and no proper commercial jackfruit orchards are available throughout the world. The numerous health benefits of jackfruit include antioxidant, anti-cariogenic, anti-inflammatory, antimicrobial, wound healing, hypoglycaemic and melanin biosynthesis inhibitory properties. Very few studies have been conducted to date focusing on shelf-life extension and the value addition of jackfruit and management of the waste generated after fresh consumption or processing by converting it to value-added products or renewable energy sources, or utilization for the extraction of pectin. Standardized processed products of jackfruit can offer the consumers another way of reaping the various health benefits. As a giant tropical compound fruit with excellent flavour and texture, jackfruit has the potential to be exploited commercially in the global market. There is an urgent need to develop appropriate vegetative propagation techniques based on micropropagation methods and to bring information about soil requirements, agronomic practices, pest and disease management, and postharvest management to a wider audience. It should not be considered only as a 'poor man's food', as its nutritional and sensory qualities are competitive enough to earn it a place in the global market. Future thrust on nutrigenomics along with more proven clinical trials on animals and humans could potentially help in improving its quality for the satisfaction of different consumer sectors around the world and convince the people about its medicinal properties. Globally, jackfruit can be considered a functionally, nutritionally and medicinally important fruit.

References

Ahmmed, R., Inam, A.K.M.S., Alim, M.A., Sobhan, M.M. and Haque, M.A. (2017) Extraction, characterization and utilization of pectin from jackfruit (*Artocarpus heterophyllus L.*) waste. *Bangladesh Journal of Environmental Science* 32, 57–66.

Ajiboye, B.O., Ojo, O.A., Adeyonu, O., Imiere, O., Olayide, I. *et al.* (2016) Inhibitory effect on key enzymes relevant to acute type-2 diabetes and antioxidative activity of ethanolic extract of *Artocarpus heterophyllus* stem bark. *Journal of Acute Disease* 5(5), 423–429. DOI: 10.1016/j.joad.2016.08.011.

Amadi, J.A.C., Ihemeje, A. and Afam-Anene, O.C. (2018) Nutrient and phytochemical composition of jackfruit (*Artocarpus heterophyllus*) pulp, seeds and leaves. *International Journal of Innovative Food, Nutrition & Sustainable Agriculture* 6, 27–32.

Anderson, J.W., Smith, B.M. and Gustafson, N.J. (1994) Health benefits and practical aspects of high-fiber diets. *The American Journal of Clinical Nutrition* 59(5 Suppl), 1242S–1247S. DOI: 10.1093/ajcn/59.5.1242S.

Anon (2017) World leaders in jackfruit production. Available at: hwww.worldatlas.com/articles/world-leaders-in-jackfruit-production.html (accessed 1 July 2022).

Arollado, E.C., Ponsaran, K.M.G. and Loquias, M.M. (2018) Isolation and characterization of pectin from the ripe fruit peels of jackfruit (*Artocarpus heterophyllus Lam.*). *Acta Medica Philippina* 52(5), 453–457. DOI: 10.47895/amp.v52i5.322.

Arora, T. and Parle, A. (2016) Jackfruit: a health boon. *International Journal of Research in Ayurveda & Pharmacy* 7(3), 59–64. DOI: 10.7897/2277-4343.073113.

Arung, E.T., Shimizu, K. and Kondo, R. (2006a) Inhibitory effect of artocarpanone from *Artocarpus heterophyllus* on melanin biosynthesis. *Biological & Pharmaceutical Bulletin* 29(9), 1966–1969. DOI: 10.1248/bpb.29.1966.

Arung, E.T., Shimizu, K. and Kondo, R. (2006b) Inhibitory effect of isoprenoid-substituted flavonoids isolated from *Artocarpus heterophyllus* on melanin biosynthesis. *Planta Medica* 72(9), 847–850. DOI: 10.1055/s-2006-931606.

Arung, E.T., Shimizu, K. and Kondo, R. (2007) Structure-activity relationship of prenyl-substituted polyphenols from *Artocarpus heterophyllus* as inhibitors of melanin biosynthesis in cultured melanoma cells. *Chemistry & Biodiversity* 4(9), 2166–2171. DOI: 10.1002/cbdv.200790173.

Arung, E.T., Shimizu, K., Tanaka, H. and Kondo, R. (2010a) 3-Prenyl luteolin, a new prenylated flavone with melanin biosynthesis inhibitory activity from wood of *Artocarpus heterophyllus*. *Fitoterapia* 81(6), 640–643. DOI: 10.1016/j.fitote.2010.03.011.

Arung, E.T., Wicaksono, B.D., Handoko, Y.A., Kusuma, I.W., Shimizu, K. *et al.* (2010b) Cytotoxic effect of artocarpin on T47D cells. *Journal of Natural Medicines* 64(4), 423–429. DOI: 10.1007/s11418-010-0425-6.

Aseef, R.M., Manikandan, K., Kavino, M., Vijayakumar, R.M. and Meenakshi Ganesan, N. (2017) Biochemical evaluation of local genotypes of jackfruit (*Artocarpus heterophyllus Lam.*) in Pudukkotai district. *Journal of Pharmacognosy and Phytochemistry* 6, 2533–2536.

Awal, H.M.A. and Gheyasuddin, S. (1991) Biochemical parameters of jackfruit seed meal. *Bangladesh Journal of Agricultural Research* 16, 17–22.

Azad, K. (2000) Genetic diversity of jackfruit in Bangladesh and development of propagation methods. PhD thesis, University of Southampton, UK.

Balakrishna, K.J. and Seshadri, T.R. (1947) Chemical examination of jack tree latex and jack fruit gum. *Proceedings of the Indian Academy of Sciences - Section A26* 46–51.

Balamaze, J., Muyonga, J.H. and Byaruhanga, Y.B. (2019) Physico-chemical characteristics of selected jackfruit (*Artocarpus heterophyllus* Lam) varieties. *Journal of Food Research* 8(4), 11. DOI: 10.5539/jfr.v8n4p11.

Baliga, M.S., Shivashankara, A.R., Haniadka, R., Dsouza, J. and Bhat, H.P. (2011) Phytochemistry, nutritional and pharmacological properties of *Artocarpus heterophyllus* Lam (jackfruit): a review. *Food Research International* 44(7), 1800–1811. DOI: 10.1016/j.foodres.2011.02.035.

Barik, B.R., Bhaumik, T., Dey, A.K. and Kundu, A.B. (1994) Triterpenoids from *Artocarpus heterophyllus*. *Phytochemistry* 35(4), 1001–1004. DOI: 10.1016/S0031-9422(00)90655-1.

Barros-Castillo, J.C., Calderón-Santoyo, M., Cuevas-Glory, L.F., Pino, J.A. and Ragazzo-Sánchez, J.A. (2021) Volatile profiles of five jackfruit (*Artocarpus heterophyllus* Lam.) cultivars grown in the Mexican Pacific area. *Food Research International* 139, 109961. DOI: 10.1016/j.foodres.2020.109961.

Beaulieu, J.C. and Baldwin, E.A. (2002) Flavour and aroma of fresh-cut fruits and vegetables. In: Lamikanra, O. (ed.) *Fresh-Cut Fruits and Vegetables: Science, Technology and Market*. CRC Press, Boca Raton, Florida, pp. 391–415. DOI: 10.1201/9781420031874.

Begum, R., Aziz, M.G., Uddin, M.B. and Yusof, Y.A. (2014) Characterization of jackfruit (*Artocarpus heterophyllus*) waste pectin as influenced by various extraction conditions. *Agriculture and Agricultural Science Procedia* 2, 244–251. DOI: 10.1016/j.aaspro.2014.11.035.

Begum, R., Yusof, Y.A., Aziz, M.G. and Uddin, M.B. (2017a) Structural and functional properties of pectin extracted from jackfruit (*Artocarpus heterophyllus*) waste: effects of drying. *International Journal of Food Properties* 20(sup1), S190–S201. DOI: 10.1080/10942912.2017.1295054.

Begum, R., Yusof, Y.A., Aziz, M.G. and Uddin, M.B. (2017b) Extraction and characterization of pectin from jackfruit (*Artocarpus heterophyllus* Lam) waste. *IOSR Journal of Pharmacy and Biological Sciences* 12, 42–49.

Benkeblia, N. (2019) Variation of reducing and total sugars starch, total phenolic contents in unripe and ripe jackfruit (*Artocarpus heterophyllus*) and West Indian locust (*Hymenaea courbaril*) fruits. *International Journal of Plant Science and Horticulture* 49–55. DOI: 10.36811/ijpsh.2019.110005.

Berry, S.K. and Kalra, C.L. (1988) Chemistry and technology of jackfruit (*Artocarpus heterophyllus*) – a review. *Indian Food Packer* 42, 62–76.

Bhat, V., Mutha, A. and Dsouza, M.R. (2017) Pharmacognostic and physiochemical studies of *Artocarpus heterorphyllus* seeds. *International Journal of ChemTech Research* 10, 525–536.

Bhadra, S., Mohan, N., Parikh, G. and Nair, S. (2019) Possibility of *Artocarpus heterophyllus* latex as an alternative source for natural rubber. *Polymer Testing* 79, 106066. DOI: 10.1016/j.polymertesting.2019.106066.

Bhatia, B.S., Siddappa, G.S. and Lal, G. (1955) Composition and nutritive value of jackfruit. *Indian Journal of Agricultural Sciences* 25, 303–306.

Bobbio, F.O., El-Dash, A.A., Bobbio, P.A. and Rodrigues, L.R. (1978) Isolation and characterization of the physicochemical properties of the starch of jackfruit seeds (*Artocarpus heterorphyllus*). *Cereal Chemistry* 55, 505–511.

Brown, W.H. (1951) *Useful Plants of Philippines*, Vol. 1. Technical bulletin.10. Philippines Department of Agriculture and Natural Resources, Manila, pp. 463–467.

Burci, L.M., da Silva, C.B., de Oliveira, M., Dalarmi, D., Zanin, S.M.W. *et al.* (2015) Determination of antioxidant radical scavenging activity and total phenolic compounds of *Artocarpus heterophyllus* (jackfruit) seeds extracts. *Journal of Medicinal Plants Research* 9(40), 1013–1020. DOI: 10.5897/JMPR2015.5926.

Burns, J.K. (1991) The polygalacturonases and lyases. In: Walter, R.H. (ed.) *The Chemistry and Technology of Pectin, Food Science and Technology*. Academic Press, New York, New York, pp. 165–188.

Chanda, I., Chanda, S.R. and Dutta, S.K. (2009) Anti-inflammatory activity of a protease extracted from the fruit stem latex of the plant *Artocarpus heterophyllus* Lam. *Research Journal of Pharmacology and Pharmacodynamics* 1, 70–72.

Chandrika, U.G., Jansz, E.R. and Warnasuriya, N.D. (2004) Analysis of carotenoids in ripe jackfruit (*Artocarpus heterophyllus*) kernel and study of their bioconversion in rats. *Journal of the Science of Food and Agriculture* 85, 186–190. DOI: 10.1002/jsfa.1918.

Chatterjee, B.P., Vaith, P., Chatterjee, S., Karduck, D. and Uhlenbruck, G. (1979) Comparative studies of new marker lectins for alkali-labile and alkali-stable carbohydrate chains in glycoproteins. *The International Journal of Biochemistry* 10, 321–327. DOI: 10.1016/0020-711x(79)90097-1.

Chatterjee, B.P., Ahmed, H. and Chowdhury, S. (1988) Further characterization of *Artocarpus lakoocha* lectin (artocarpin) purified using rivanol. *Carbohydrate Research* 180, 97–110. DOI: 10.1016/0008-6215(88)80067-3.

Chowdhury, F.A., Azizur Raman, Md. and Jabbar Mian, A. (1997) Distribution of free sugars and fatty acids in jackfruit (*Artocarpus heterophyllus*). *Food Chemistry* 25, 25–28. DOI: 10.1016/S0308-8146(96)00294-4.

Chrips, N.R., Balasingh, R.G.S. and Kingston, C. (2008) Nutrient constituents of neglected varieties of *Artocarpus heterophyllus* Lam. from Kanyakumari district, South India. *Journal of Basic and Applied Biology* 2, 36–37.

Coronel, R.E. (1983) Jack. In: *Promising Fruits of the Philippines.* College of Agriculture, University of Philippines at Los Banôs, Laguna, Philippines, pp. 267.

Dayal, R. and Seshadri, T.R. (1974) Colourless compounds of the roots of *Artocarpus heterophyllus*. Isolation of new compound artoflavone. *Indian Journal of Chemistry* 12, 895–896.

de Faria, A.F., de Rosso, V.V. and Mercadante, A.Z. (2009) Carotenoid composition of jackfruit (*Artocarpus heterophyllus*), determined by HPLC-PDA-MS/MS. *Plant Foods for Human Nutrition (Dordrecht, Netherlands)* 64(2), 108–115. DOI: 10.1007/s11130-009-0111-6.

Devi, P.S.S., Kumar, N.S. and Sabu, K.K. (2021) Phytochemical profiling and antioxidant activities of different parts of *Artocarpus heterophyllus* Lam. (Moraceae): a review on current status of knowledge. *Future Journal of Pharmaceutical Sciences* 7(1), 30. DOI: 10.1186/s43094-021-00178-7.

Duthie, G.G., Gardner, P.T. and Kyle, J.A.M. (2003) Plant polyphenols: are they the new magic bullet? *The Proceedings of the Nutrition Society* 62(3), 599–603. DOI: 10.1079/PNS2003275.

Elevitch, C.R. and Manner, H.I. (2006) *Artocarpus heterophyllus* (jackfruit), ver.I.IV. In: Elevitch, C.R. (ed.) *Species Profiles for Pacific Island Agroforestry*. Permanent Agriculture Resources, Holualoa, Hawaii, p. 16.

Fang, S.C., Hsu, C.L. and Yen, G.C. (2008) Anti-inflammatory effects of phenolic compounds isolated from the fruits of *Artocarpus heterophyllus*. *Journal of Agricultural and Food Chemistry* 56(12), 4463–4468. DOI: 10.1021/jf800444g.

Faruque, A.J.M. and Omar. (1985) The nutri-value of mango, pineapple and jackfruit. *Krishi-Khata, Bangladesh* 45, 114–118.

Fernandes, F., Ferreres, F., Gil-Izquierdo, A., Oliveira, A.P., Valentão, P. *et al.* (2017) Accumulation of primary and secondary metabolites in edible jackfruit seed tissues and scavenging of reactive nitrogen species. *Food Chemistry* 233, 85–95. DOI: 10.1016/j.foodchem.2017.04.068.

Fernando, M.R., Thabrew, M.I. and Karunanayake, E.H. (1990) Hypoglycaemic activity of some medicinal plants in Sri-Lanka. *General Pharmacology* 21(5), 779–782. DOI: 10.1016/0306-3623(90)91033-n.

Fernando, M.R., Wickramasinghe, N., Thabrew, M.I., Ariyananda, P.L. and Karunanayake, E.H. (1991) Effect of *Artocarpus heterophyllus* and *Asteracanthus longifolia* on glucose tolerance in normal human subjects and in maturity-onset diabetic patients. *Journal of Ethnopharmacology* 31(3), 277–282. DOI: 10.1016/0378-8741(91)90012-3.

Fraga, S.R.G. (2005) Investigação dos voláteis e precursores de voláteis glicosilados da jaca (*Artocarpus heterophyllus* Lam.) e do Muruci (*Byrsonima crassifolia* Lam. Rich). PhD thesis, Institute of Chemistry, Federal University of Rio de Janeiro, Rio de Janeiro, Brazil.

Fu, Y.H., Guo, J.M., Xie, Y.T., Yu, X.M., Su, Q.T. *et al.* (2020) Prenylated chromones from the fruits of *Artocarpus heterophyllus* and their potential anti-HIV-1 activities. *Journal of Agricultural and Food Chemistry* 68(7), 2024–2030. DOI: 10.1021/acs.jafc.9b06417.

Galdón, B.R., Rodríguez, C.T., Rodríguez, E.R. and Romero, C.D.. (2008) Organic acid contents in onion cultivars (*Allium cepa* L.). *Journal of Agricultural and Food Chemistry* 56(15), 6512–6519. DOI: 10.1021/jf800282h.

Gao, X., Shi, R., Gu, F., Zhang, Y., Wu, G. *et al.* (2021) Effect of pasteurization on volatile aroma components of jackfruit (*Artocarpus heterophyllus* Lam) juice from different cultivars. *Food Science* 35, 63–68.

Goswami, C. and Chacrabati, R. (2016) Jackfruit (Artocarpus heterophyllus). In: Simmonds, M.S.J. and Preedy, V.R. (eds) *Nutritional Composition of Fruit Cultivars*. Elsevier, Amsterdam, pp. 317–335.

Grimm, J.E. and Steinhaus, M. (2019) Characterization of the major odor-active compounds in jackfruit pulp. *Journal of Agricultural and Food Chemistry* 67, 5838–5846.

Gunasena, H.P.M., Ariyadasa, K.P., Wikramasinghe, A., Herath, H.M.W., Wikramasinghe, P. *et al.* (1996) *Manual of Jack Cultivation in Sri Lanka*. Department of Forest Publication, Forest Information Service, Sri Lanka.

Gupta, D., Mann, S., Sood, A. and Gupta, R.K. (2011) Phytochemical, nutritional and antioxidant activity evaluation of seeds of jackfruit (*Artocarpus heterophyllus* Lam). *International Journal of Pharma and Bio Sciences* 2, 336–345.

Gupta, N., Sood, M., Hameed, F. and Singh, J. (2020) Uses and health benefits of jackfruit. *Rashtriya Krishi* 15, 103–104.

Hasan, M.K. (2002) Biochemical content of flesh and seed of two jackfruit (*Artocarpus heterophyllus* Lam) germplasm from two seasons. MSc thesis, Department of Biochemistry, Bangladesh Agricultural University, Mymensingh, Bangladesh.

He, S., Wu, G., Zhang, Y., Xu, F., Zhu, K. *et al.* (2019) Analysis of volatile aroma compounds in jackfruit (*Artocarpus heterophyllus*) of different genotypes. *Chinese Journal of Tropical Crops* 40, 1304–1311.

Hettiaratchi, U.P.K., Ekanayake, S. and Welihinda, J. (2011) Nutritional assessment of a jackfruit (*Artocarpus heterophyllus*) meal. *Ceylon Medical Journal* 56(2), 54–58. DOI: 10.4038/cmj.v56i2.3109.

Jagadeesh, S.L., Reddy, B.S., Swamy, G.S.K., Gorbal, K., Hegde, L. *et al.* (2007) Chemical composition of jackfruit (*Artocarpus heterophyllus* Lam.) selections of Western Ghats of India. *Food Chemistry* 102(1), 361–365. DOI: 10.1016/j.foodchem.2006.05.027.

Jagtap, U.B., Panaskar, S.N. and Bapat, V.A. (2010) Evaluation of antioxidant capacity and phenol content in jackfruit (*Artocarpus heterophyllus* Lam.) fruit pulp. *Plant Foods for Human Nutrition* 65, 99–104. DOI: 10.1007/s11130-010-0155-7.

Jagtap, U.B., Waghmare, S.R., Lokhande, V.H., Suprasanna, P. and Bapat, V.A. (2011) Preparation and evaluation of antioxidant capacity of jackfruit (*Artocarpus heterophyllus* Lam.) wine and its protective role against radiation induced DNA damage. *Industrial Crops and Products* 34(3), 1595–1601. DOI: 10.1016/j.indcrop.2011.05.025.

Jha, P.K., Choudhary, R.S. and Choudhary, S.K. (1997) Studies of medicinal plants of Palamau (Bihar) – 2nd Part. *Biojournal* 9, 21–38.

Kabir, S. (1998) Jacalin: a jackfruit (*Artocarpus heterophyllus*) seed-derived lectin of versatile applications in immunobiological research. *Journal of Immunological Methods* 212(2), 193–211. DOI: 10.1016/s0022-1759(98)00021-0.

Kader, A.A. (2008) Flavor quality of fruits and vegetables. *Journal of the Science of Food and Agriculture* 88(11), 1863–1868. DOI: 10.1002/jsfa.3293.

Karthy, E.S., Ranjitha, P. and Mohankumar, A. (2009) Antimicrobial potential of plant seed extracts against multidrug resistant methicillin resistant *Staphylococcus aureus* (MDR-MRSA). *International Journal of Biology* 1(1), 34–40. DOI: 10.5539/ijb.v1n1p34.

Kays, S.J. (1991) Development of plants and plant parts. In: *Postharvest Physiology of Perishable Plant Products*. AVI Publishing, Van Nostrand Reinhold, New York, New York, pp. 257–333.

Kelebek, H. (2010) Sugars, organic acids, phenolic compositions and antioxidant activity of grapefruit (*Citrus paradisi*) cultivars grown in Turkey. *Industrial Crops and Products* 32(3), 269–274. DOI: 10.1016/j.indcrop.2010.04.023.

Kennedy, D.O. (2016) B vitamins and the brain: mechanisms, dose and efficacy--a review. *Nutrients* 8(2), 68. DOI: 10.3390/nu8020068.

Khan, M.R., Omoloso, A.D. and Kihara, M. (2003) Antibacterial activity of *Artocarpus heterophyllus*. *Fitoterapia* 74(5), 501–505. DOI: 10.1016/s0367-326x(03)00120-5.

Kim, H.-J., Hyun, E.-K., Kim, Y.-S., Lee, Y.-J. and Oh, D.-K. (2006) Characterization of an Agrobacterium tumefaciens D-psicose 3-epimerase that converts D-fructose to D-psicose. *Applied and Environmental Microbiology* 72(2), 981–985. DOI: 10.1128/AEM.72.2.981-985.2006.

Ko, F.N., Cheng, Z.J., Lin, C.N. and Teng, C.M. (1998) Scavenger and antioxidant properties of prenylflavones isolated from *Artocarpus heterophyllus*. *Free Radical Biology & Medicine* 25(2), 160–168. DOI: 10.1016/s0891-5849(98)00031-8.

Koh, P.C., Leong, C.M. and Noranizan, M.A. (2014) Microwave-assisted extraction of pectin from jackfruit rinds using different power levels. *International Food Research Journal* 21, 2091–2097.

Kotowaroo, M.I., Mahomoodally, M.F., Gurib-Fakim, A. and Subratty, A.H. (2006) Screening of traditional antidiabetic medicinal plants of Mauritius for possible α-amylase inhibitory effects *in vitro*. *Phytotherapy Research* 20, 228–231.

Kumar, S., Singh, A.B., Abidi, A.B., Upadhyay, R.G. and Singh, A. (1988) Proximate composition of jackfruit seeds. *Journal of Food Science and Technology* 25, 308–309.

Kurian, S., Joseph, L. and Josekumar, V.S. (2018) Phytochemical evaluation, GC-MS analysis and antimicrobial activity of the leaves petiole of *Artocarpus heterophyllus* Lam. *Asian Journal of Pharmacy and Pharmacology* 4(3), 280–287. DOI: 10.31024/ajpp.2018.4.3.7.

Lagouge, M., Argmann, C., Zachary, G., Meziane, H., Lerin, C. *et al*. (2006) Resveratrol improves mitochondrial function and protects against metabolic disease by activating SIRT1 and PGC-1a. *Cell* 127, 1109–1122.

Lauzon, R.D., Galvez, L.A., Vinculado, J.F., Amestoso, F.J. and Diczbalis, Y. (2017) Effect of maturity, fruit and pulp location and pulp size and thickness on the quality of vacuum-fried jackfruit (*Artocarpus heterophyllus*) pulp from Eviarc sweet variety. In: *International Symposium on Tropical Fruits, Nadi, Fiji, 23-25 October 2017*, pp. 103–115.

Lee, P., Tan, R., Yu, B., Curran, P. and Liu, S. (2013) Sugars, organic acids, and phenolic acids of exotic seasonable tropical fruits. *Nutrition & Food Science* 43(3), 267–276. DOI: 10.1108/00346651311327927.

Li, Y., Liu, S., Zhu, Z. and Yang, Y. (2014) Determination of glucose, fructose and sucrose in jackfruit by HPLC. *Food Science* 35, 84–87.

Li, Y., Duan, X., Liu, S., Li, Y. and Zhang, X. (2017) Changes in soluble sugar accumulation and activities of sucrose-metabolizing enzymes during fruit ripening of jackfruit. *Journal of Agricultural Science* 9(8), 155. DOI: 10.5539/jas.v9n8p155.

Lin, C.N. and Lu, C.M. (1993) Heterophylol, a phenolic compound with novel skeleton from *Artocarpus heterophyllus*. *Tetrahedron Letters* 34, 8249–8250.

Lin, C.N., Lu, C.M. and Huang, P.L. (1995) Flavonoids from *Artocarpus heterophyllus*. *Phytochemistry* 39(6), 1447–1451. DOI: 10.1016/0031-9422(95)00135-T.

Liu, Y.P., Yu, X.M., Zhang, W., Wang, T., Jiang, B. *et al*. (2020) Prenylated chromones and flavonoids from *Artocarpus heterophyllus* with their potential antiproliferative and anti-inflammatory activities. *Bioorganic Chemistry* 101, 104030. DOI: 10.1016/j.bioorg.2020.104030.

Loizzo, M.R., Tundis, R., Chandrika, U.G., Abeysekera, A.M., Menichini, F. *et al*. (2010) Antioxidant and antibacterial activities on food-borne pathogens of *Artocarpus heterophyllus* Lam (Moraceae) leaves extracts. *Journal of Food Science* 75, M291–M295.

Lu, C.M. and Lin, C.N. (1993) Two 2′,4′,6′-trioxygenated flavanones from *Artocarpus heterophyllus*. *Natural Products Research Centre* 33, 909–911.

Madhav, A. and Pushpalatha, P.B. (2002) Characterization of pectin extracted from different fruit wastes. *Journal of Tropical Agriculture* 40, 53–55.

Maia, J.G.S., Andrade, E.H.A. and Zoghbi, M. das G.B. (2004) Aroma volatiles from two fruit varieties of jackfruit (*Artocarpus heterophyllus* Lam.). *Food Chemistry* 85(2), 195–197. DOI: 10.1016/S0308-8146(03)00292-9.

Meera, M., Ruckmani, A., Saravanan, R. and Lakshmipathy Prabhu, R. (2018) Anti-inflammatory effect of ethanolic extract of spine, skin and rind of jack fruit peel - a comparative study. *Natural Product Research* 32(22), 2740–2744. DOI: 10.1080/14786419.2017.1378200.

Miah, M.Y., Bhattacharjee, S., Sultana, A., Bhowmik, S., Sarker, A.K. *et al*. (2017) Evaluation of amino acid profile of jackfruit (*Artocarpus heterophyllus*) seed and its utilization for development of protein enriched supplementary food. *Journal of Noakhali Science and Technology University* 1, 77–84.

Mohamed, S. and Hasan, Z. (1995) Extraction and characterization of pectin from various tropical agro-wastes. *ASEAN Food Journal* 10, 43–50.

Mohnen, D. (2008) Pectin structure and biosynthesis. *Current Opinion in Plant Biology* 11(3), 266–277. DOI: 10.1016/j.pbi.2008.03.006.

Morton, J. (1987) Jackfruit (*Artocarpus heterophyllus*). In: *Fruits of Warm Climates*. Morton, J.F., Miami, Florida, pp. 58–64.

Muangthai, P. and Katinted, A. (2014) Trehalose and psicose sugar in jackfruit. *International Journal of Advance Research* 2, 1–6.

Nahar, N., Mosihuzzaman, M. and Dey, S.K. (1993) Analysis of free sugar and dietary fibre of some vegetables of Bangladesh. *Food Chemistry* 46(4), 397–400. DOI: 10.1016/0308-8146(93)90011-4.

Neha, C.P. and Abdussalam, A.K. (2019) GC-MS analysis of bioactive components on the fruit extracts of *Artocarpus hirsutus* Lam.: a potential wild edible plant. *International Journal of Pharmaceutical Sciences and Research* 10, 2981–2984.

Ojwang, A.R., Muge, E.K., Mbatia, B., Mwanza, B. and Ogoyi, D.O. (2017) Comparative analysis of phytochemical composition and antioxidant activities of methanolic extracts of leaves, roots and bark of jackfruit (*Artocapus heterophyllus*) from selected regions in Kenya and Uganda. *Journal of Advances in Biology & Biotechnology* 16(1), 1–13. DOI: 10.9734/JABB/2017/37355.

Ojwang, R.A., Muge, E.K., Mbatia, B.N., Mwanza, B.K. and Ogoyi, D.O. (2018) Compositional, elemental, phytochemical and antioxidant characterization of jackfruit (*Artocarpus heterophyllus*) pulps and seeds from selected regions in Kenya and Uganda. *European Journal of Medicinal Plants* 23(3), 1–12. DOI: 10.9734/EJMP/2018/40967.

Olias, J.M., Sanz, C., Rios, J.J. and Perez, A.G. (1995) Substrates specificity of alcohol acyl transferases from strawberry and banana fruits. In: *Fruit Flavours*, Vol. 596. ACS Symposium, Oxford University Press, Oxford, UK, pp. 135–141.

Ong, B.T., Nazimah, S.A.H., Osman, A., Quek, S.Y., Voon, Y.Y. *et al.* (2006) Chemical and flavour changes in jackfruit (*Artocarpus heterophyllus* Lam.) cultivar J3 during ripening. *Postharvest Biology and Technology* 40(3), 279–286. DOI: 10.1016/j.postharvbio.2006.01.015.

Ong, B.T., Nazimah, S.A.H., Tan, C.P., Mirhosseini, H., Osman, A. *et al.* (2008) Analysis of volatile compounds in five jackfruit (*Artocarpus heterophyllus* L.) cultivars using solid-phase microextraction (SPME) and gas chromatography-time-of-flight mass spectrometry (GC-TOFMS). *Journal of Food Composition and Analysis* 21(5), 416–422. DOI: 10.1016/j.jfca.2008.03.002.

Patil, K.S., Jadhav, A.G. and Joshi, V.S. (2005) Wound healing activity of leaves of *Artocarpus heterophyllus*. *Indian Journal of Pharmaceutical Sciences* 67, 629–632.

Patil, R., Joshi, G.D., Haldankar, P.M. and More, M. (2011) Estimation of pectin content in jackfruit (*Artocarpus heterophyllus*). *Asian Journal of Horticulture* 6, 536–537.

Pavanasasivam, G., Uvais, M. and Sultanbawa, S. (1973) Cycloartenyl acetate, cycloartenol and cycloartenone in the bark of *Artocarpus* species. *Phytochemistry* 12(11), 2725–2726. DOI: 10.1016/0031-9422(73)85088-5.

Pebbuli, A. and Bauri, F.K. (2018) Assessment of different quality characters of twenty jackfruit genotypes under new alluvial zone of West Bengal. *International Journal of Current Microbiology and Applied Sciences* 7(12), 2621–2626. DOI: 10.20546/ijcmas.2018.712.297.

Peng, S.D., Lin, L.J., Ouyang, L.J., Zhu, B.Q., Yuan, Y. *et al.* (2013) Comparative analysis of volatile compounds between jackfruit (*Artocarpus heterophyllus L.*) peel and its pulp. *Advanced Materials Research* 781–784, 1413–1418. DOI: 10.4028/www.scientific.net/AMR.781-784.1413.

Pereira-da-Silva, G., Moreno, A.N., Marques, F., Oliver, C., Jamur, M.C. *et al.* (2006) Neutrophil activation induced by the lectin KM+ involves binding to CXCR2. *Biochimica et Biophysica Acta* 1760(1), 86–94. DOI: 10.1016/j.bbagen.2005.09.011.

Prakash, O., Kumar, R., Mishra, A. and Gupta, R. (2009) *Artocarpus heterophyllus* (jackfruit): an overview. *Pharmacognosy Review* 3, 353–358.

Pullar, J.M., Carr, A.C. and Vissers, M.C.M. (2017) The roles of vitamin C in skin health. *Nutrients* 9(8), E866. DOI: 10.3390/nu9080866.

Rahman, A.K.M.M., Huq, E., Mian, A.J. and Chesson, A. (1995) Microscopic and chemical changes occurring during the ripening of two forms of jackfruit (*Artocarpus heterophyllus* L.). *Food Chemistry* 52(4), 405–410. DOI: 10.1016/0308-8146(95)93290-8.

Rahman, M., Nahar, N., Jabbar, M.A. and Mosihuzzaman, M. (1999) Variation of carbohydrate composition of two forms of fruit from jack tree (*Artocarpus heterophyllus* L.) with maturity and climatic conditions. *Food Chemistry* 65(1), 91–97. DOI: 10.1016/S0308-8146(98)00175-7.

Rama Rao, A.V. and Venkataraman, M.V. (1973) Colouring matter of the *A. hetetophyllus*. *Indian Journal of Chemistry* 11, 298–299.

Rao, J., Singh, K., Singh, S., Mishra, S.K. and Bajpai, M. (2014) *Artocarpus heterophyllus* (jackfruit) potential unexplored in dentistry – an overview. *Universal Journal of Pharmacy* 3, 50–55.

Rasmusson, P. (1983) Identification of volatile components of jackfruit by gas chromatography/mass spectrometry with two different columns. *Analytical Chemistry* 55(8), 1331–1335. DOI: 10.1021/ac00259a033.

Ratnasooriya, W.D. and Jayakody, J.R. (2002) *Artocarpus heterophyllus* seeds inhibits sexual competence but not fertility of male rats. *Indian Journal of Experimental Biology* 40, 304–308.

Rees, T.A.P., Dixon, W.L., Pollock, C.J. and Franks, F. (1981) Low temperature sweetening of higher plants. In: Friend, J. and Robert, M.J.C. (eds) *Recent Advances in the Biochemistry of Fruits and Vegetables*. Academic Press, London/New York, pp. 41–60.

Rosa, J.C., Greene, L.J., De Oliveira, P.Sér.L., Garratt, R., Beltramini, L. *et al.* (1999) KM+, a mannose-binding lectin from *Artocarpus integrifolia*: amino acid sequence, predicted tertiary structure, carbohydrate recognition, and analysis of the β-prism fold. *Protein Science* 8(1), 13–24. DOI: 10.1110/ps.8.1.13.

Ruiz-Montañez, G., Burgos Hernández, A., Calderón-Santoyo, M., López-Saiz, C.M., Velázquez-Contreras, C.A. *et al.* (2014) Screening antimutagenic and antiproliferative properties of extracts isolated from Jackfruit pulp (*Artocarpus heterophyllus* Lam). *Food Chemistry* 175, 409–416. DOI: 10.1016/j.foodchem.2014.11.122.

Samaddar, H.M. (1985) Jackfruit. In: Bose, T.K. (ed.) *Fruits of India Tropical and Subtropica*. Naya Prokash, Kolkata, India, pp. 638–649.

Sampath, W.D.M., Sandaruwan, P.D., Edirisinghe, D.G. and Sudusingha, Y.C.Y. (2020) Effect of *Artocarpus heterophyllus* latex on properties of calcium carbonate filled natural rubber/low density polyethylene/waste polyethylene composites. *Physical Science & Biophysics Journal* 4: 000140.

Sato, M., Fujiwara, S., Tsuchiya, H., Fujii, T., Iinuma, M. *et al.* (1996) Flavones with antibacterial activity against cariogenic bacteria. *Journal of Ethnopharmacology* 54(2–3), 171–176. DOI: 10.1016/s0378-8741(96)01464-x.

Saxena, A., Bawa, A.S. and Raju, P.S. (2009) Phytochemical changes in fresh-cut jackfruit (*Artocarpus heterophyllus* L.) bulbs during modified atmosphere storage. *Food Chemistry* 25, 1443–1449. DOI: 10.1016/j.foodchem.2009.01.080.

Saxena, A., Bawa, A.S. and Raju, P.S. (2011) Jackfruit (*Artocarpus heterophyllus* Lam). In: Yahia, E.M. (ed.) *Postharvest Biology and Technology of Tropical and Subtropical Fruits*, Vol. 3. Woodhead Publishing, Cambridge, UK, pp. 275–298.

Seddon, S.A. and Lennox, G.W. (1980) *Trees of the Caribbean*. Macmillan Education, London.

Selvaraj, Y. and Pal, D.K. (1989) Biochemical changes during the ripening of jackfruit (*Artocarpus heterophyllus* L). *Journal of Food Science and Technology* 25, 304–307.

Sengupta, U.K. and Rao, C.V.N. (1963) The pectic acid from the pulp of jackfruit (*Artocarpus integrifolia*). I. Methylation studies. *Bulletin of the Chemical Society of Japan* 25, 1683–1688. DOI: 10.1246/bcsj.36.1683.

Septama, A., Jantan, I., Panichayupakaranant, P., Aluwi, M.F.M. and Rahmi, E. (2020) Immunosuppressive and antibacterial activities of dihydromorin and norartocarpetin isolated from *Artocarpus heterophyllus* heartwoods. *Asian Pacific Journal of Tropical Biomedicine* 10(8), 361. DOI: 10.4103/2221-1691.287162.

Setiawan, B., Sulaeman, A., Giraud, D.W. and Driskell, J.A. (2001) Carotenoid content of selected Indonesian fruits. *Journal of Food Composition and Analysis* 14(2), 169–176. DOI: 10.1006/jfca.2000.0969.

Shanmugapriya, K., Saravana, P.S., Payal, H., Mohammed, S.P. and Binnie, W. (2011) Antioxidant activity, total phenolic and flavonoid contents of *Artocarpus heterophyllus* and *Manilkara zapota* seeds and its reduction potential. *International Journal of Pharmacy and Pharmaceutical Sciences* 3, 256–260.

Sharaf, A. and El-Saadany, S.S. (1987) Biochemical studies on guava fruits during different maturity stages. *Chemie, Mikrobiologie, Technologie der Lebensmittel* 25, 145–149.

Sharma, W. and Trivedi, P.C. (1995) Nematocidal and nematostostatic response of aqueous extract of certain plants of semi-arid niche. *Current Nematology* 6, 45–53.

Shen, T., Wang, X.N. and Lou, H.X. (2009) Natural stilbenes: an overview. *Natural Product Reports* 26(7), 916–935. DOI: 10.1039/b905960a.

Shirajum, M. (2014) Antioxidant and antidiarrhoeal activities of methanolic extract of *Artocarpus heterophyllus* seed. *International Journal of Pharmaceutics and Drug Analysis* 2, 890–895.

Singh, A., Kumar, S. and Singh, I.S. (1991) Functional properties of jackfruit seed flour. *Lebensmittel-Wissenschaft Technologia* 24, 373–374.

Singh, A., Maurya, S., Singh, M. and Singh, U.P. (2015) Studies on the phenolic acid contents in different parts of raw and ripe jackfruit and their importance in human health. *International Journal of Applied Science – Research and Review* 2, 69–73.

Singh, J., Chauhan, P.S., Kavita, A., Bhatnagar, P. and Meena, R.R. (2015) Jackfruit: a food of promise. *HortFlora Research Spectrum* 4, 277–281.

Soetan, K.O., Olaiya, C.O. and Oyewole, O.E. (2010) The importance of mineral elements for humans, domestic animals and plants – a review. *African Journal of Food Science* 4, 200–222.

Soobrattee, M.A., Neergheen, V.S., Luximon-Ramma, A., Aruoma, O.I. and Bahorun, T. (2005) Phenolics as potential antioxidant therapeutic agents: mechanism and actions. *Mutation Research* 579(1–2), 200–213. DOI: 10.1016/j.mrfmmm.2005.03.023.

Soong, Y.-Y. and Barlow, P.J. (2004) Antioxidant activity and phenolic content of selected fruit seeds. *Food Chemistry* 88(3), 411–417. DOI: 10.1016/j.foodchem.2004.02.003.

Srinivasan, K. and Kumaravel, S. (2016) Mass spectrometry analysis of volatile constituents of jackfruit powder. *Indo American Journal of Pharmaceutical Sciences* 3, 331–339.

Srivastava, R. and Singh, A. (2020) Jackfruit (*Artocarpus heterophyllus* Lam) biggest fruit with high nutritional and pharmacological values: a review. *International Journal of Current Microbiology and Applied Sciences* 9(8), 764–774. DOI: 10.20546/ijcmas.2020.908.082.

Sultana, A., Amin, M.N., Miah, M.Y., Sarker, A.K. and Rasel, M.M.A. (2017) Determination of proximate composition and amino acid profile of jackfruit seed and utilization of its seed flour for development of protein enriched supplementary food. *Cell Biology* 5(6), 57. DOI: 10.11648/j.cb.20170506.11.

Sundarraj, A.A. and Ranganathan, T.V. (2017a) Physicochemical characterization of jackfruit (*Artocarpus integer* (thumb.).) peel. *Research Journal of Pharmaceutical, Biological and Chemical Sciences* 8, 2285–2295. DOI: 10.7897/2230-8407.089171.

Sundarraj, A.A. and Ranganathan, T.V. (2017b) Phytochemical screening and spectroscopy analysis of jackfruit. *International Research Journal of Pharmacy* 8(9), 151–159. DOI: 10.7897/2230-8407.089171.

Sundarraj, A.A., Ranganathan, T.V. and Sriramulu, G. (2018) Optimized extraction and characterization of pectin from jackfruit (*Artocarpus integer*) wastes using response surface methodology. *International Journal of Biological Macromolecules* 106, 698–703. DOI: 10.1016/j.ijbiomac.2017.08.065.

Suresh Kumar, G., Appukttan, P.S. and Basu, D.K. (1982) D-Galactose – specific lectin from jackfruit seed. *Journal of Bioscience* 4, 257–261.

Susana, and Gunawan, D.H. (2018) The characteristic of minimally processed jackfruit with wrapping composite film from hydroxypropilcelullose which synthesized from pineapple crown during storage. *International Journal of Modern Engineering Research* 8, 23–28.

Swami, S.B. and Kalse, S.B. (2019) Jackfruit (*Artocarpus heterophyllus*): biodiversity, nutritional contents, and health. In: Merillon, J.M. and Ramawat, K.G. (eds) *Bioactive Molecules in Food*. Reference Series in Phytochemistry, Springer Nature, Cham, Switzerland, pp. 2237–2259. DOI: 10.1007/978-3-319-78030-6.

Swami, S.B., Thakor, N.J., Haldankar, P.M. and Kalse, S.B. (2012) Jackfruit and its many functional components as related to human health: a review. *Comprehensive Reviews in Food Science and Food Safety* 11(6), 565–576. DOI: 10.1111/j.1541-4337.2012.00210.x.

Swords, G., Bobbio, P.A. and Hunter, G.L.K. (1978) Volatile constituents of jack fruit (*Arthocarpus heterophyllus*). *Journal of Food Science* 43(2), 639–640. DOI: 10.1111/j.1365-2621.1978.tb02375.x.

Sy Mohamad, S.F., Mohd Said, F., Abdul Munaim, M.S., Mohamad, S. and Wan Sulaiman, W.M.A. (2019) Proximate composition, minerals contents, functional properties of Mastura variety jackfruit (*Artocarpus heterophyllus*) seeds and lethal effects of its crude extract on zebrafish (*Danio rerio*) embryos. *Food Research* 3(5), 546–555. DOI: 10.26656/fr.2017.3(5).095.

Takaya, J., Higashino, H. and Kobayashi, Y. (2004) Intracellular magnesium and insulin resistance. *Magnesium Research* 17, 126–136.

Thakur, B.R., Singh, R.K. and Handa, A.K. (1997). Chemistry and uses of pectin – a review. *Critical Reviews in Food Science and Nutrition* 37, 47–73.

Tiwari, K. and Vidyarthi, A.S. (2015) Nutritional evaluation of various edible fruit parts of jackfruit (*Artocarpus heterophyllus*) at different maturity stages. *International Journal of Chemical and Pharmaceutical Review and Research* 1, 21–26.

Tramontin, D., Cadena-Carrera, S.E., Assreuy, J., Nunes, R., Santin, J.R. *et al.* (2021) Response surface methodology (RSM) to evaluate both the extraction of triterpenes and sterols from jackfruit seed with supercritical CO_2 and the biological activity of the extracts. *Journal of Food Science and Technology* 58, 3303–3313.

Trindade, M.B., Lopes, J.L.S., Soares-Costa, A., Monteiro-Moreira, A.C., Moreira, R.A. *et al.* (2006) Structural characterization of novel chitin binding lectins from the genes *Artocarpus* and their fungal activity. *Biochimica et Biophysica Acta* 1764(1), 146–152. DOI: 10.1016/j.bbapap.2005.09.011.

Tyagi, S., Sahay, S., Nanher, A.H. and Kumar, V. (2015) Medicinal uses of jackfruit (*Artocarpus heterophyllus*). In: *XII Agricultural Science Congress*, ICAR-NDRI, February 3–6, 2015, p. 227.

Uddin, M.T., Rukanuzzaman, M., Khan, M.M.R. and Islam, M.A. (2009) Adsorption of methylene blue from aqueous solution by jackfruit (*Artocarpus heteropyllus*) leaf powder: a fixed-bed column study. *Journal of Environmental Management* 90(11), 3443–3450. DOI: 10.1016/j.jenvman.2009.05.030.

USDA (United States Department of Agriculture) (2016) The PLANTS database. Natural Resource Conservation Service (NRCS), National Plant Data Team, Greensboro, North Carolina. Available at: https://plants.sc.egov.usda.gov/home (accessed 1 July 2022).

Venkataraman, K. (2001) Wood phenolics in the chemotaxonomy of the Moraceae. *Phytochemistry* 11(5), 1571–1586. DOI: 10.1016/0031-9422(72)85002-7.

Wang, X.L., Di, X.-X., Shen, T., Wang, S.Q. and Wang, X.-N. (2017) New phenolic compounds from the leaves of *Artocarpus heterophyllus*. *Chinese Chemical Letters* 28(1), 37–40. DOI: 10.1016/j.cclet.2016.06.024.

Wealth of India (1985) *The Wealth of India: A Dictionary of Indian Raw Materials & Industrial Products*, Vol. 1A. National Institute of Science Communication and Information Resources (NISCAIR), Council of Scientific & Industrial Research (CSIR), New Delhi.

Wei, B.L., Weng, J.R., Chiu, P.H., Hung, C.F., Wang, J.P. *et al.* (2005) Antiinflammatory flavonoids from *Artocarpus heterophyllus* and *Artocarpus communis*. *Journal of Agricultural and Food Chemistry* 53(10), 3867–3871. DOI: 10.1021/jf047873n.

Wong, K.C., Lim, C.L. and Wong, L.L. (1992) Volatile flavour constituents of chempedak (*Artocarpus polyphema* Pers.) fruit and jackfruit (*Artocarpus heterophyllus* Lam.) from Malaysia. *Flavour and Fragrance Journal* 7(6), 307–311. DOI: 10.1002/ffj.2730070604.

Wong, K.T., Poh, G.Y.Y., Goh, K.K.T., Wee, M.S.M. and Jeyakumar Henry, C. (2021) Comparison of physicochemical properties of jackfruit seed starch with potato and rice starches. *International Journal of Food Properties* 24(1), 364–379. DOI: 10.1080/10942912.2021.1885439.

Wongsa, P. and Zamaluddien, A. (2005) Total phenolic content, antioxidant activity and inhibitory potential against α-amylase and α-glucosidase of fifteen tropical fruits. In: *Proceedings of the 37th Congress on Science and Technology of Thailand, Bangkok, Thailand, 2005*.

Xiao, J., Li, C., Xu, S., Xing, L., Xu, Y. *et al.* (2015) JACALIN-LECTIN LIKE1 regulates the nuclear accumulation of GLYCINE-RICH RNA-BINDING PROTEIN7, influencing the RNA processing of FLOWERING LOCUS C antisense transcripts and flowering time in Arabidopsis. *Plant Physiology* 169, 2102–2117.

Xu, S.Y., Liu, J.P., Huang, X., Du, L.P., Shi, F.L. *et al.* (2018) Ultrasonic-microwave assisted extraction, characterization and biological activity of pectin from jackfruit peel. *LWT – Food Science and Technology* 90, 577–582.

Yamin, R., Mistriyani, S., Ihsan, S., Armadany, F.I., Sahumena, M.H. and Fatimah, W.O.N. (2021) Determination of total phenolic and flavonoid contents of jackfruit peel and *in vitro* antiradical test. *Food Research* 5, 84–90.

Yuan, W.-J., Yuan, J.-B., Peng, J.-B., Ding, Y.-Q., Zhu, J.-X. *et al.* (2017) Flavonoids from the roots of *Artocarpus heterophyllus*. *Fitoterapia* 117, 133–137. DOI: 10.1016/j.fitote.2017.01.016.

Zhang, L., Tu, Z.-C., Xie, X., Wang, H., Wang, H. *et al.* (2017) Jackfruit (*Artocarpus heterophyllus* Lam.) peel: a better source of antioxidants and α-glucosidase inhibitors than pulp, flake and seed, and phytochemical profile by HPLC-QTOF-MS/MS. *Food Chemistry* 234, 303–313. DOI: 10.1016/j.foodchem.2017.05.003.

Zheng, Z.-P., Cheng, K.-W., To, J.T.-K., Li, H. and Wang, M. (2008) Isolation of tyrosinase inhibitors from *Artocarpus heterophyllus* and use of its extract as antibrowning agent. *Molecular Nutrition & Food Research* 52(12), 1530–1538. DOI: 10.1002/mnfr.200700481.

Zheng, Z.P., Chen, S., Wang, S., Wang, X.C., Cheng, K.W. *et al.* (2009) Chemical components and tyrosinase inhibitors from the twigs of *Artocarpus heterophyllus*. *Journal of Agricultural and Food Chemistry* 57(15), 6649–6655. DOI: 10.1021/jf9014685.

Zheng, Z.-P., Xu, Y., Qin, C., Zhang, S., Gu, X. *et al.* (2014) Characterization of antiproliferative activity constituents from *Artocarpus heterophyllus*. *Journal of Agricultural and Food Chemistry* 62(24), 5519–5527. DOI: 10.1021/jf500159z.

Zuraidah, M.A., John, B.A. and Kamaruzzaman, Y. (2017) Cytotoxicity on MCF7 cell lines exposed to an extract of the jacalin from jackfruit seed. *Science Heritage Journal* 1(2), 16–18. DOI: 10.26480/gws.02.2017.16.18.

Zuwariah, I., Noor Fadilah, M.B., Hadijah, H. and Rodhiah, R. (2018) Comparison of amino acid and chemical composition of jackfruit seed flour treatment. *Food Research* 2, 539–545.

3 Processing and Products

Sisir Mitra[1]*, Mathala J. Gupta[2] and Pradyot K. Pathak[3]
[1]Former Professor, Faculty of Horticulture, Bidhan Chandra Krishi Viswavidyalaya, Mohanpur, West Bengal, India; [2]ICAR-Central Coastal Agricultural Research Institute, Ela, Goa, India; [3]KVK, 24 Pargana (N), West Bengal University of Animal and Fishery Science, West Bengal, India

3.1 Introduction

The present status of postharvest handling of jackfruit in most of the major growing countries is very poor. There is high wastage because of seasonal production causing gluts in the market and low prices. The ripe fruit is highly perishable and its shelf life is very poor under ambient conditions. The fruits could better be utilized by employing adequate processing technologies and preserving the fruit as different value-added products.

Immature green fruits are widely used as a vegetable and are cooked and used like potato, often sautéed with fish, chicken or egg. The tender fruits are also canned to use out of season. However, the bulk of fruits are allowed to mature to be harvested for dessert fruits, despite a somewhat pungent odour. As well as use as a dessert fruit, ripe pulp is used to flavour products such as ice cream and beverages. The fresh pulp can also be used to prepare chips, jams, jellies, candies and leathers. It can be canned in syrup and is sometimes fermented and distilled to produce liquor. The seeds are also eaten after boiling or roasting, or are dried and salted as table nuts, or ground to make flour, which is blended with wheat flour for baking. The inedible parts of jackfruit include the rind/peel, non-edible perianth and the central core. The non-edible waste may be utilized for extraction of pectin, clarified juice, and production of biofuel, biochar and some other products. The various products developed from jackfruit are shown in Fig. 3.1.

3.2 Tender Jackfruit

Tender jackfruit has a mild, pleasant flavour and distinctive texture, and is rich in vitamins and minerals. It also contains isoflavones, phytonutrients and antioxidants (Mukherjee, 1993). Tender jackfruit curry is exotically delicious and makes a delicious vegetable dish, similar to chicken in consistency, and is even referred to as 'vegetable meat' in many parts of South-east Asia (Lakshmana *et al.*, 2013). A number of recipes and ready-to-eat packs are available for preparation of curry, biryani, fried, pickle, papad and so on (Fig. 3.2).

The use of hurdle processing in stabilizing curry prepared from tender jackfruit extends the storage life. The hurdles applied include acidification, vacuum packaging and in-pack pasteurization. The hurdles were found to be adequate to render the product microbiologically safe, giving an overall shelf life of 120, 270

*Corresponding author: sisirm55@gmail.com

 Jackfruit : Botany, Production and Uses (ed. S. Mitra)
DOI: 10.1079/9781800622319.0003

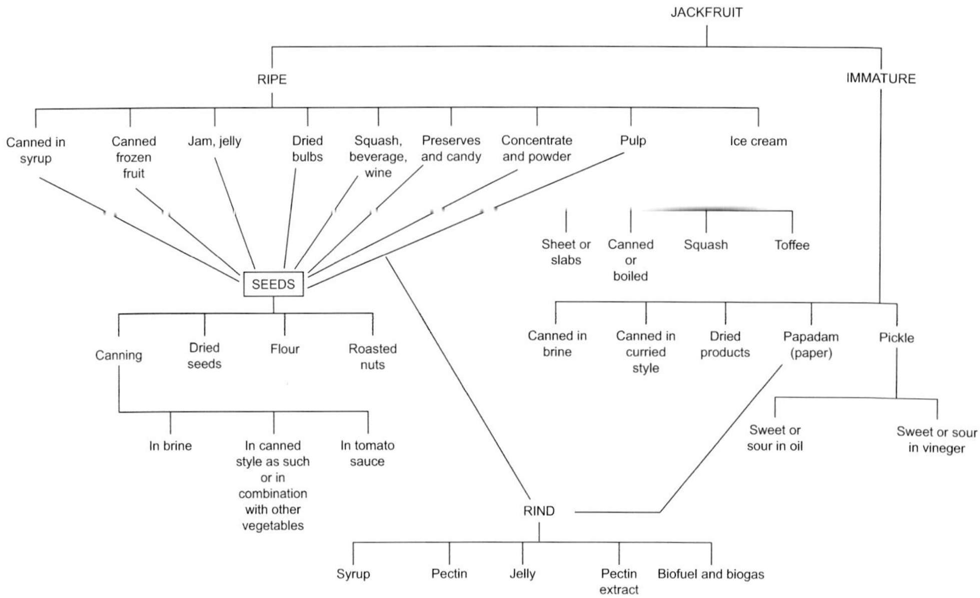

Fig. 3.1. The various products produced from jackfruit.

and 360 days at storage temperatures of 28°C, 4°C and −18°C, respectively (John *et al.*, 1993). Lakshmana *et al.* (2013) developed technology to keep tender jackfruit curry for 12 months under ambient conditions with good texture and sensory characteristics in a shelf-stable retort pouch. Prefabricated multilayer retortable pouches consisting of 12 µm polyester/12 µm aluminium foil, 75 µm cast polypropylene of dimension 15 × 20 cm were filled with the product (Fig. 3.3).

Some examples of ways in which jackfruit can be cooked are given below.

3.2.1 Curry

Tender jackfruit curry is very popular in India, Bangladesh, Vietnam, the Philippines, Sri Lanka and some other countries. Gunasena *et al.* (1996) from Sri Lanka suggested several recipes and methods of curry preparation from tender jackfruit and some of the preparation methods are as follows:

- *Ingredients*: 1 medium-sized tender jackfruit, 1 dessertspoon coriander powder, 2 teaspoons cumin powder, 2 teaspoons sweet cumin powder, 8 peppercorns, curry leaves, 2–3 dessert spoons chilli powder, 2–3 cardamoms, 2–3 cloves, 6 pieces of goraka (*Garcinia cambogia*) fruit, 7 g garlic, 4 teaspoons salt, 850 ml second extract of coconut milk, 1/2 teaspoon fenugreek, 60 g red onions and 250 ml first extract of coconut milk.
  Method: Peel jackfruit and cut into small pieces. Score deep on skin side of pieces and put into water. Chop the onion. Roast the spices except the chilli powder to golden-brown. Put the jackfruit pieces with all other ingredients, and the second extract of coconut milk into a pan. Bring to boil and simmer until the sauce has reduced and the jackfruit is soft. Add the first extract of coconut milk, bring to the boil and simmer until the sauce is reduced (use a small pan so that the coconut milk covers the fruit). Remove the goraka
- *Ingredients*: 450 g young jackfruit, ¼ teaspoon turmeric powder, 4½ teaspoons salt, 40 g onions, thin slice of ginger, 2½ dessert spoons chilli powder, 4 teaspoons cumin powder, 2 teaspoons fenugreek, 2 teaspoons sweet cumin powder, 2 pieces goraka fruit, 15 g garlic, 4 cardamoms, 2

Fig. 3.2. Jackfruit products. Photograph courtesy of Wakao Foods, Panjim, Goa, India.

cloves, 2.5 cm cinnamon, coconut milk (extract of one coconut).

Method: Peel the fruit and cut into large pieces. Chop the onions. Put all the ingredients together and mix well. Bring to boil and simmer until the fruit is soft and the sauce is thick.

- The recipe and method suggested by Best (2015) from Australia is as follows:

  Ingredients: young jackfruit (400 g) after skin removed) 40 g bird chillies, 5 cm turmeric root, 255 ml coconut milk, 50 g anchovies (small marine fish), 2 pieces dried tamarind slices and salt to taste.

  Method: Peel the young jackfruit, cut into 2 × 2 cm cubes and boil in water with salt and one piece of dried tamarind slice until soft. Grind the chillies and turmeric root finely, pound the anchovies coarsely and then cook anchovies, ground ingredients, jackfruit, tamarind slices, coconut milk and a dash of water in a pot. Add salt to taste, and simmer for 20–30 min over a moderate heat with a little more water if necessary. It is now ready to serve.

- Ginataang langka with munggo (jackfruit and mung bean (*Vigna radiata*) in coconut cream). This is a Filipino dish is made of sautéed, young, immature jackfruit, mung beans, dried fish, spices (garlic, onion, ginger and peppercorn) with coconut cream (locally known as 'gata'). The ingredients and method of preparation (courtesy of Jigzcel Basoy, personal communication) are as follows:

  Ingredients: 300 g young, immature jackfruit, 100 g mung beans (boiled), 2 pieces regular-size dried fish, 5 cloves of garlic, 1 piece onion, 1 thumb-size piece ginger and 2 g peppercorns.

  Method: Slice the peeled, young, immature jackfruit into cubes of about 3 × 3 cm. Add

Fig. 3.3. Tender jackfruit, minimally processed and frozen. Photograph courtesy of Dr M.J. Gupta.

to a pot and add water enough to submerge the cubes and let it boil. Upon boiling, sprinkle in some salt. Boil until the jackfruit becomes tender, then drain the water and set aside the cooked jackfruit. Slice the garlic, onion and ginger. In a separate pot, sauté the garlic, onion, ginger and dried fish. Add the boiled jackfruit and mung beans. Add water enough to cover the jackfruit and mung beans and let it boil. Upon boiling, taste the soup to check the salinity of the dish. If it is not salty enough, add extra salt to taste. Add the peppercorn and the coconut cream and boil for 5 min. Serve as a viand while hot (Fig. 3.4)

- Vegetarian curry: this is an Indian jackfruit curry dish (courtesy of I. Bandyopadhyaya, personal communication). The ingredients and method of preparation are as follows:

Ingredients: 500 g peeled tender jackfruit, 20 ml cooking oil, ½ teaspoon turmeric powder, ¼ teaspoon chilli powder, ground and mixed powder of 6 cloves, 4 small

Fig. 3.4. Ginataang langka with munggo. Photograph courtesy of Dr J.D.F. Bassoy.

cardamoms, 5 cm of cinnamon stick, 5 g raisins, asafetida (pinch), 2 bay leaves, and 1 piece onion (80–100 g).
Method: Soak the raisins in water and smash them up. Slice the peeled jackfruit in to pieces of 5–6 cm each and boil in a pressure cooker with two teacups of water up to two whistles. Drain out the water and smash the pieces gently. Gently heat a cooking pan, add the oil, a pinch of asafoetida and ¼ teaspoon of the ground mixed powder of clove, cardamom and cinnamon and two bay leaves. Add the chopped onion and fry it in low heat until it turns light brown. Take out the fried onion. Add the boiled and smashed jackfruit pieces in the pan and cook slowly in a medium heat until it turns brown. Add the fried onion and salt to taste and cook for 2 min. Add the smashed raisins, mix well and cook again until the the jackfruit turns dark brown. Add the mixed powder and remove from the heat. It is now ready to serve (Fig. 3.5).

3.2.2 Biriyani

Biriyani is dish that usually prepared with meat; however, vegetarian biriyani can be prepared by replacing the meat with tender jackfruit. In the preparation of biriyani, the main ingredients are a mixture of spices and aromatic substances that provide the special taste and biriyani flavour. The following method for making jackfruit biryani is courtesy of U. Banerjee, personal communication):

- In a frying pan add 2 teaspoons coriander seed, 2 teaspoons cumin seed, 2 teaspoons fennel seeds, 3–4 cm cinnamon stick, 4 green cardamoms and 1 black cardamom, 4 cloves, ½ a mace, 2 dried red chillis, 8 peppercorns. Roast until fragrant and then cool and grind finely. The Biriyani spice is ready.
- In a large cooking pan, add ¾ cup of cooking oil and heat. Fry three finely chopped large

Fig. 3.5. Tender jackfruit curry. Photograph courtesy of Dr I. Bandyopadhyaya.

onions (300 g) until brown and drain on absorbant paper.

- Marinate 500 g peeled and sliced (cubes) tender jackfruit with ½ cup of curd, 1 teaspoon ginger paste, 1 teaspoon garlic paste, paste of 10 cashew nuts, 1 teaspoon biriyani spices mixture, ½ the fried onion and 1 tablespoon oil taken from the pan after frying the onion for 10 min. Mix ½ cup milk with a pinch of saffron and 1 teaspoon of kewra water (an extract distilled from pandanus flower). Cover it and put aside.
- Wash and soak 2 cups of basmati (scented, long grain) rice. Heat sufficient water in a large pan. Add 2 teaspoons salt, 1 teaspoon rose water, 1 tablespoon fried onion oil, and 5 pieces of cinnamon (2–3 cm each), 5 green cardamoms and 10 cloves tied in a small, clean cloth. Bring to the boil and then add the soaked, scented basmati rice. Cook until the rice is 80% cooked. Drain the excess water but collect ½ cup of the drained water.
- In the large pan, heat the rest of the fried onion oil. Add the marinated tender jackfruit cubes. Cook over a medium heat for 10 min. Then reduce the heat to a simmer, add the drained water, spread the rice on the tender jackfruit cubes and add the rest of the fried onion, 2 teaspoons butter or clarified butter, and saffron milk. Spread these over the rice. Cover tightly and cook

on a medium heat for 5 min. Then reduce the heat to a simmer for 10 min, switch off the heat and keep the biriyani for 20 min with the lid on. It is now ready to serve.

3.2.3 Fritters

The following is a method for preparing jackfruit fritters:

- *Ingredients*: 250 g young jackfruit, 125 g lentils, ½ tablespoon cumin seeds, ½ tablespoon black peppercorns, ¼ tablespoon cloves, pinch of cinnamon, ½ stick ginger, 1 piece garlic cloves, 2 large onions, 1 medium-sized dry red chilli, 1–2 green chillies, 1–2 tablespoons cilantro, 1 cup water and salt to taste.
- *Method*: Peel and chop fresh jackfruit. In a saucepan, boil the jackfruit pieces in the water along with half an onion and the green and red chillies, lentils, ginger, garlic, cumin seeds, cloves, peppercorns, cinnamon and salt. Simmer until tender and then let all the liquid evaporate completely. Grind to a fine paste in a blender. Remove the spices that are difficult to grind. Mix well and then make small flat patties. These can either be deep fried in oil or baked in a pre-heated oven at 375°F till they are golden-brown. Serve with mint chutney (Best, 2015).

3.2.4 Fry and chips

Jackfruit fry is a delicious snack prepared by deep frying tender jackfruit flakes in oil after mixing with gram flour, ginger and garlic paste, chilli powder, asafetida, baking powder and salt to taste.

3.2.5 Canning

Canning is widely practised to extend the period of availability of fruits. Jackfruit bulbs, both tender and ripe, can be successfully canned for subsequent use in vegetable curries and also for table purposes.

Tender jackfruit cubes can be canned in brine. Marak *et al.* (2019) reported that tender jackfruit can be brined at 8% salt strength in glass jars or standing pouches of 75 μm for a period of at least 6 months. The brined tender jackfruit can be used as a vegetable in curries (Fig. 3.6) during the off-season and can easily be transported to places where it is not available.

Fig. 3.6. Tender jackfruit pieces in brine. Photograph courtesy of Dr M.J. Gupta.

Thermal processing is a promising preservation method to ensure year-round availability of tender jackfruit. The identification of suitable time/temperature combinations is an essential prerequisite for thermal processing. Pritty and Sudheer (2019) standardized the thermal processing parameters for canned tender jackfruit. Tender jackfruits were washed, peeled and cut into pieces of uniform size. The pieces were dipped in 0.1% potassium metabisulfite solution (2 kg solution kg^{-1} sample) for 15 min to prevent browning. The fruit pieces were then blanched in boiling water (100°C) for 1 min with 0.3% citric acid. The product was immediately cooled to room temperature. Thermal processing was performed in cans (900 ml volume, dimensions 11.5 × 10.0 cm). About 250 g blanched sample was added to a can and the remaining portion was filled with about 550 ml brine solution (2%) with provision of suitable headspace (7 mm). Residual air was removed and the cans were cooled to room temperature after double seaming. The cans were then subjected to pasteurization and a sterilized temperature in a retort. Based on the results of both microbiological and quality attributes (texture, colour, pH, total soluble solids (TSS) and crude fibre content), the study advocated the use of pasteurization at 90°C for 19 min or sterilization at 121°C for 38 min for thermal processing of canned tender jackfruit.

3.3 Processing of Ripe Fruit

3.3.1 Chips

Jackfruit chips are a popular delicacy in South Asia (Yi *et al.*, 2016a). Many processing technologies have been applied to produce fruit chips, such as deep-fat frying, vacuum frying, freeze drying, steam puff drying and microwave vacuum drying (Yi *et al.*, 2016b). Vacuum frying improves the quality of fried food and can reduce the final oil uptake in the product. Vacuum frying is a deep-fat frying process that is carried out in a closed system, below atmospheric pressure, and thus at a lower frying temperature. The low frying temperatures and minimal exposure to oxygen are responsible for most of the benefits of the fried products, which include nutrient preservation (da Silva and Moreira, 2008), oil quality protection (Shyu *et al.*, 1998) and a reduction in toxic compound generation (Granda *et al.*, 2004).

Both half-ripe and fully ripe jackfruit bulbs can be vacuum fried. However, vacuum-fried chips from half-ripe fruit lack the aroma and taste of those from ripe jackfruit. The ripe fruit pulps are processed into vacuum-fried and dehydrated products, and both products are highly acceptable by consumers (Fig. 3.7).

The production of vacuum-fried jackfruit described by Barga *et al.* (2019) is as follows. The fruit is washed, brushed and rinsed with 10 p.p.m. sanitizing solution, cut in half, and depulped and deseeded. The pulp (4 kg) is blanched in boiling water for 1 min or until the pulp colour changes to bright yellow and kept in a chest freezer for 24 h. Frozen jackfruit pulps are then vacuum fried at 89°C at 30 mmHg vacuum for 2 h and 10 min. After this, the vacuum-fried jackfruit pulps are sealed and packed in foil pouches. Coconut oil is used for frying. The oil can be recycled for at least ten frying cycles to produce high-quality and safe vacuum-fried jackfruit product for consumers. The reusing of oil provides cost-effectiveness in the vacuum-fried jackfruit process.

Maity *et al.* (2017) suggested preconditioning of bulbs by freezing overnight for 16 h at 20°C before vacuum frying. Prefreezing produces fried chips that have less oil content as well as a very high porosity and less surface shrinkage. Prefreezing bulbs also retains a high carotenoid content after vacuum frying. Yi *et al.* (2016a) evaluated three drying methods; instant controlled pressure drop-assisted freeze drying (FD-DIC), instant controlled pressure drop-assisted hot air drying and freeze drying. The FD-DIC-dried jackfruit chips showed the highest expansion ratio (119%) and the best colour and texture (hardness 42 N, crispness 19). FD-DIC-finished jackfruit chips exhibited stronger antioxidant capacities (2,2-diphenyl-1-picrylhydrazyl (DPPH)-free, radical scavenging activity, ferric reducing power and ABTS (2,2′-azino-bis(3-ethylbenzothiazoline-6-sulfonic acid))-positive radical cation scavenging activity) and higher retention of phenolics (1.2 mg caffeic acid equivalents g^{-1}) and carotenoids (3.1 μg

Fig. 3.7. Packaged jackfruit chips. Photograph courtesy of Dr M.J. Gupta.

β-carotene equivalents g^{-1}). This combination of FD-DIC could be an alternative method for obtaining high-quality fruit chips.

3.3.2 Dehydrated bulbs

The method for preparing dehydrated bulbs is as follows. Cut the deseeded bulb into two to four pieces. Blanch the fruit segments by plunging into boiling water for 2 min and cool rapidly under clean cold water. Place the blanched segments in a single layer on mesh dryer trays. Load the trays into a drying cabinet and dry at 55°C for 6–7 h until the moisture content is reduced to 5%. Remove the trays and pack the dried bulbs immediately in moisture-proof containers (e.g. 400-gauge polythene or polypropylene pouches) and heat seal them. The dried product has a shelf life of 4–6 months. It can be soaked in water for 30 min and used as fresh fruit, in baby food, as a health mix or in soup.

Dehydrated jackfruit sometimes develops browning after 1–2 months, even if the product is properly packed. Jackfruit bulbs immersed in 0.1% sodium metabisulfite prior to dehydration and those that were steam blanched for 3 min and then immersed in 0.1% sodium metabisulfite did not differ significantly in terms of sensory qualities. The browning of stored dehydrate jackfruit is caused by a non-enzymatic reaction (Galvez *et al.*, 2013).

Better sensory and nutritional characteristics can be retained by a combination of osmotic dehydration and convective drying. Souza *et al.* (2014) suggested dehydration of bulbs in sucrose with a soluble solids content of 40°Bx and then air drying at 60±2°C until they reached a constant weight. The water activity of dried bulbs was 0.56 and they had a good aroma, flavour and colour. Li *et al.* (2017) developed the technology for processing light-sugar dried jackfruit that have a natural flavour and are a non-sulfur product. Fully ripened bulbs were boiled in water after adding 0.6% D-sodium erythorbate at 80°C for 3 min. After drying, the bulbs were osmodehydrated using intermittent forced sugar solution cycling technology (initial sugar solution of 25°Bx and adding a 15% proportion of 75% maltose syrup and 0.3% glycerine) and the bulbs were drained when the sugar solution remained stable at 35°Bx. This technology is reported to preserve natural flavour, contain low total sugar levels and have a sulfur dioxide residue of 0.02 g kg^{-1} or less.

3.3.3 Canning

Chin and Nushirwan (1989) suggested the following method for canning of ripe bulbs in syrup. Remove the bulbs from ripe fruit, deseed and slice into suitable sizes. Wash thoroughly in clean water and fill a can (250 g, dimensions 7.3 × 11.7 cm). Pour in hot syrup (at 90°C; 30°Bx with 0.1% each of citric acid and calcium chloride, pH 3.5–3.7). Exhaust the can with steam for 3 min and seal the can at about 75°C. Retort at 100°C for 7 min and then cool to below 50°C. Label the can. The processed product can be kept up to 12 months without deterioration of quality.

3.3.4 Squash

Squash is a non-alcoholic concentrated juice that is usually fruit flavoured made from fruit juice, water and sugar or sugar substitute. Shwetha and Ranganna (2018) described squash preparation from jackfruit bulb. Well-matured and ripe, deseeded jackfruit bulbs were washed in hot water, cut into pieces and mashed in a blender to produce pulp. Sugar (1200 g) citric acid (8–10 g) and mashed pulp (400–450 g) were added to 1 l water to get 45% pulps, with 40°Bx and 1.2% acidity. The mixture was then boiled at 80–85°C for 15–20 min. After cooling, preservative potassium metabisulfite was added at 350 p.p.m. and filtered through a stainless steel filter. Plain heated glass bottles were filled with filtrate concentrate and cork sealed by a cork-sealing machine. The cork-sealed squash bottles were pasteurized in a water bath at 90–95°C for 15–20 min and allowed to stand overnight in ambient conditions. The shelf life of the squash was found to be about 45–60 days at ambient temperature and 90 days in refrigerated conditions (Fig. 3.8).

3.3.5 Nectar

Nectar is a juice or non-carbonated, non-alcoholic beverage made from pulp by mixing with water, sugar and citric acid. Jackfruit nectar is a palatable beverage with a pleasant taste and aroma. Deseeded ripe bulbs are cut into pieces and boiled in water with stirring. The mashed and boiled softened pieces are crushed and rubbed through a fine sieve to remove any large pieces. Sugar syrups of 15–18°Bx are prepared and added to the blended juice. The acidity of the juice (around 0.30%) is maintained by adding citric acid and preservative (sodium benzoate or potassium metabisulfite). The prepared product is filtered through muslin cloth into presterilized bottles and sealed with crown caps. The filled bottles are pasteurized for about 30 min in boiling water. Thermo-ultrasound processing conditions (80% amplitude at 45±1°C for 20±1 min) allow 68% physical stability, 60% pectin methylesterase active, 420 mg ascorbic acid l^{-1} and a total phenolic content of 134 mg gallic acid equivalents l^{-1}. The thermo-ultrasound-treated jackfruit nectar had a lower microbial load and higher TSS, ascorbic acid, total phenolic content and antioxidant activity than pasteurized nectar (Cruz-Cansino *et al.*, 2020).

3.3.6 Beverages

The ingredients required for preparation of beverage from 1 kg jackfruit are: 150 g pulp,

Fig. 3.8. Jackfruit squash. Photograph courtesy of Dr M.J. Gupta.

140 g sugar, 710 ml water and 140 mg potassium metabisulfite (ICUC, 2005). The method is as follows. Cut the ripe fruit in half lengthwise, carve out the core of the fruit, scoop out the bulbs, remove the seeds, homogenize the pulp using a pulper or blender, boil the pulp, add pectin-degrading enzyme and keep the mixture at room temperature overnight. Filter the extract. Prepare a 50% sugar syrup solution at 90°C, make the volume up to 1 l and combine the fruit juice (30%) and sugar syrup (70%). Add preservative (0.05% sodium metabisulfite), pour into presterilized bottles and seal. Pasteurize the sealed bottle at 80–95°C for 10–20 min and cool the bottle to room temperature by immersing in cool water.

Dushyantha *et al.* (2011) isolated lactic acid bacteria from the jackfruit phyllosphere, perianth lobes and juice. They found that lactic acid bacteria strain JFL1 was superior to other isolates in fermenting the jackfruit juice. This strain was further screened for beverage making. Casein and soy protein were supplemented as a nitrogen source to enhance the fermentation efficiency of the strains.

In Vietnam, the pulp of jackfruit is usually used in the food industry as a vacuum-dried product but the 'rags' (fibrous strands of perianth that surrounded the segments), which comprise about 25% of the fruit weight, are waste. Dam and Nguyen (2013) developed a new value-added product of fermented beverages from fruit rags of jackfruit. The rags were blanched in boiling water for 3 min and then quickly cooled and blended with water (1:5) for 5 min. The resultant juice was adjusted to pH 4.5 using saturated citric acid and incubated with 0.2% Pectinex Ultra SP at 45–55°C for 1.5 h. The enzyme-treated juice was extracted and then adjusted to pH 4.2. The juice was boiled at 100°C for 15 min and then quickly cooled. After being left for 30 min to deposit the sediments, the juice obtained from the supernatant was supplemented with syrup (65°Bx). The optimal conditions for maximum juice yield were the use of 0.3% pectinase at 90°C with a time of 2.5 min. The fermentation process was studied using $10–10^6$ cells of *Saccharomyces oviformis* ml^{-1}. The optimum conditions for juice making were 21°Bx and pH 4.6. The optimum fermentation time was 3.5 days at 25°C. The final product was 11°Bx, pH 4.35, with 1.28% total acids and 5.5% (v/v) ethanol. It was evaluated to have acceptable sensory quality.

3.3.7 Jam

Jackfruit jam is prepared from the pulp of ripe fruits. Deseeded bulbs from a fully ripe jackfruit are blended and boiled for 5–7 min to extract the juice, and 700 g sugar and 10 g pectin is then added to 1 kg jackfruit pulp and cooked until the TSS reach 64°Bx. Citric acid (0.25%) is then added. The end point is determined using the flake test and the jam is poured while hot into sterilized jars and stored at room temperature (Bhuyan *et al.*, 2013). Mixed fruit jam containing the pulps of jackfruit (65%), avocado (*Persea americana*) (15%) and kokum (*Garcinia indica*) (20%) was developed by Kushala *et al.* (2012). The product was found to be acceptable with good organoleptic scores for appearance, aroma, flavour and taste. It can be stored for 120 days at room temperature.

3.3.8 Jelly

Jelly is a product of gelatinous consistency prepared by boiling strained fruit extract with sugar. It is transparent with an attractive colour (Haq, 2006). Jackfruit bulb jelly can be prepared using 450 g juice extracted from ripe bulbs, 120 g sugar, 2.5 g citric acid, 5 g pectin, 1 g agar, 0.3 g potassium metabisulfite and 450 ml water (Mondal *et al.*, 2016). The end product should have a TSS of 70–75°Bx.

Jackfruit contains a large amount of waste material comprising the outer prickly rind and the inner perigones (e.g. underdeveloped bulbs). The waste material contains about 2% pectin of high quality. It can be used to prepare a clear, light yellowish jelly with a good taste and aroma.

3.3.9 Pickle

Tender jackfruits are used for preparing pickle. Small, tender, green peeled jackfruit are cut into several pieces and the bulbs are removed by hand; the seeds may be removed and the bulbs are sliced into pieces (1 kg). The slices need to be cured in brine solution (5%) for 8–10 days prior to pickling. The slices are drained from the brine solution. The following spices are ground and mixed: 2.5 g turmeric powder, 25 g coriander seeds, 10–20 g chilli powder, 10 g salt and 150 g sugar. The spice mix and vinegar (10 ml kg^{-1}) are added to the jackfruit slices and the mixture is cooked in a stainless steel boiling pan for 30 min with stirring. The pickle is then poured into presterilized jars and sealed. Jackfruit pickle in oil is also possible (ICUC, 2005). Rudrawar and Kale (2017) suggested the use of jaggery to replace sugar, which has several health benefits.

Sweet pickle in oil can be prepared using the following: pieces of peeled tender jackfruit (500 g), sugar (60 g), ginger (9 g), mustard powder (5.6 g), turmeric (2.5 g), cumin (1.5 g), aniseed (1.5 g), fenugreek (1.7 g), black cumin (1.5 g), cinnamon (1.5 g), cardamom (1 pod), vinegar (100 ml), 4 cloves and mustard oil (100 ml). This sweet pickle can be stored for 12 months (Mondal *et al.*, 2016).

3.3.10 Leather

Fruit leather is a dehydrated product prepared from fresh fruit pulp or fruit juice concentrate (Fig. 3.9). It is a leather-like sheet and is also known as fruit bar or fruit slab (Babu *et al.*, 2020). The preparation method suggested by ICUC (2005) is as follows: add sugar (10–15% of the weight of fully ripe deseeded bulb = 100–150 g kg^{-1}) according to the variety used and taste, and blend the bulbs. Dissolve preservative (e.g. potassium or sodium metabisulfite (0.1 g kg^{-1}) in water and add to the product. Concentrate the mixture in a steam-jacketed pan, and spread the concentrate on stainless steel trays lined with greaseproof paper at 3 mm thickness. Dry the leather using a solar (2 days) or mechanized (18–20 h) dryer; turn it over after 1 day in the solar dryer or 5 h in the artificial dryer until the moisture content is 9–12%.

Babu *et al.* (2020) used 64.9 g jackfruit bulb, 3.2 g maize flour, 2.6 g honey, 29.2 g water, 0.03 g potassium metabisulfite and 0.01 g sodium benzoate for preparation of 100 g of jackfruit leather. Honey was added to the cooked bulbs and they were ground to a smooth pulp. Potassium metabisulfite and sodium benzoate were added and the pulp was

Fig. 3.9. Jackfruit leather. Photograph courtesy of Dr M.J. Gupta.

spread on trays coated with butter and dried at 52°C and 1.5 m s^{-1} air velocity for 24 h in a tray drier. Jackfruit leather is rich in minerals such as potassium (7442 μg g^{-1}) and sodium (1842 μg g^{-1}). It has a critical moisture content of 48.39±0.44% with respect to 96% relative humidity and a water activity of 0.47. The product is packed in 220-gauge polyethylene laminated with 50-gauge metalized polyester and 150-gauge biaxially oriented polypropylene laminated with 50-gauge polyester. The physical and chemical parameters of the product were evaluated during storage at room temperature (28±2°C) and the results revealed a shelf-life stability of 90 days in either of the packaging materials.

The effect of solar drying on the quality and consumer acceptability of jackfruit leather was compared by Okilya *et al.* (2010) with cabinet and convection oven drying methods commonly used in the preparation of good-quality fruit leathers. The moisture content of solar-dried leather (18.50%) was not significantly different ($p > 0.05$) from that of cabinet-dried leather (18.85%). However, solar-dried leather had significant lower colour readings compared with cabinet-dried leather, and was disliked and received significantly lower scores ($p < 0.05$) on all sensory attributes evaluated.

3.3.11 Papad

The shelf life of jackfruit papad is much longer than that of ordinary papad and can be stored for longer periods. It can be utilized as an accompaniment to lunch as well as an evening snack. Papad can be prepared by steaming the deseeded ripe bulb for 20 min; when cool, add salt to taste, mix and mash well with the hands. Do not use a mixer. The dough should be very soft and should be made into small balls. Two small plastic sheets are smeared with oil on one side, and one ball is placed on one oiled sheet and covered with the other sheet, oil-side down. The plastic sheet-covered ball is pressed with a manual wooden press or metal press (a device with two flat surfaces and a handle) to form round the papad. The papads are dried in an oven on a low heat, in batches if necessary.

Fig. 3.10. Jackfruit candy. Photograph courtesy of Dr J.D.F. Bassoy.

When completely dried, they can be stored in an air-tight container (Best, 2015).

3.3.12 Candy

Jackfruit candies (Fig. 3.10) are naturally very nutritious as they possess most of the nutraceuticals of the fruit. Candy can be prepared from jackfruit pulp by adding it to sugar syrup and gradually raising its TSS to 70°Bx. The ingredients required for preparation of candy are: 500 g deseeded pulp, 1250 ml water, 750 g sugar, 2 g potassium metabisulfite and 3 g citric acid. The method is as follows. Prepare a 15% brine solution with 1% calcium chloride and soak the bulbs in the brine solution for 2 days. Remove and wash thoroughly to remove the salt and drain. Prepare a sugar syrup of 40°Bx, boil the deseeded bulbs for 5 min in the syrup and cool at room temperature (28–31°C) for 24 h. Remove the bulbs from the syrup and add sugar to the syrup until it is 50°Bx; add the citric acid and immerse the bulbs at room temperature for 24 h. Remove the bulbs from the syrup again and add more sugar to the syrup until it reaches 62°Bx. Re-immerse the bulbs in the syrup and keep at room temperature for 24 h, then remove the bulbs and quickly rinse in water to remove the surface syrup; drain off the syrup by spreading the bulbs on wire trays. Dry the drained bulbs for about 24 h in a solar

Fig. 3.11. Jackfruit-flavoured ice cream. Photograph courtesy of Dr M.J. Gupta.

or tray drier, and pack the candies in jars, tins, cardboard cartons or polythene pouches and seal them (ICUC, 2005). Sreedevi *et al.* (2012) suggested drying the slices for about 28 h in a convective tray dryer at 70°C to obtain a shelf-stable product.

3.3.13 Ice cream

Consumer demand for value-added ice creams incorporating different types of fruits and nut is increasing. Ice creams flavoured with fruits such as strawberry, mango, guava and pineapple are available in the market. Jackfruit pulp, which is rich in nutrients and has a distinct flavour and colour, may be utilized for preparation of ice cream (Fig. 3.11). Ice cream made with 110 g jackfruit pulp, 150 ml milk, 300 ml fresh cream and 100 g sugar was reported to be highly acceptable in terms of body and texture. It has also a good colour and appearance score (Safia and Simi, 2019). Gaikwad *et al.* (2020) observed maximum scores for flavour, colour, appearance, texture, sweetness and overall acceptability in ice cream containing 15% jackfruit pulp, 15% sugar and 0.1% carrageen.

Jackfruit seed powder (Shinde *et al.*, 2021) and seed milk (Lumbantobing *et al.*, 2019) can also be used in preparing jackfruit-flavoured ice cream. The use of plant-based non-dairy milks is increasing in developing food products, particularly for people with lactose intolerance. Substituting dairy milk with jackfruit seed-based milk would provide a vegan, low-fat and crude fibre-rich ice cream (Lumbantobing *et al.*, 2019).

3.3.14 Wine

Wine is defined as an alcoholic beverage with an alcohol content of 8–15% (v/v), usually produced from fruit juice. The fruit juice should have a sugar content of at least 14% (w/w) to convert it to alcohol. If the sugar content is less than 14% (w/w), additional sugar needs to be added (Kumoro *et al.*, 2012). The initial sugar concentrations, fermentation temperature, sulfur dioxide concentrations and specific yeast strains are key factors in determining a successful fermentative process for fruit wine (Sibounnavong *et al.*, 2010).

Fermentation of jackfruit juice with 14% (w/w) sugar concentration using 0.5% (w/v) yeast (*Saccharomyces cerevisiae*) under anaerobic conditions at 30°C for 9 days produced a good-quality wine with 12.13% (v/v) ethanol and a specific jackfruit aroma (Kumoro *et al.*, 2012). Jackfruit wine has good antioxidant activity owing to its gradual increase in phenolic content as the fermentation progresses. The fermentation was completed on day 14, with a maximum alcohol content of 18% (Sharma *et al.*, 2013).

Wine can also be made from jackfruit coproducts such as seedcoat pith, rags, peel and overripe pulp (Cagasan *et al.*, 2021).

3.4 Products from Seed

A ripe jackfruit contains 8–15% seed weight out of the total weight (Amornrat and Kamontip, 2004). Jackfruit seeds are rich in protein (13.5%), fibre (3.19%) and carbohydrate (79.34%) (Ocloo *et al.*, 2010). In jackfruit seed flour, sucrose, glucose and fructose were found to be the predominant soluble sugars, and non-digestible oligosaccharides such as 1-ketose were also found. The total polyphenol content was 2.42 mg gallic acid equivalents g^{-1}. The flour also had a protective and preventative effect associated with a potential prebiotic activity of *Lactobacillus casei* and *Bifidobacterium longum* (Trejo Rodríguez *et al.*, 2021). Jackfruit seeds have been utilized for processing products such as flour for bakery, extruded products, chapatis, starch extraction, confectionery, etc. (Fig. 3.12). To use the seed, the thin, slimy coating of pulp around the seed should be removed by thoroughly rinsing in water and the seeds dried in the shade. Boiled seeds (Fig. 3.13) can be used to prepare curry. Sun-dried or oven-dried (cabinet-dried) seeds can be stored in air-tight containers for over 1 year. The dehydrated seeds are soaked in water for about 18 h and can then be used as fresh seeds for canning or for cooking and consumption (Bhatia *et al.*, 1956).

3.4.1 Flour

Jackfruit flour is prepared from the seeds. The seeds are first sorted and cleaned. Cleaning is followed by roasting at 160°C for about 60 min. The roasted seeds are then dried in a tray drier or cabinet drier and milled in a flour mill (Mandave *et al.*, 2018). Roasting of the seeds prior to flour production makes the protein more available and increases the water-absorption capacity with reduced oil absorption. Roasting also positively affects amylase content, which is an important factor with regard to the end-use properties of the product (Eik Ejiofor *et al.*, 2014). However, compared with wheat flour, jackfruit seed flour has limited cooking properties, particularly for baking (Cuadra *et al.*, 2018). Incubation of pre-ground seeds with lactic acid at 1–3% solution for 16–32 h at room temperature improves the majority of the quality parameters and the fermented seed powder had outstanding potential in food systems (Cuadra *et al.*, 2018). Jackfruit seed flour (Fig. 3.14) is a rich source of minerals such as potassium (6466 p.p.m.), magnesium (4582 p.p.m.) and sodium (8906 p.p.m.). It has a high water-absorption capacity (2.91 ml g^{-1}), oil absorption capacity (0.884 ml g^{-1}) and bulk density (0.873 g ml^{-1}). It was found to have a low gelation capacity of 17% (Abraham and Jayamuthunagai, 2014). The gluten-free seed flour can be used as a substitute for wheat flour or mixed with wheat or

Fig. 3.12. Products from jackfruit seeds. Photograph courtesy of Mr Shree Padre.

Fig. 3.13. Boiled jackfruit seeds. Photograph courtesy of Dr M.J. Gupta.

Fig. 3.14. Jackfruit seed flour. Photograph courtesy of Dr M.J. Gupta.

rice flour, and can be used in making bread, chapattis, cakes, biscuits, noodles, pasta, etc. (Hema, 2015).

Jackfruit seed flour packed in polyethylene pouches can be stored for 6 months in a refrigerator without any unwanted changes (Ariani, 2007).

3.4.2 Bakery products

Jackfruit seed powder can be used to fortify together with functional ingredients in bakery products such as bread, biscuits, cakes and muffins. Supplementation with 5–25% seed powder makes the bread nutritionally rich with a higher carbohydrate, fat, protein and crude fibre content (Tulyathan *et al.*, 2002; Hossain, 2014). In the preparation of chapattis, 20–25% of the wheat flour can be replaced with jackfruit seed flour (Sultana *et al.*, 2014). In biscuits, 10–40% supplementation of seed powder makes them nutritionally rich (Islam *et al.*, 2015), while 5–15% supplementation of seed flour increases the protein content and reduces the fat content in cakes (Arpit and John, 2015). Replacing 16% of wheat flour with jackfruit flour and 11% of sugar with polydextrose reduced the calories by 34% in cake compared with normal chocolate cake (Siti-Faridah and Noor-Aziah, 2012).

3.4.3 Coffee

Roasted jackfruit seeds (after drying and/or fermentation) produced changes in the aroma sensory profiles that resulted in an agreeable chocolate aroma (Papa Spada *et al.*, 2018). The price of cocoa beans has increased significantly in recent years following an increase in demand, but production is not growing to match the demand. Thus, there is a need to substitute cocoa with other products. Papa Spada *et al.* (2018) stated that the use of jackfruit seed powder replacing cocoa powder in a cappuccino formulation is possible. It was suggested to use dry seed powder and the ideal level of cocoa substitution in cappuccino formulations was found to be 50–75%. The use of jackfruit seed

Fig. 3.15. Coffee from jackfruit seed powder. Photograph courtesy of Mr Sree Padre.

powder has advantages such as moisture reduction, high wettability and solubility, sensory acceptation in terms of the chocolate aroma and other similarities, such as colour, density and appearance (Fig. 3.15).

3.4.4 Extruded products

Jackfruit seed powder is used to improve the nutritional and nutraceutical properties of extruded products such as breakfast cereals, snacks, pellets, crispy flat bread, pre-cooked flours and cereal-based baby foods (Waghmare *et al.*, 2019). Nandkule *et al.* (2015) reported an increase in protein, ash, fat and crude fibre contents and reduced carbohydrate content in noodles fortified with jackfruit seed flour. Incorporation of 30–40% jackfruit seed meal increased the fibre content of cereal bars (Santos *et al.*, 2011), and incorporation of 10% jackfruit seed flour increased the nutrient content and firmness of pasta (Abraham and Jayamuthunagai, 2014).

3.5 Products from Jackfruit Waste

The inedible parts of jackfruit include the prickly rind/peel, the non-edible perianth and the central core (Xu *et al.*, 2018). Jackfruit peel consists around 40–60% of total fruit weight depending on the varieties. This non-edible waste may be utilized as animal feed (Ndyomugyenyi *et al.*, 2014), or for the production of bioethanol (Yuvarani and Dhas, 2017), pectin (Xu *et al.*, 2018) and other valuable products. The peel is rich in cellulose (27.75%), pectin (7.52%), protein (6.2%) and starch (4%) (Sundarraj and Ranganathan, 2017).

Bioethanol from jackfruit peel was extracted by fermentation using *S. cerevisiae* yeast (Yuvarani and Dhas, 2017). Xu *et al.* (2018) extracted pectin from peel using different organic and mineral acids and ultrasonic microwave-assisted extraction.

The inedible jackfruit parts are potential raw materials for bioenergy and biochar production (Nsubuga *et al.*, 2020). The high sugar content in jackfruit waste can be converted anaerobically into biogas (Chakravarty, 2016). Jackfruit waste is a potential raw material for biochar production through pyrolysis given its carbon content, and this biochar can be used for the production of briquettes, adsorption of heavy metals in water and soil amendment (Nsubuga *et al.*, 2020).

A process for the preparation of clarified juice from waste has been developed, which involves the treatment of jackfruit waste with pectic enzyme at 0.03% concentration. Incubation for 2 h at 40°C and subsequent filtration yielded 60% clarified juice. Thus, an economic way to obtain a beverage from waste brings utilization of the jackfruit to over 80% (John and Narasimhan, 2007)

3.6 Conclusion

Jackfruit is a rich source of minerals and vitamins, and is useful in food and many other industrial applications. Because of its flavour and nutritional quality, jackfruit has the potential to become a commercially important tropical fruit crop not only for fresh consumption but also for processing into value-added products. Despite this, utilization of jackfruit has not been exploited to its full potential. Recently, several advanced technologies have been developed for value addition, and there is immense scope for value-added products of jackfruit including blending with other fruits. Products such as chips, canned fruit, jam, jelly, beverages, leather and toffee/candy have great potential to attract consumers and for marketability of the fruit. Minimally processed ready-to-eat tender jackfruit for cooking and in cans also has great potential. The development of low-cost processing technologies, value addition through extension of the shelf life and processing of marketable surplus into value-added products, as well as utilization of food industry waste/by-products are the key areas on which to focus in order to increase the share of these value-added products in the market. The policy environment is also important. Agro-processing sectors involved with major fruit crops receive government support in the form of subsidies, foreign-exchange allowances, price stabilization or guarantees, and access to specialists and consultants. Small-scale processors involved with underutilized fruits such as jackfruit do not have such advantages.

References

Abraham, A. and Jayamuthunagai, J. (2014) An analytical study on jackfruit seed flour and its incorporation in pasta. *Research Journal of Pharmaceutical, Biological and Chemical Sciences* 5, 1597–1610.

Amornrat, M. and Kamontip, S. (2004) Physico-chemical properties of flour and starch from jackfruit seeds (*Artocarpus heterophyllus* Lam.) compared with modified starches. *International Journal of Food Science and Technology* 39, 271–276.

Ariani, S. (2007) Nutritional quality and value addition to jackfruit seed flour. PhD thesis, University of Agricultural Science, Dharwad, India.

Arpit, S. and John, D. (2015) Effect of different levels of jackfruit seed flour on quality characteristics of chocolate cake. *Research Journal of Agriculture and Forestry Sciences* 3, 1597–1610.

Babu, N., Dinakaran, A., Raman, M., Ravindran, A., Sankar, T.V. *et al.* (2020) Development of jackfruit leather and its shelf-life analysis in different packaging materials. *Food Science Research Journal* 11(2), 158–168. DOI: 10.15740/HAS/FSRJ/11.2/158-168.

Barga, J., Galvez, L., Lauzon, R. and Diczbalis, Y. (2019) Chemical quality of extracted oil and sensory quality of vacuum fried jackfruit (*Artocarpus heterophyllus* Lam.) pulp as influenced by frying cycle. *Annals of Tropical Research* 130–141. DOI: 10.32945/atr41210.2019.

Best, A. (2015) *Value-Adding Options for Tropical Fruit Using Jackfruit as a Case Study*. Rural industries research and development corporation Publication No. 15/042. Australian Government, Wagga, NSW, Australia.

Bhatia, B.S., Siddappa, G.S. and Lal, G. (1956) Development of products from jackfruit. V. Dried jack seeds, flour, roasted nut and jack papad and toffee. *Indian Food Packer* 10, 9–16.

Bhuyan, M.A.J., Saha, M.G. and Rahman, M.A. (2013) Value addition to jackfruit (*Artocarpus heterophyllus* Lam.) through integrated processing and preservation. In: *Workshop on Volarisation of Traditional Processing of Indigenous and Underutilized Fruits, 14-16 January, 2013*, Institute of Technology of Cambodia, Phom Penh, Cambodia.

Cagasan, C.U., L'ingatong, C.A., Pore, K.M., Ramada, R., Restor, C.D. *et al.* (2021) Production and quality evaluation of wine from jackfruit co-products. *International Journal of Life Sciences and Biotechnology* 4, 340–352. DOI: 10.38001/ijlsb.827739.

Chakravarty, G. (2016) Evaluation of fruit waste as a substrate for the production of biogas. *Annals of Biological Research* 7, 25–28.

Chin, A.H.G. and Nushirwan, Z. (1989) Jackfruit quality for canning in syrup. *MARDI Research Journal* 17, 266–274.

Cruz-Cansino, N.S., Ariza-Ortega, J.A., Alanis-Garcia, E., Ramírez-Moreno, E., Velázquez-Estrada, R.M. *et al.* (2020) Optimal thermoultrasound processing of jackfruit (*Artocarpus heterophyllus* Lam.) nectar: physicochemical characteristics antioxidant properties, microbial quality, and fatty acid profile comparison with pasteurized nectar. *Journal of Food Processing and Preservation* 45: e15029.

Cuadra, M.N.J., Galvez, L.A. and Amestoso, F.J. (2018) Evaluation of physicochemical, functional and microbiological properties of fermented EVIARC sweet jackfruit (*Artocarpus heterophyllus* Lam.) seeds flour. *Annals of Tropical Research* 40, 100–113. DOI: 10.32945/atr4018.2018.

Dam, S.M. and Nguyen, N.T. (2013) Production of fermented beverage from fruit rags of jackfruit (*Artocarpus heterophyllus*). *Acta Horticulturae* 989, 285–292. DOI: 10.17660/ActaHortic.2013.989.37.

da Silva, P.F. and Moreira, R.G. (2008) Vacuum frying of high-quality fruit and vegetable-based snacks. *LWT - Food Science and Technology* 41(10), 1758–1767. DOI: 10.1016/j.lwt.2008.01.016.

Dushyantha, D.K., Raghavendra Kumar, M. and Suvarna, V.C. (2011) Study on fermentation of jackfruit (*Artocarpus heterophyllus* L.) juice by beneficial lactic acid bacteria. *Research Journal of Pharmaceutical, Biological and Chemical Science* 2, 732–739.

Eik Ejiofor, J., Beleya, E.A. and Onyenorah, N.I. (2014) The effect of processing methods on the functional and compositional properties of jackfruit seed flour. *International Journal of Nutrition and Food Sciences* 3(3), 166. DOI: 10.11648/j.ijnfs.20140303.15.

Gaikwad, S.B., Kamble, D.K. and Jaybhay, V.B. (2020) Development of ice cream using jackfruit pulp. *Pharma Innovation Journal* 9, 533–539.

Galvez, L.A., Patindol, R.A. and Mabesa, L.B. (2013) Influence of variety and sulfite on controlling browning within four months of storage of dehydrated jackfruit (*Artocarpus heterophyllus* Lam.) pulp. *Annals of Tropical Research* 35, 40–59. DOI: 10.32945/atr3523.2013.

Granda, C., Moreira, R.G. and Tichy, S.E. (2004) Reduction of acrylamide formation in potato chips by low-temperature vacuum frying. *Journal of Food Science* 69(8), E405–E411. DOI: 10.1111/j.1365-2621.2004.tb09903.x.

Gunasena, H.P.M., Ariyadasa, K.P., Wikramasinghe, A., Hearth, H.M.W., Wikramasinghe, P. *et al.* (1996) *Manual of Jack Cultivation in Sri Lanka*, Vol. 48. Forest Information Service, Forest Department, Sri Lanka.

Haq, N. (2006) *Jackfruit, Artocarpus heterophyllus*. Southampton Centre for Underutilized Crops, University of Southampton, Southampton, UK.

Hema, J. (2015) Development of nutritious instant dried powder by mixing bulb and seeds of the jackfruit. MSc thesis, Department of Agroprocessing, Bangabandhu Sheikh Mujibur Rahman Agricultural University, Gazipur, Bangladesh.

Hossain, M.T. (2014) Development and quality evaluation of bread supplemented with jackfruit seed flour. *International Journal of Nutrition and Food Sciences* 3(5), 484. DOI: 10.11648/j.ijnfs.20140305.28.

ICUC (International Centre for Underutilised Crops) (2005) *Training Manual on Processing and Small Business Development*. ICUC, University of Southampton, UK.

Islam, S., Begum, R. and Khatun, M. (2015) A study on nutritional and functional properties analysis of jackfruit seed flour and value addition to biscuits. *International Journal of Engineering Research and Technology* 4, 139–147.

John, P.J., Balasubramanyam, N. and Narasimham, P. (1993) Effect of packaging, processing and storage conditions on the quality of raw jackfruit curry in flexible pouches. *Journal of Food Processing and Preservation* 17(2), 109–118. DOI: 10.1111/j.1745-4549.1993.tb00228.x.

Kumoro, A.C., Sari, D.R., Pinadita, P.P., Retnowati, D.S. and Bidiyati, C.S. (2012) Preparation of wine from jackfruit (*Artocarpus heterophyllus* Lam.) juice using baker yeast and initial sugar concentration. *World Applied Science Journal* 16, 1262–1268.

Kushala, G., Sreenvas, K.N. and Siddappa, R. (2012) Product development acceptability and cost effectiveness of jackfruit jam blended with avocado and kokum. *Food Science Research Journal* 3, 78–80.

Lakshmana, J.H., Jayaprahash, C., Kumar, R., Kumaraswamy, M.R., Kathiravan, T. *et al.* (2013) Development and evaluation of shelf stable retort pouch processed ready-to-eat tender jackfruit (*Artocurpus heterophyllus*) curry. *Journal of Food Processing & Technology* 04(10), 1000274. DOI: 10.4172/2157-7110.1000274.

Li, Q.J., Feng, C.M., Li, X.R., Wen, L.X., Huang, H. *et al.* (2017) Optimization for processing technology of light-sugar dried jackfruit with natural flavour. *Journal of Southern Agriculture* 48, 889–895.

Lumbantobing, E., Tanardi, S. and Putra, A. (2019) Development of vegan ice cream from jackfruit (*Artocarpus heterophyllus*) seed-based milk. In: *Proceedings of the 16th ASEAN Food Conference - Outlook and Opportunities of Food Technology and Culinary for Tourism Industry. Science and Technology Publications*, Indonesia, pp. 66–71. DOI: 10.5220/0009983700660071.

Maity, T., Bawa, A.S. and Raju, P.S. (2017) Effect of preconditioning on physicochemical, microstructural, and sensory quality of vacuum-fried jackfruit chips. *Drying Technology* 36(1), 63–71. DOI: 10.1080/07373937.2017.1300590.

Mandave, P., Bobade, H. and Patil, S. (2018) Jackfruit seed flour. Processing technologies and applications. *International Journal of Agricultural Engineering* 11(Special issue), 149–154. DOI: 10.15740/HAS/IJAE/11.Sp.Issue/149-154.

Marak, N.R., Nganthoibi, R.K. and Momin, C.W. (2019) Process development for brining of tender jackfruit. *International Journal of Current Microbiology and Applied Sciences* 8(4), 2408–2414. DOI: 10.20546/ijcmas.2019.804.280.

Mondal, C., Sultana, S., Mannan, M.A. and Khan, S.A.K.U. (2016) Preparation and sensorial evaluation of pickles, jam, jelly and squash developed from jackfruit (*Artocarpus heterophyllus*). *Journal of Environmental Science and Natural Resources* 9(2), 35–41. DOI: 10.3329/jesnr.v9i2.32149.

Mukherjee, B. (1993) *Traditional Medicine*. Oxford and IBH Publishing Co, New Delhi.

Nandkule, V.D., Masih, D., Sonkar, C. and Patil, D.D. (2015) Development and quality evaluation of jackfruit seed and soy flour noodles. *International Journal of Science Engineering and Technology* 3, 802–806.

Ndyomugyenyi, E.K., Okot, M.W. and Muteikka, D. (2014) Production and availability of jackfruit (*Artocarpus heterophyllus*) and Java plum (*Syzygium cumini*) beans for livestock feeding in eastern and central regions of Uganda. *Livestock Research for Rural Development* 26, 73. DOI: 10.5455/jasa.20141113124747.

Nsubuga, D., Banadda, N., Kabenge, I. and Wydra, K.D. (2020) Potential of jackfruit waste for biogas, briquettes and as a carbondioxide sink-a review. *Journal of Sustainable Development* 13(4), 60. DOI: 10.5539/jsd.v13n4p60.

Ocloo, F., Bansa, D., Boatin, R., Adom, T. and Agbemavor, W. (2010) Physico-chemical, functional and pasting characteristics of flour produced from jackfruits (*Artocarpus heterophyllus*) seeds. *Agriculture and Biology Journal of North America* 1(5), 903–908. DOI: 10.5251/abjna.2010.1.5.903.908.

Okilya, S., Mukisa, I.M. and Kaaya, A.N. (2010) Effect of solar drying on the quality and acceptability of jackfruit leather. *Electronic Journal of Environmental, Agricultural and Food Chemistry* 9, 101–111.

Papa Spada, F., da Silva, P.P.M., Mandro, G.F., Margiotta, G.B., Spoto, M.H.F. *et al.* (2018) Physicochemical characteristics and high sensory acceptability in cappuccinos made with jackfruit seeds replacing cocoa powder. *PLOS ONE* 13(8), e0197654. DOI: 10.1371/journal.pone.0197654.

Pritty, S.B. and Sudheer, K.P. (2019) Standardization of thermal process parameters of canned tender jackfruit. *Journal of Postharvest Technology* 7, 6–13.

Rudrawar, B.D. and Kale, V.S. (2017) Development and standardization of jackfruit pickle and study of sensory analysis. *Life Science Leaflets* 84, 57–65.

Safia, N.E. and Simi, S. (2019) Sensory evaluation of ice creams prepared with different inclusion levels of jackfruit pulp. *Asian Journal Of Home Science* 14(2), 411–414. DOI: 10.15740/HAS/AJHS/14.2/411-414.

Santos, C.T., Bonomo, R.F., Fontan, R.D.C.I., Bonomo, P., Veloso, C.M. *et al.* (2011) Characterization and sensorial evaluation of cereal bars with jackfruit. *ActaScientiarum Technology* 33(1). DOI: 10.4025/actascitechnol.v33i1.6425.

Sharma, N., Bhutia, S.P. and Aradhya, D. (2013) Process optimization for fermentation of wine from jackfruit (*Artocarpus heterophyllus* Lam.). *Journal of Food Processing & Technology* 04(2). DOI: 10.4172/2157-7110.1000204.

Shinde, V.L., Pawar, C.D., Warang, O.S., Dandekar, V.S., Kulkarni, M.M. *et al.* (2021) Studies on preparation of ice-cream from jackfruit (*Artocarpus heterophyllus*) seed powder. *International Journal of Chemical Studies* 9(1), 2710–2712. DOI: 10.22271/chemi.2021.v9.i1al.11636.

Shwetha, M.S. and Ranganna, B. (2018) Incidence of bacteria-yeast-moulds on processed jackfruit squash from different genotypes. *International Journal of Current Microbiology and Applied Sciences* 7(2), 775–779. DOI: 10.20546/ijcmas.2018.702.098.

Shyu, S.L., Hau, L.B. and Hwang, L.S. (1998) Effect of vacuum frying on the oxidative stability of oils. *Journal of the American Oil Chemists Society* 75(10), 1393–1398. DOI: 10.1007/s11746-998-0188-3.

Sibounnavong, P., Daungpanya, S., Sidtiphanthong, S., Keoudone, C. and Sayavong, M.J. (2010) Application of *Saccharomyces cerevisiae* for wine production from star gooseberry and carambola. *Journal of Agricultural Technology* 6, 99–105.

Siti-Faridah, M.A. and Noor-Aziah, A.A. (2012) Development of reduced calorie chocolate cake with jackfruit seed (*Artocarpus heterophyllus* Lam.) flour and polydextrose using response surface methodology (RSM). *International Food Research Journal* 19, 515–519.

Souza, L.C., Souza, V.C., Silva, L.T.S., Pereira, A.S. and Silva, A.G. (2014) Sensory evaluation of jaca (*Artocarpus heterophyllus*) processed by osmoconvective dehydration. *Magistra* 26, 141–145.

Sreedevi, M.S., Palanimuthu, V., Nagaraja, G. and Rajhupati, C.J. (2012) Development and standardization of recipe for jackfruit candy. *International Journal of Processing and Postharvest Technology* 3, 73–76.

Sultana, A., Rahman, Md.R.T., Islam, M., Rahman, M. and Alim, Md.A. (2014) Evaluation of quality of chapaties enriched with jackfruit seed flour and Bengal gram flour. *IOSR Journal of Environmental Science, Toxicology and Food Technology* 8(5), 73–78. DOI: 10.9790/2402-08537378.

Sundarraj, A.A. and Ranganathan, T.V. (2017) Physiochemical characterization of jackfruit (*Artocarpus integer* Thumb.) peel. *Research Journal of Pharmaceutical, Biological and Chemical Sciences* 93, 2285–2295.

Trejo Rodríguez, I.S., Alcántara Quintana, L.E., Algara Suarez, P., Ruiz Cabrera, M.A. and Grajales Lagunes, A. (2021) Physicochemical properties, antioxidant capacity, prebiotic activity and anticancer potential in human cells of jackfruit *(Artocarpus heterophyllus)* seed flour. *Molecules* 26(16), 4854. DOI: 10.3390/molecules26164854.

Tulyathan, V., Tananuwong, K., Songjinda, P. and Jaiboon, N. (2002) Some phytochemical properties of jackfruit (*Artocarpus heterophyllus* Lam.) seed flour and starch. *Science Asia* 28(1), 37. DOI: 10.2306/scienceasia1513-1874.2002.28.037.

Waghmare, R., Memon, N., Gat, Y., Gandhi, S., Kumar, V. *et al.* (2019) Jackfruit seed: an accompaniment to functional foods. *Brazilian Journal of Food Technology* 22, e2018207. DOI: 10.1590/1981-6723.20718.

Xu, S.-Y., Liu, J.-P., Huang, X., Du, L.-P., Shi, F.-L. *et al.* (2018) Ultrasonic-microwave assisted extraction, characterization and biological activity of pectin from jackfruit peel. *LWT - Food Science and Technology* 90, 577–582. DOI: 10.1016/j.lwt.2018.01.007.

Yi, J., Wang, P., Bi, J., Liu, X., Wu, X. *et al.* (2016a) Developing novel combination drying method for jackfruit bulb chips: instant controlled pressure drop (DIC)-assisted freeze drying. *Food Bioprocess Technology* 9(3), 452–462. DOI: 10.1007/s11947-015-1643-4.

Yi, J., Zhou, L., Bi, J., Chen, Q., Liu, X. *et al.* (2016b) Influence of pre-drying treatments on physicochemical and organoleptic properties of explosion puff dried jackfruit chips. *Journal of Food Science and Technology* 53(2), 1120–1129. DOI: 10.1007/s13197-015-2127-2.

Yuvarani, M. and Dhas, C.S.I. (2017) Synthesis of bioethanol from *Artocarpus heterophyllus* peel by fermentation using *Saccharomyces cerevisiae* at low cost. *GRD Journal for Engineering* 2, 1–8.

4 Breeding and Varieties

Sisir Mitra*
Former Professor, Faculty of Horticulture, Bidhan Chandra Krishi Viswavidyalaya, Mohanpur, West Bengal, India

4.1 Introduction

Jackfruit (*Artocarpus heterophyllus*) is a native of India that grows wild in the rainforests of the Western Ghats. It has been cultivated since time immemorial. It was introduced and became naturalized in many tropical countries, particularly in South and South-east Asia (Mitra, 2000). The genus *Artocarpus* is a member of the family Moraceae. The genus contains about 50–60 species, distributed throughout the Indo-Malayan region and in southern China. All species are monoecious, the male flower being borne in the pseudo-catkins and the female in pseudo-heads (Mitra, 2017).

Most of the jackfruit trees grown have seedling origins and, as they are cross-pollinated, they have produced highly variable and diversified heterozygous trees. The improvement of jackfruit through breeding has not yet attracted serious attention from research workers as for other major commercial fruit crops because the jackfruit is not a priority listed fruit for most of the countries where it is grown.

As it is protandrous, cross-pollinated and mostly propagated from seed, jackfruit has innumerable types and forms. It differs in age of first bearing, season of bearing (early, medium and late, and off-season), and size and shape of fruits. There is variation in the density, size and shape of the spines on the rind, and in maturity period of the fruits. The fruits also vary in bulb/flake colour, consistency, sweetness, flavour and taste. Fruits of all jackfruit trees in any locality are different from one another in a number of characteristics. Such variabilities between and among the trees offers ample scope for improvement of the crop by clonal selection (Amzad Hossain, 1996). The jackfruit varieties available in different countries are listed in Table 4.1.

The seedling progenies exhibit a wide range of morpho-agronomic characteristics (Table 4.2). Variation exists in fruit shape, colour, aroma, flesh colour and seed numbers (Table 4.3), as well as in fruit characteristics (Table 4.4). A few attempts have been made to understand the extent of genetic diversity in jackfruit from the study of morphological characteristics, and to select superior types.

Jackfruit is indigenous to India, and there is a wide range of variability in the germplasm of jackfruit in this country. Cultivated types fall broadly into two groups: soft- and firm-fleshed types. Soft-fleshed types have a pulp that is highly juicy and soft, and are popular in eastern India; firm-fleshed types are found mostly in southern India, particularly in Kerala state. The varieties of this type are good for freezing and drying. Local selections have been named 'Gulabi' (rose-scented), 'Champa' (flavour like that of *Michelia* spp.) and 'Hazari' (bearing a large number of fruits in a tree). As

*sisirm55@gmail.com

DOI: 10.1079/9781800622319.0004

Table 4.1. Jackfruit varieties growing in different countries. Data from: Morton (1987); Campbell and McNaughton (1994); Mitra (2000, 2017); Wasielewski and Campbell (2000); Mitra and Maity (2002); Haq (2006); Magdalita *et al.* (2011); Wangchu *et al.*, 2011Wangchu *et al.*, 2013; Wang *et al.* (2012); and Karunakaran *et al.* (2017).

Country	Varieties
Australia	Golden Nugget, Black Gold, Honey Gold, Lemon Gold, Rajang, Amber, Galaxy, Fitzory, Nahen, Kapa
Bangladesh	Topa, Hazari, Chola, Goal, Koa, Khaja, Shariatpur, Sreepur, Comilla, Savar, BAU Kanthal 1, BARI Kanthal 1
Brazil	Jaca mole, Jaca-dura, Jack-manteiga
China	Malaixiya No. 1, Changyou, Siji, Xiangmi No. 17, Xiangmi No. 1, Hongbao, Sijimition, Red Flesh Jackfruit
India	Champa, Hazari, Gulabi, Safeda, Khaja, Ghila, Varikka, Koozha, Navarikka, Palar 1 Jack, PPI-1 Jack, PLR (J)-2, KT-7, Swarna Halsu, NSP, Konkan Prolific, Janagare, Byrachandra, Kantanukunatae, Sharath, Kachahalli, JT5, JT40, JT42, Swarna Manohar, Swarna Poorti, Siddu, Shankara, CHESJF3
Indonesia	Kandel, Mini, Tabouey
Malaysia	J29, J32 (Mantin), J32 (NS1), J33 (Tekam Yellow), J35 (Crystal Jackfruit 1 (CJ1)), J37 (Mastura)
Myanmar	Talaing, Kala
Philippines	Torres, EVIARC Sweet, Sinapelo, Burabod, Cerventes Gold, MMSU-SRO Sweet, Mabini
Sri Lanka	Father Long, Kothmale, Maharagama, Mandoor, Arthur V. Dias, Tellippalai, Ganegoda, Kurucos, Rosacos, Hirosa, Singaporejack
Thailand	Dangs Rasimi, Kun Wi Chan, Kha-num Nang, Kha-num Lamoud
USA (Florida)	Black Gold, Cheena, Cochin, Golden Nugget, Honey Gold, Lemon Gold, Dang Rasimi, Kun Wi Chan, J-31, NS1, Tabouey
Vietnam	Sieu Som, La Bang, Vien Linh, Red Flesh, MDN06, Mit Wghe, Mit Dua, Mit Uot

a result of local surveys, some improved types have been collected. As raw jackfruit is in high demand in India as a vegetable for culinary purposes, emphasis is also given to fruit characteristics such as the thickness of the rind and softness of the flesh at premature stages of fruit development (Mitra, 2000).

4.2 Breeding Objectives

The first breeding objective was to develop a variety that was regular and prolific in bearing, as jackfruit yield is not similar every year and the trees are often erratic in bearing (Ray, 2002).

As mentioned earlier, in terms of quality, two types of jackfruit are commonly recognized: one with firm, sweet pulp and the other with tender (or softer), mushy pulp. The later type is considered inferior. The varieties developed through breeding should bear fruits containing larger quantity of edible pulp (bulbs) and small seeds, and should be good-quality fruits with acceptable pulp colour and texture, and good flavour and sweetness.

Usually, large-sized fruits are inconvenient to handle and are also not preferred by the consumer, so new varieties should be an acceptable size and have a symmetrical form.

The jackfruit tree has a long juvenile phase (7–8 years). Reducing this period is another important objective. A few varieties such as 'Singapore' in Sri Lanka (Richards, 1950) and 'Nangka Salak' in Java (Yaacob and Subhadrabandhu, 1995) have a short juvenile phase (1.5–3 years). Such varieties can be used as one of the parents for inducing precocity.

Table 4.2. Variation in morpho-agronomic characteristics of jackfruit.

Tree habit	Open, spreading, low spreading, sparse upright
Tree growth rate	Fast, moderate, slow
Canopy	Dense, mostly dome-shaped, slightly pyramidal or flat topped; ranges from 3.5 to 6.7 m
Leaf shape	Elliptic, elliptic-obovate, obovate, oblong, lanceolate, oval
Leaf size	4–25 cm long; 2–12 cm wide
Leaf petiole	1.2–4.0 cm long
Fruit maturity	Variable
Fruit seasons	Variable
Fruit shape	Oblong, ellipsoid, triangular, spheroid, claviform, round
Fruit numbers per tree	15–1450
Fruit weight	1.2–22.0 kg
Fruit texture	Fibrous, firm, coarse, melting, crisp
Seed shape	Oblong, ellipsoid, irregular, reniform, elongate, spheroid
Seed weight	250–1230 g

Table 4.4. Variation in the fruit characteristics of jackfruit.

Characters	Range
Fruit weight (kg)	1.2–22.0
Fruit length (cm)	20.5–60.6
Fruit diameter (cm)	16.4–29.5
Fruit girth (cm)	50.5–95.8
No. of bulbs per fruit	24.2–580.2
Pulp (%)	18.3–60.9
Seed (%)	2.6–23.1
Rachis (%)	1.5–21.4
Rind (%)	20.6–72.0
Brix (°Bx)	13.8–25.2

Tender jackfruits (about 45–60 days old) are used as a vegetable in India and in South Asian countries. Lately, the demand has been increasing from western countries as a substitute for meat. One breeding objective should be to develop varieties that bear year-round or twice a year. The other objectives should be to develop varieties that are resistant to borers, dieback and flooding, those with wider environmental adaptations, dwarf trees and plants with a long postharvest life of the fruits.

As a cross-pollinated and mostly seed-propagated species, jackfruit population diversity results from the breeding system and natural selection associated with local environmental differences (evolution) or from human selection and the preference of the local community that is cultivating them (domestication). As an underutilized crop, jackfruit has escaped intensive selection and cultivation (APAARI, 2012). For these reasons, a wide range of genetic and morphological variation have been reported for jackfruit (Mitra and Mani, 2000; Mitra and Maity, 2002; Azad *et al.*, 2007; Wangchu *et al.*, 2013; Shyamalamma *et al.*, 2017). The variations in morpho-agronomic characteristic and in the fruits have been documented. These studies have, however, been limited in geographical scope and with limited number of individuals. Azad *et al.* (2007) evaluated the diversity of jackfruit in five regions of Bangladesh and selected 10 trees as potential tree types based

Table 4.3. Variation in the fruit flesh characteristics of jackfruit.

Flesh aroma	Mild, strong
Flesh colour	Creamy white, light yellow, deep yellow, yellow, reddish, red-golden
Flesh texture	Crisp, coarse, fibrous/coarse, fibrous, smooth
Quantity of fibre	Scarce, medium, abundant
Juiciness of pulp	Very juicy, juicy, medium juicy, less juicy, dry

on farmers' criteria, which included high yield, fruit quality, sweetness and early-fruiting types. These types were recommended for cloning. Maity *et al.* (2002) and Wangchu *et al.* (2013) selected 44 genotypes from six districts (Nadia, South 24 Parganas, North 24 Parganas, Midnapore, Purulia and Cooch Behar) of West Bengal, India. The criteria of selection was to identify: (i) suitable genotypes to use as table fruit (small to medium in size); (ii) fruits suitable to use as a vegetable (less fibre content and medium to large in size); (iii) fruits suitable for processing (large fruit, soft flesh and juicy); (iv) trees suitable for commercial orcharding (early bearing, precocious, medium-sized fruit and sweet); and (v) clones bearing twice a year. Farmers have also propagated clones from selections taken from natural populations for their desirable characteristics, but the selection has not been rigorous. Effective selection of clones with desirable characteristics and vegetative propagation of these clones is practicable (Haq, 2006). Superior jackfruit mother trees (or plus trees) can be used for improvement of the species as the population of jackfruit in the field is composed of variable genotypes (Gunasena *et al.*, 1996).

4.3 Germplasm Conservation

The wealth of genetic diversity of jackfruit is the basis for present and future use. Any erosion of germplasm resources will not only result in loss of genetic material but also hinder improvement of the crop. Conservation of jackfruit and related gene pool diversity thus needs more attention (Haq, 2006).

Moderate genetic erosion of jackfruit was indicated by Sastrapradja (1975) to be occurring in South-east Asia, with replacement of seedlings by clonal cultivars. In Bangladesh, a shift from direct seedlings by tree owners to the planting of nursery-raised, seed-propagated saplings started in the early 1980s (Haq, 2006). Reports of a decline in jackfruit morphological variation have raised concerns about a possible decline in genetic diversity (Khan *et al.*, 2010). However, Witherup *et al.* (2019) reported that sapling propagation does not appear to have reduced the genetic diversity of jackfruit in Bangladesh. It is important to bear in mind that Bangladeshi jackfruit has not yet undergone a transition to clonally propagated modern varieties, and now is the time to conserve existing diversity (Witherup *et al.*, 2019).

4.3.1 *In situ* conservation and field gene banks

There is a need to identify/locate precise areas of diversity and possible natural wild populations of jackfruit in forest habitats such as the Western Ghats of India, the Andaman Islands and the south-eastern part of India for *in situ* conservation (Reddy *et al.*, 2001). Germplasm conservation, out of necessity, needs to be in the form of field gene banks, which are expensive to maintain. Moreover, jackfruit is not a priority crop for research in most of the countries where it is grown, and there is lack of long-term financial support to develop and maintain field gene banks. In order to develop on-farm conservation, farmers will need to be persuaded to conserve trees that they find useful.

Field gene banks have recently been established by government institutions and some agricultural universities in India, Bangladesh, Sri Lanka, Malaysia, Vietnam, Thailand, the Philippines and Australia. In India, a large collection of jackfruit germplasm (242) is being maintained in the field gene banks at the Indian Council of Agricultural Research (ICAR) National Bureau of Plant Genetic Resources (NBPGR) Regional Station Ranchi (Jharkhand, India). Some accessions are also being maintained at the NBPGR Regional Station, Thrissur, Kerala, and the Horticulture Experiment and Training Centre, Saharanpur, Uttar Pradesh, in India (Gupta *et al.*, 2019). These collections are a valuable source of genetic materials to initiate improvement programmes and/or testing. Jackfruit germplasms were planted in the forestry and plantation area of Gunungkidul Regency, Yogyakarta, Indonesia, in 2011–2012. The collection area was utilized as an *ex situ* area for the genetic conservation of jackfruit in Indonesia (Palupi and Daryono, 2021)

4.3.2 *Ex situ* conservation

Jackfruit seeds are recalcitrant. They cannot be dried and stored for a long. As the seeds lose viability quickly, seed storage is not possible. *In vitro* storage of vegetatively propagated clonal material can be done by using slow-growth techniques for medium-term conservation (Haq, 2006). Cryogenic storage of embryos has been investigated for jackfruit. This technique involves the selection of fruits of correct maturity and subsequent processing of the seeds and embryos prior to treatment. A lower moisture content is essential for cryogenic storage (Fu *et al.*, 1993). Changes in physiological characteristics, and desiccation and freezing sensitivity of the embryonic axis with increasing seed maturity were studied (Chandel *et al.*, 1995). Embryos measuring 4–5 mm from mature, unripe fruits were found to be best for survival and could be regenerated repeatedly. The selected fresh embryos were partially desiccated (to 60% moisture content) prior to treatment with a dimethyl sulfoxide and 0.5% proline mixture. Freezing was carried out in stages: the embryos were pre-frozen slowly at 10°C min^{-1} down to –40°C, after which they were plunged directly into liquid nitrogen at –196°C (Chandel *et al.*, 1995).

Rajasekharan and Ganeshan (2019) developed the technology for cryopreservation of pollen grains. Pollen grains were extracted on a butter paper cover by rubbing the male spike and tapping, and the collected pollens were dried using zeolite granules. Pre-and post-storage germination assessment was carried out in Brewbakers and Kwack medium containing 5% sucrose using the hanging drop technique (Fig. 4.1). The pollen was then packed with the butter paper in an aluminium pouch. The packed pollen was rapidly plunged into the cryobiological system containing liquid nitrogen. Post-storage germination assessment was done as described above. The *in vitro* germination percentage of fresh pollen was 72.0% (Fig. 4.1A), while that of cryopreserved pollen was 69.0% after 2 months of cryopreservation, showing that there was no decline in the viability of pollen. Scanning electron microscopy of both fresh (Fig. 4.1B) and cryopreserved (Fig. 4.1C) pollen found no changes.

In vivo pollination was examined by bagging the female spike before opening to avoid contamination (Rajasekharan and Ganeshan, 2019). Fresh pollen from the male spikes was collected and rubbed against the female spike at 10.00–10.30 a.m. (when there is higher stigma receptivity); three such spikes were pollinated. The procedure was repeated for 3–4 days and bagging was done after every pollination. Similarly, pollen that had been cryopreserved for 2 months was pollinated on the pre-bagged female spike; again, three female spikes were pollinated. As a control, a female spike was bagged before opening and was not pollinated. Fruit set was observed in all spikes pollinated with either fresh or cryopreserved pollen, while the control spike that was bagged without pollination dropped after 2 weeks without setting fruit.

Jackfruit pollen can be cryopreserved successfully and can be used in breeding programmes for the transfer of quality traits and to broaden the genetic base by excluding the limitations of time and season.

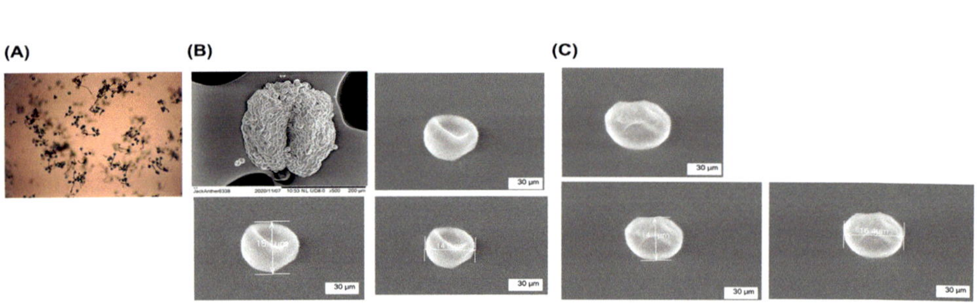

Fig. 4.1. (A) Pollen germination in Brewbakers and Kwack medium. (B, C) Scanning electron microscope images of fresh pollen (B) and cryopreserved pollen (C). Photographs courtesy of Dr P.E. Rajasekharan.

In theory, seeds under cryogenic storage are in a state of suspended animation and should survive indefinitely (Haq, 2006). Thammasiri (1998) successfully cryopreserved embryonic axes of jackfruit by vitrification. Embryonic axes from fully mature seeds were excised and pre-cultured on woody plant medium (WPM) supplemented with 0.3 M sucrose and 0.5 M glycerol at 25°C for 16 h. Subsequently, the axes were transferred to a 1.8 ml cryotube filled with a highly concentrated vitrification solution at 25°C for 50 min. The axes were immediately plunged into liquid nitrogen. After rapid warming, the vitrification solution was replaced with 0.5 ml of 1.2 M sucrose in WPM solution and kept at 25°C for 20 min prior to culture on WPM agar medium. The survival rate from cryopreservation was about 50% and the cryopreserved axes were able to develop into whole plantlets.

For *in vitro* conservation, the most widely applied technique is temperature reduction, which can be combined with a decrease in light intensity or culture in the dark. Ganeshan *et al.* (2012) developed a protocol for *in vitro* conservation of seedling-derived *in vitro* plantlets and *in vitro* plantlets derived from mature jackfruit trees using low temperature and low light (2.97 $\mu M\ m^{-2}\ s^{-1}$). The *in vitro* plantlet could be maintained beyond 1 year and up to 4 years or more with an annual subculture schedule. After 1 year of conserving seedling-derived *in vitro* plantlets, there was significant conservation gain at 10°C in terms of reduction in shoot length and number of shoots, indicating that lowering the temperature and reducing the light intensity favour slow growth.

4.4 Genetic Diversity Analysis

Heritability and genetic advance are important selection parameters. Heritability is the ratio of genetic variance to phenotypic variance and is expressed as a percentage. It is a good index to understand the transmission of any characteristic from parents to their offspring in the next generation. Genetic advance is the genotypic value of selected plants over the base population and is the major genetic gain under selection. Path coefficient analysis measures the contribution of an independent variable to a dependent variable and helps in determining the yield components. Genetic diversity in selecting desirable parents for a further breeding programme is measured by the important technique of genetic divergence, which is measured by D^2 statistics. D^2 statistics measures the degree of differentiation and thus helps in the selection of genetically divergent parents. This technique also determines the relative contribution of each component trait to the total divergence.

Working with 44 selected genotypes, Mitra and Maity (2002), Maity *et al.* (2002) and Wangchu *et al.* (2013) observed a wide range of variation in fruit weight, fruit rind weight, peel weight and yield per tree. Among the different traits, maximum heritability was recorded for the weight of fresh flakes/ bulbs without seed and yield per tree. Most of the fruit quality characteristics, however, showed very little genetic advance, revealing that these characteristics may not be effective for selection to improve the fruit quality of jackfruit. The fruit weight showed a highly significant positive association with rind weight, rachis length, fruit length and flake length (Wangchu *et al.*, 2013). Sharma *et al.* (2006) reported higher genotypic correlation coefficients than the corresponding phenotypic correlation coefficients in most of the fruit characteristics. They observed a significant and positive association of fruit yield with average fruit weight at the phenotypic and genotypic level, and fruit weight had a highly significant positive association with fruit length and the weight of the bulbs. Thus, these characteristics are considered to be important components for fruit yield per tree in jackfruit. Path analysis revealed that the number of fruits per tree and the number of bulbs per fruit, followed by average fruit weight, exhibited the most important direct effect on fruit yield per tree. Multivariate analysis of genetic divergence among the 44 genotypes selected by Wangchu *et al.* (2013) revealed that the weight of fresh flakes without seed showed the highest percentage contribution (45.35%), followed by flake width (28.86%), and that fruit diameter and total sugar showed the lowest percentage contribution (both 6.98%) (Table 4.5).

Table 4.5. Contribution of 26 physico-chemical characters to genetic divergence in 44 superior genotypes of jackfruit. Modified from Wangchu *et al.* (2013).

Characteristic	Contribution (%)
Stalk length (cm)	13.85
Stalk diameter (cm)	9.62
Fruit length (cm)	9.51
Fruit diameter (cm)	6.98
Fruit weight (g)	7.82
Fruit rind weight (g)	7.72
Shelf life (days)	8.03
Number of flakes kg^{-1} fruit	12.90
Weight of flakes kg^{-1} fruit (g)	7.82
Weight of fresh flakes with stone (g)	11.94
Weight of fresh flakes without stone (g)	45.35
Flake weight/fruit weight ratio	8.46
Flake length (cm)	8.03
Flake width (cm)	28.86
Rachis length (cm)	7.61
Rachis diameter (cm)	7.82
Number of stones per fruit	7.61
100-Stone weight (g)	7.61
Yield per tree	13.53
Citric acid (%)	8.35
Reducing sugar (%)	8.03
Total sugar (%)	6.98
TSS (°Bx)	7.93
Vitamin C (mg 100 g^{-1})	7.82
TSS/acid ratio	10.57
Sugar/acid ratio	17.12

TSS: total soluble solids.

A study of the seed characteristics of yellow and orange bulb types showed that light yellow types had the highest seed weight (7.66 g), seed length (3.23 cm), seed breadth (2.10 cm) and average total weight of seeds per fruit (913.2 g). The yellow types had the maximum number of seeds per fruit (124.6) and the highest pulp:seed ratio (4.24) (Guruprasad and Thimmaraju, 1989).

Several studies have been conducted to investigate genetic diversity among jackfruit germplasms using biotechnological tools. Schnell *et al.* (2001) analysed the genetic diversity of 26 accessions from different parts of the world, using amplified fragment length polymorphism (AFLP) markers. In their study, accessions from India were found to be different from the south-coast accessions. The genetic diversity of Indonesian jackfruit was studied by Palupi *et al.* (2019) using random amplified polymorphic DNA markers. In this study, H-15 primers produced more polymorphic fragments (91.7%) than the other six primers (A-13, A-27, B-4, L-1, OPB-1 and OPD-19). Jackfruit tree accessions from the 11 provinces were divided into two clusters with a similarity coefficient ranging from 63.33% to 85.71%. Provinces in East Java and West Java were the most closely similar (85.71%) based on molecular characteristics. Ye *et al.* (2009) used inter-simple sequence repeat markers to analyse the genetic diversity of 76 Chinese jackfruit accessions. They reported low genetic diversity in the studied germplasm, and the genetic similarity coefficient ranged from 0.626 to 0.945, with an average of 0.775. Cluster analysis showed that four groups could be clustered at a genetic distance coefficient of 0.752. The results concluded that two accessions from the Tropical Plant Garden, WJ2 and GSYWJ1, were genetically different from the other accessions. Accessions of soft- and firm-fleshed types could not be discriminated.

Witherup *et al.* (2013) isolated and characterized 25 microsatellite loci from enriched genomic libraries of *Artocarpus altilis* and tested them in four *Artocarpus* spp. and one hybrid. Twenty-one microsatellite loci were evaluated on *A. altilis* (n = 241), *Artocarpus camansi* (n = 34), *Artocarpus mariannensis* (n = 15) and *A. altilis* × *Artocarpus mariannensis* (n = 64) samples. Nine of these loci plus four additional loci were evaluated on *Artocarpus heterophyllus* (n = 426) samples. All loci were found to be polymorphic for at least one species. The average number of alleles ranges from two to nine within taxa. They concluded that these microsatellite primers will facilitate further studies on the genetic structure and evolutionary and domestication history of

Artocarpus spp. They will aid in variety identification and establishing germplasm conservation strategies for breadfruit and jackfruit (this is discussed further in Chapter 5, this volume).

4.5 Hybridization

There has been very little research on hybridization in jackfruit. In the Fairchild Tropical Garden, Florida, breeding work was initiated in 1995 (Campbell *et al.*, 2004). Reciprocal crosses were made among a subset of the core genetic collections. Varieties were selected for inclusion in the project based on precocity and productivity and the fruit quality traits of aroma, edible percentage, flesh firmness, colour and flavour. 'Black Gold', 'Cheena', 'Dang Rasmi', 'Galaxy', 'Golden Nugget', 'Honey Gold', 'Lemon Gold', 'J-30', 'J-31', 'NS-1' and 'Taboliey' were used. Controlled crosses were made over two fruiting seasons, seed was collected and 15 seedlings of each cross were planted at the US Department of Agriculture-Agricultural Research Service (USDA-ARS) Subtropical Research Station (Chapman Field). Evaluations began in 2003 with the objective of selecting superior jackfruit varieties. Fruiting was observed in less than 5 years, and often earlier. For the controlled crosses in this study, flowering began on three seedling trees within 2 years of planting. Within 3 years, there were 15 trees producing fruits and in the fourth year, 70 of the 400 trees were in production. External appearance of the fruit was similar among the seedling population; however, differences were observed among the seedlings in the overall size of the fruit, its shape and the colour of the spines. Colour differences among the seedlings were in the shades of green and the presence or absence of a russet overcolour. Preliminary results showed excellent internal quality characteristics of many of the crosses.

The Fruit Experimental Station at Kallar, India, started breeding work in 1952 with a view to developing short, compact, multi-branched trees that were precocious and productive, bearing large, yellow, high-quality fruits, and bearing twice (early and late in the season). The variety 'Singapore Jack' was chosen as the female parent because of its early and late crops and 'Vilipala', a local selection from the forest having large fruits with large carpels of superior quality and borne regularly in the main summer season, was selected as the male parent. After 25 years of testing, one hybrid was rated outstanding for its precocity, fruit size, and off-season as well as main-season production, with its yield excelling that of its parents (Chellappan and Roche, 1982). The Department of Agriculture, Malaysia, reported a hybrid named 'J37' ('Mastura'), which is a hybrid of 'J35' and 'J36' producing large, good-quality fruit.

As wide genetic variability exists in this crop, there is a need to exploit heterosis for yield gain and improvement in fruit quality. Interspecific hybridization also deserve attention in order to determine the species that are compatible with jackfruit and their contribution in a particular direction such as tree size (dwarfness), precocity, salt hardiness, or low water stress (Ray, 2002)

4.6 Genetic Resources in Different Countries

4.6.1 India

In southern India (Kerala, Tamil Nadu and Karnataka states), different forms of jackfruit ('Varikka', 'Koozha' and 'Navarikka') are available. The Tamil Nadu Agricultural University released 'Palur 1 Jack' ('PLR-1') in 1992, 'PPI-1 Jack' in 1999 and 'PLR (J)-2' in 2007. The University of Agricultural Science, Bengaluru, developed the varieties 'Swarna', 'NSP' 'Janagare', 'Byrachandra', 'Kantanukunatae' and 'Sharath' and conserved 105 of the best genotypes in the germplasm block of the university (Shyamalamma *et al.*, 2017).

The Central Horticultural Experimental Station, Hirehelli, Karnataka, surveyed the jackfruit-growing tracts of southern Karnataka and identified three superior accessions namely CHESHJF 1 ('Siddu'), CHESHJF 2 ('Shankara') and CHESHJF 3 (yield of 250 fruits per tree, average fruit weight 7.35 kg, yellowish-orange flakes, 28.5°Bx) (Karunakaran *et al.*, 2017).

Survey, selection and characterization of jackfruit in West Bengal was initiated in 1991 and continued until 2005, resulting in the selection of 44 superior genotypes (Mitra and Maity, 2002; Maity *et al.*, 2002; Wangchu *et al.*, 2011, 2013). The selected genotypes differ widely among themselves in average fruit weight (1.6–16.17 kg), number of flakes kg^{-1} fruit (6–60.3), flake colour (yellowish-white, yellow, orange, coppery-red), Brix (20–28.67°Bx) and total soluble solids (TSS)/acid ratio (35.69–178.02). Correlation studies have shown that fruit weight exhibited a highly significant positive association with rind weight, rachis length, fruit length and flake length. Studies with biochemical markers (peroxidase and esterase isozymes) showed that factors other than geographical origin, such as exchange of genetic stocks, genetic drift, spontaneous variations, natural and artificial selection, cross-pollination and seedling origin of the plant, are responsible for genetic diversity (Wangchu *et al.*, 2011). Rai *et al.* (2003) evaluated 21 genotypes of jackfruit from Bihar, West Bengal and Uttar Pradesh. Based on overall performance with respect to the bearing potential, maturity period, and fruit and flake characteristics, genotypes HPJS-5/8 and HPJS-3/10 were selected for table purposes and HPJS-4/5 and HPJS-2/6 for culinary purposes.

A total of 204 trees belonging to 67 accessions including wild and cultivated types from Kerala, Karnataka and Tamil Nadu are maintained at the NBPGR Regional Station at Thrissur, Kerala. Seedlings generated through seed progeny are maintained in the field along with the standard varieties 'Singapore Jack' and 'Muttam Varikka'. Results have shown that all except one belonged to the 'Varikka' type with firm flesh. Trees numbered V-504-1-A, IC96152-1-A, IC9/624-1A and IC-91737-1A were found to be excellent table fruit types. IC93388-2-A was the only 'Koozha' type with soft bulbs that can be used for fruit puddings. IC-99627-1-A had very thin flakes that are good for preparation of chips. OBL-8 was extra-early (off-season) with fruit initiation in August. IC-97624-1A had early fruit ripening (March). IC91665-2A and OBL-1 ('Undachakka') had small round fruits, with single fruit weight ranging from 0.75 to 2.0 kg. These types exhibiting early fruit maturity are suitable to use as vegetables. IC96148-3-A has coppery-red flakes that are thick and crisp with a good aroma.

A survey conducted in Assam showed that both soft-fleshed ('Pakikhua') and firm-fleshed ('Khoja') varieties are available. Commonly, jackfruit ripens during March–May in the Barak Valley, while in the Assam Valley, fruits ripen during May–July. Certain trees giving two or three crops (January–February, April–May and June–July) known as 'Baromahia' were identified during the survey. In southern India, maximum diversity has been reported from the Wayanad plateau of the Western Ghats of Kerala (Mitra, 2017).

'Siddu'

This variety was identified from Karnataka. The tree shape is broadly pyramidal. It bears solitary fruits on primary as well as secondary branches with small-sized fruits (2–5 kg) (Fig. 4.2). The fruiting season is May–July. The fruit shape is irregular, and around 450 fruits are produced per tree. It is highly suitable for homesteads and commercial cultivation. Each fruit contains an average of 30 bulbs. The flakes are attractive, sweet (31.0°Bx), firm and coppery-red in colour. The fruit is rich in phytochemicals such as carotenoids (4.43 mg 100 g^{-1}), lycopene (1.12 mg 100 g^{-1}), flavonoids (3.74 mg catechin equivalents 100 g^{-1}), phenols (31.76 mg gallic acid equivalents 100 g^{-1}) and vitamin C (6.48 mg 100 g^{-1}). It is a high yielder, with an average yield of 1098 kg per tree (Karunakaran and Singh, 2019).

'Shankara'

'Shankara' was a selection from a farmer's field in Karnataka. The tree is medium height with a semi-spreading nature. It bears solitary fruits on primary as well as secondary branches. The fruits are small (2–5 kg), and the skin colour is dark green when mature, the tip of the spine becoming brown to black when ripe. The fruit contains about 60 flakes with an average flake weight of 17.85 g. The bulbs are medium crispy with coppery-red coloured flakes (Fig. 4.3). The TSS content is quite high (28°Bx) with high amounts of carotenoids (5.83 mg 100

Fig. 4.2. 'Siddu' fruit. Photograph courtesy of Dr G.Karunakaran.

g^{-1}) and lycopene (2.2 mg 100 g^{-1}). The crop is generally harvested during April–June, and a well-managed tree produces about 460 kg fruit annually (Karunakaran and Singh, 2019).

'Swarna Halsu'

The University of Agricultural Science (UAS) in Bengaluru, selected and released this variety in 2010. 'Swarna Halsu' is a high yielder that bears fruit on the trunk, primary branches and secondary branches. The fruits are medium-sized (6–8 kg) and the flakes are golden-brown in colour (Fig. 4.4). The fruit is juicy and sweet (25–26°Bx), with thick flakes and a very thin rind (<1.0 cm).

'Konkan Prolific'

This variety was released in 2004 by the Regional Research Station, Dapoli, Maharashtra. It is a seedling selection with a potential yield of 450–500 kg per tree and an average yield of 420.6 kg per tree, producing around 73 fruits per tree each year with an individual fruit weight of 5.7 kg, fruit length 34.4 cm and fruit width 15.2 cm. The fruits are oblong in shape, with firm flesh, and

Fig. 4.3. 'Shankara' fruit. Photograph courtesy of Dr G. Karunakaran.

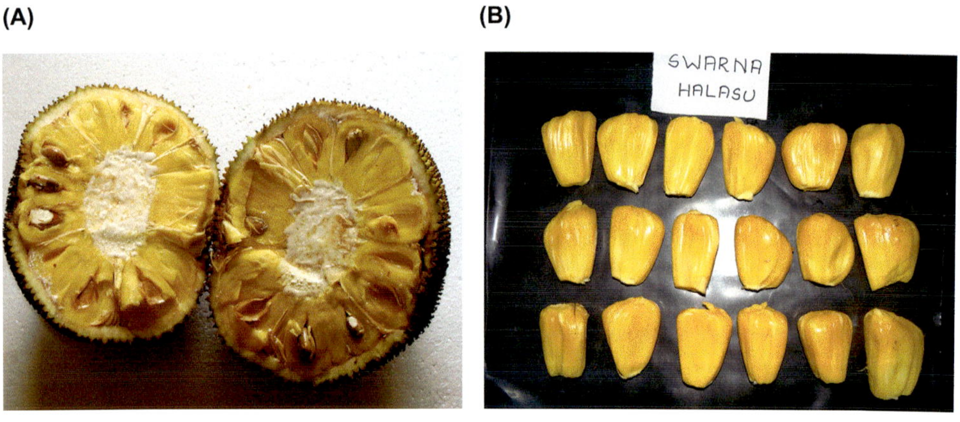

Fig. 4.4. 'Swarna Halsu' fruit (A) and bulbs (B). Photographs courtesy of Dr S. Shyamalamma.

golden-yellow in colour with a spiny surface. Flowering is in January–February and the trees bear fruit 6–7 years after planting. The fruits have an average TSS of 25°Bx.

'Palar 1 Jack' ('PLR-1')

This improved variety was selected in 1992 by the Tamil Nadu Agricultural University (TNAU) in India. The trees are medium in height, less spreading than other vaieties and suitable for high-density planting. In addition to the regular bearing season (March–June), it produces fruits during the off-season (October–December). The annual yield per tree is about 80 fruits, weighing around 900 kg in total. The average fruit weight is 12 kg, and each contains around 115–120 flakes.

Fig. 4.5. 'Palar 1 Jack' fruit. Photograph courtesy of Dr G. Karunakaran.

The fruit quality is good with attractive golden-yellow, firm flakes. The flakes are sweet (19°Bx) and the fruit has a good keeping quality (Fig. 4.5).

'PPI-1 Jack'

This clonal selection was made by TNAU in 1996. It is a medium-height tree, producing an average of 105 fruits per tree, each weighing about 17 kg. It bears fruits twice a year (April–June and November–December). The flakes are crispy and sweet with a pleasant aroma.

'PLR (J)-2'

This clonal selection was identified from Pathirakottai by TNAU. The tree yields about 95–100 fruits year^{-1}. The fruits are of good quality and have very good keeping quality.

'KT-7'

This selection was made at UAS. It produces 50–70 fruits year^{-1}, each weighing about 5.0 kg. The flakes are yellow and firm, with a sweet acid taste; they are not juicy and the TSS content is about 25°Bx.

'Lalbaugh Madhura'

This is an early-bearing, high-yielding variety (650–800 kg fruits per tree year^{-1}) released from UAS and the State Department of Horticulture, Government of Karnataka, in 2018. The fruits are 6–18 kg in weight. The flakes are large (75–100 g, 9–10 bulbs kg^{-1}), crisp, thick, and very sweet (29–30°Bx), and are suitable for processing into different products (Fig. 4.6).

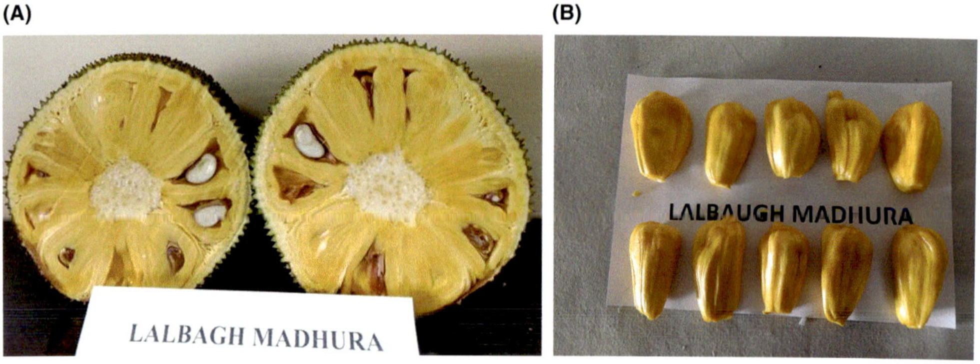

Fig. 4.6. 'Lalbaugh Madhura' fruit (A) and bulbs (B). Photographs courtesy of Dr S. Shyamalamma.

Fig. 4.7. 'Ramachandra' fruit. Photograph courtesy of Dr S. Shyamalamma.

'Nelugudigae'

The fruits of this variety are borne on the trunk, primary branches and secondary branches. The fruits weigh 80–10 kg and have a thick rind (2 cm). The thick orange bulbs are suitable to use as table fruit, as well as for processing.

'Ramachandra'

The fruits of 'Ramachandra' are round, medium-sized (7–9 kg) with a thick rind (2 cm), and are suitable for fresh fruit. The pulp is pink to light orange in colour (Fig. 4.7), thick and firm with a TSS of 23°Bx. It has a distinct flavour and aroma.

Fig. 4.8. 'JT5' fruit. Photograph courtesy of Dr D. Maji.

'Kantanukuntae'

This was selected at UAS. The tree is very large and a high yielder, producing 100–150 fruits year^{-1}, each weighing about 500–600 g. The fruits are ellipsoid in shape. The flakes are orange, firm, sweet and juicy, and the TSS content is about 30°Bx.

'NSP'

This variety was selected from the GKVK campus at UAS. The tree produces 50–100 fruits, mainly on the trunk. The fruits are large (average weight 22.4 kg) and ellipsoid in shape, and the flakes are yellow, soft, sweet (TSS 25°Bx) and juicy.

'JT5'

This was selected at Bidhan Chandra Krishi Vishwavidyalaya (BCKV), West Bengal. The trees are vigorous with dense branches. It is regular in bearing, with fruit clusters on the main trunk and primary branches (Fig. 4.8), yielding about 180 fruits per tree each year, each weighing 4.0–5.2 kg. The flakes are yellow in colour, low fibre, soft and medium sweet (16.5°Bx) with a good aroma. It is suitable to use as a vegetable when immature (Majhi, 2012).

'JT40'

'JT40' was selected at BCKV. The tree is dwarf in height with a spreading crown. It bears fruits twice a year, with clusters on the main trunk and primary branches, yields around 400 fruit per tree each year, each weighing 1.5–2.0 kg. The flakes are yellow in colour, low fibre and medium sweet (19.6°Bx) with a good aroma. It is also suitable to use as a vegetable when immature (Wangchu, 2005; Wangchu *et al.*, 2013).

'JT42'

This was also selected at BCKV in 2002 (Wangchu, 2005). The tree is dwarf and erect. It is regular in bearing, and bears solitary fruits on the main trunk and primary branches. The fruits are oblong in shape, each weighing 7.5–8.5 kg. The flakes are thick, firm, juicy, dark yellow and sweet (26°Bx) and have a mild aroma. The fruit has a good shelf life.

'Swarna Manohar'

This variety was selected by the ICAR Regional Centre, Ranchi, in 2004. It is an early variety suitable for table purposes. The fruits have an attractive skin colour, are uniform in shape and are of high quality (TSS 18°Brix, acidity 0.09%) with a fragrant pulp, and sweet, firm flakes. The average fruit weight is 13.7 kg.

'Swarna Poorti'

This variety was also selected by the ICAR Regional Centre in 2004. It is a late variety suitable to use as a vegetable at the early stage of fruit development (30–60 days after fruit set). The fruits have an attractive light green skin colour and are roundish in shape. The fruits are medium sweet (TSS 18°Bx) with an average weight of 4.2 kg.

4.6.2 Sri Lanka

Jackfruit has a long history in Sri Lanka. The existence of this fruit is mentioned in the Mahavamsa as far back as the 4th century BC (Medagoda and Jayawardena, 1997). *Artocarpus* is represented in Sri Lanka mainly by four species. *A. heterophyllus* and *A. incisus* are the most commonly and widely grown. The other species, *A. nobilis* and *A. gomenzianus*, grow wild and are mostly used for timber.

The genetic diversity of *A. heterophyllus* and *A. incisus* is conserved mainly through the home garden system. The University of Peradeniya has a collection of 441 accessions of jackfruit at Dodangolla. The field gene bank site for jackfruit trees is located at the Horticultural Crop Research and Development Institute, Peradeniya (Medagoda and Jayawardena, 1997).

Cultivated jackfruit has been divided into two groups according to the quality of the fruit: (i) 'Waraka' with firm fleshy pericarp; and (ii) 'Wela' with a soft fleshy pericarp. Ten varieties of 'Waraka' types have been identified. The important varieties are 'Father Long', 'Maharagama', 'Kothmale', 'Hirosa' and 'Mandoor'. The other varieties include 'Arthur V. Dias', 'Tellippalai', 'Ganegoda', 'Kurucos' and 'Singapore Jack'.

'Father Long'

This has a large fruit (10–15 kg) that is oblong in shape (Fig. 4.9). The flesh is dark yellow, thick, crispy and sweet in taste (20–23°Bx). The edible portion comprises about 48% of the fruit.

'Maharagama'

The fruit is large (10–15 kg) and oblong in shape (Fig. 4.10). The flesh is yellowish-orange, crispy and sweet (20–23°Bx). The edible portion comprises around 53% of the fruit.

'Kothmale'

This has a medium-sized fruit (8–10 kg) that is oblong in shape (Fig. 4.11). The flesh is yellowish-orange, crispy and sweet (20–23°Bx). The edible portion comprises about 52% of the fruit.

'Hirosa' ('Rosa Kos')

This has a large, oblong fruit (10–15 kg) (Fig. 4.12). The flesh is yellowish-orange, crispy and sweet (20–30°Bx). The edible portion comprises around 47% of the fruit.

'Mandoor'

The medium-sized fruit (8–10 kg) of this variety are round in shape (Fig. 4.13). The flesh is dark yellow, crispy and sweet (23–25°Bx). The edible portion comprises about 46% of the fruit.

4.6.3 Bangladesh

Jackfruit is a major fruit crop of Bangladesh. Jackfruit accessions are maintained at the Fruit Research Stations of Hathazari (Chittagong) and Jaintiapur (Sylhet). Four superior trees were selected and named 'Shariatpur' (average fruit weight 5.7 kg, soft and juicy pulp, TSS 21.5°Bx), 'Sreepur' (average fruit weight 10.6 kg, soft and juicy pulp, TSS 23.2°Bx), 'Comilla'

Fig. 4.9. 'Father Long' leaves, fruit and bulbs. Photograph courtesy of Department of Agriculture, Sri Lanka.

(average fruit weight 5.0 kg, each fruit contains about 150 bulbs, soft and juicy pulp, very high TSS 30.0°Bx, fruit harvested December–May) and 'Savar' (average fruit weight 7.3 kg, crispy pulp, TSS 23.2°Bx). Bangladesh Agricultural University at Mymensingh also has 94 accessions of jackfruit.

Almost every rural family in Bangladesh has access to a home garden in which jackfruit is the second most commonly planted fruit tree, after mango (Hocking *et al.*, 1996; Khan *et al.*, 2010). Villagers use a variety of names to describe these trees, such as *Chaila Kathal* (hard, tasteless jackfruit), *Roachhara Kathal* (jackfruit without roa or pulp), *Hazari Kathal* (bearing a thousand small fruits), *Khokra Kathal* (wavy, irregular shaped, hard fruit) and *Neisha Kathal* (very soft, sweet, small-sized pulp). These names indicate that the trees are somewhat wild or not entirely domesticated. These are considered inferior in quality and are not preferred in the market. However, such trees are still kept for timber and wood, and villagers use the fruits mainly as vegetables before they reach maturity (APAARI, 2012). Ibrahim *et al.* (2013) evaluated different varieties of jackfruit-growing in Rajshahi, Natore, Naogoan and Chapai Nawabganj. They observed wide variability in the morphological fruit characteristics and nutritional food values. The percentage of pulp varied between 38.60% and 47.37%, rind from 15.67% to 20.00%, skin from 14.86% to 23.68%, seeds from 9.46% to 19.33%, TSS from 18.80% to 27.37%, total sugar from 11.84% to 17.01%, vitamin C content from 17.82 to 31.55 mg 100 g^{-1} and acidity (as citric acid) from 0.037% to 0.075%. Established varieties include 'Topa', 'Hazari', 'Chola', 'Goal', 'Koa' and 'Khaja'.

'BARI Kanthal 1'

This variety was developed by the Horticulture Research Centre in Bangladesh in 2008. It is

Fig. 4.10. 'Maharagama' leaves, fruit and bulbs. Photograph courtesy of Department of Agriculture, Sri Lanka.

a high-yielding, regular-bearing variety. The fruits are medium-sized (9.5 kg) and the pulp is medium soft, yellow and sweet (TSS 22°Bx).

'BAU Kanthal 1'

This was developed by the Bangladesh Agricultural University in 2010. It is also a high-yielding (20–40 t ha^{-1}) and regular-bearing variety. The fruits are uniform in size and are good quality.

4.6.4 Vietnam

Jackfruit is known as 'Mit' in Vietnam and is grown in almost all provinces. The jackfruit grown has great diversity with large differences in yield, fruit size, skin colour, pulp colours, ratio of edible parts, etc. Some selected lines have fruits with an average weight of 10–12 kg with yellow, firm, brittle and sweet pulp. The edible part comprises over 35–40% of the fruit and the yield obtained is around 30–40 fruits per tree from 6–7-year-old plants. In northern provinces, jackfruit is divided into two major groups: (i) 'Mit dai', which has firm, brittle and sweet pulp; and (ii) 'Mit mat', which has soft succulent pulp when ripe. Similarly, in southern provinces, jackfruit is divided into two major groups: 'Mit Ta' and 'Mil To Nu'. There are at least 30 varieties of jackfruit recorded in production, but only a few varieties are widely known. They are 'Sieu Som', 'La Bang' ('La Lon'), 'Vien Linh', 'Red Flesh' (yellow rags), 'Red Flesh' (white rags) and 'MDNO6' (Tri and Vu, 2020).

Fig. 4.11. 'Kothmale' leaves, fruit, bulbs and seeds. Photograph courtesy of Department of Agriculture, Sri Lanka.

'Sieu Som'

This tree is fast growing and spreading. It is regular in bearing and is a high yielder (45–65 t ha^{-1}). The fruits are ovoid-oblong in shape, and small to medium in size (8–12 kg). The fruit has a lime green to yellow rind colour, with yellow-orange bulbs (Fig. 4.14) that are oblong with a smaller base, firm and crispy. The fruits are medium sweet (20–22°Bx) and the pulp portion of the fruit is about 46%.

'La Bang' ('La Lon')

This tree is fast growing and spreading, with sparse branching. It is regular in bearing and is a high yielder (50–70 t ha^{-1}). The fruits are oval-obovoid in shape (Fig. 4.15), and medium to large (10–15 kg). The rind is lime green-yellow in colour, with yellow-orange bulbs that are oblong with a curved tip, medium firm texture and medium sweet (19–21°Bx) in taste. The edible portion comprises about 44% of the fruit.

'Vien Linh'

The tree is medium growing and spreading, with medium to dense branching, and is regular in bearing and medium yielder (40–50 t ha^{-1}). The fruits are spheroid-globose in shape and small to medium in size (7–12 kg) (Fig. 4.16). The rind is green-yellow, and the bulbs are rectangular in shape, yellow-orange in colour, medium firm and sweet (22–24°Bx). The edible portion comprises around 47% of the fruit.

'Mit Wghe'

The fruits of this variety are small to medium in weight (5–12 kg). The fruit skin is yellow to dark yellow, and the pulp is yellow, very tasty and sweet (25–30°Bx).

'Mit Dua'

This is one of the popular varieties of Vietnam. The fruits are small to medium in weight (7–20

Fig. 4.12. 'Hirosa' leaves, fruit and bulbs. Photograph courtesy of Department of Agriculture, Sri Lanka.

kg), and the fruit skin is dark yellow. The flakes are white to bright yellow in colour, sweet to very sweet and have a good aroma.

'Mit Uot'

The ripe fruits of this variety are yellow to blue in colour and 6–15 kg in weight. The flakes are white to pale yellow, soft and juicy.

4.6.5 The Philippines

The National Plant Genetic Resource Laboratory, Institute of Plant Breeding, Los Banôs, Laguna, maintains 30 accessions of jackfruit. Farmer participatory breeding is being used by fruit breeders in the Philippines to develop jackfruit varieties. The breeders exploited the natural variability existing in the jackfruit population consisting of trees grown from seeds. On-site selection and evaluation of desirable jackfruit genotypes was conducted directly. The first jackfruit variety obtained in this way and released in the 1980s by the Institute of Plant Breeding (IPB) was 'Torres' (Magdalita *et al.*, 2011). Other jackfruit varieties developed are 'EVIARC Sweet', Sinapelo', 'Burabod', 'Cerventes Gold' and 'MMSU-SRO Sweet'. Valencia and Magdalita (2010) conducted on-site selection and evaluation of jackfruit trees grown from seeds in Laguna, Batangas and Cavite. One selection, named 'Mabini', released by IPB in 2005, is a dual-purpose jackfruit variety; the fruit is oblong and weighs around 15.4 kg. The flesh is bright yellow-orange, juicy, smooth, semi-crispy and sweet (20.40°Bx), and has a strong aroma, moderate latex and thick fibre (5.97 mm). The tree is tall, vigorous and a regular bearer, yielding 50–80 mature fruits per season, with a peak fruiting season from February to July.

Fig. 4.13. ‘Mandoor’ fruits and bulbs. Photograph courtesy of Department of Agriculture, Sri Lanka.

‘EVIRAC Sweet’

The fruit is ovoid in shape with an average weight of 12 kg. The flesh is golden-yellow, sweet (25°Bx) and crispy, with a strong aroma, and the flesh comprises about 45% of the fruit.

‘Burabod’

The fruit is oblong in shape and weighs 8 kg. The flesh is yellow-orange, firm and very sweet (29.7°Bx) with strong aroma and comprises about 48% of the fruit.

‘Cerventes Gold’

The fruit is ellipsoid in shape and weighs about 10 kg. The flesh is golden-yellow, medium sweet (22°Bx) and crispy, with a strong aroma, and flesh comprises around 45% of the fruit.

‘MMSU-SRO Sweet’

The fruit is ellipsoid in shape and weighs about 8 kg. The flesh is yellow-orange, very sweet (30°Bx) and firm with a mild aroma, and comprises around 43% of the fruit.

4.6.6 Australia

In Australia, jackfruit is grown in tropical Queensland and at the top end of the Northern Territory. Jackfruit production increased from 338 t in 2003 to 742 t in 2012. Australia has a limited range of varieties with the majority of material based on seedling selection introduced during the 1960s and 1970s. The fruit has either soft or firm flesh (crispy) and the rind colour is

Fig. 4.14. 'Sieu Som' fruit (A, B) and bulbs (C). Photographs courtesy of Dr Van Tri Mai.

yellow to pinkish-orange. The old varieties such as 'Black Gold', 'Honey Gold', 'Lemon Gold' and 'Golden Nugget' are now mainly in southern USA collections. Two new clonal selections made by nurserymen are 'Rajang' and 'Amber'. Both of these clones have crisp orange flesh. 'Rajang' is larger and has later-maturing fruit than 'Amber'. 'Rajang' fruit flesh has medium firmness with a slight rubbery and fibrous texture, and is juicy; it is the preferred variety among Australian consumers. 'Amber' has a mild aroma, and the fruit is fibrous and rubbery but still juicy (Diczabalis, 2012).

Fig. 4.15. 'La Bang' ('La Lon') fruit (A, B) and bulbs (C). Photographs courtesy of Dr Van Tri Mai.

4.6.7 China

Jackfruit is grown in South China. Li *et al.* (2007) evaluated jackfruit germplasm accessions from Leizhou Peninsula in Guangdong Province. The accessions were classified into soft- and firm-fleshed types based on fruit flesh texture and water content, and into annual bearing and biannual bearing based on flowering time and fruiting habit. The accessions were characterized by low genetic diversity, and were highly similar to accessions from Malaysia, suggesting that the two groups of accessions may have the same origin. The accessions classified as soft type and firm type did not show significant differences in DNA analysis. Li *et al.*

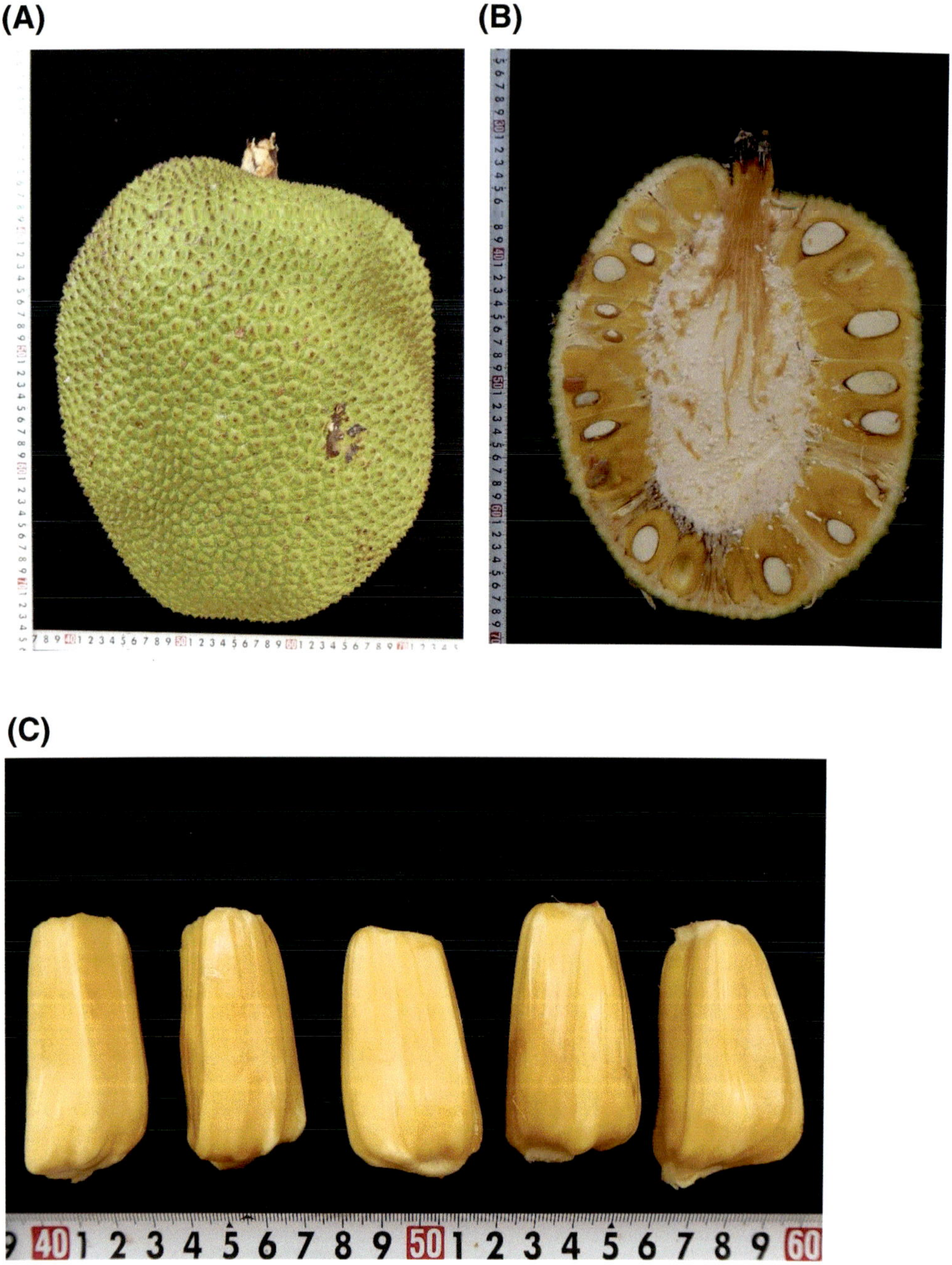

Fig. 4.16. 'Vien Linh' fruit (A, B) and bulbs (C). Photograph courtesy of Dr Van Tri Mai.

(2010) further studied the genetic diversity of 50 accessions from three provinces of China using AFLP markers. They noted a total pf 320 unambiguous bands by using eight primer combinations, and 65 (20.3%) of these were polymorphic. Genetic similarity coefficients ranged from 0 to 0.9841, with an average of 0.5000, indicating moderate genetic diversity in this collection. The dendogram derived using unweighted pair group method with arithmetic mean algorithm analysis revealed five groups, and no correlations between genetic relationship and geographical origin were found. Accessions of soft- and firm-fleshed type were not clustered into distinct groups.

'Sijimition' and 'Hongbao' are the two varieties that perform well in the Maoming area, Guangdong Province. Adult trees produce fruit crops of over 500 kg. The fruit weigh 10–20 kg and are cylindrical in shape. The flesh of the 'Hongbao' variety is crispy with TSS 13.5–16.0°Bx. The flesh of the 'Sijimition' variety is moderately juicy, with TSS 18.2–23.8°Bx and a sweet aroma. Both bear fruit year-round (Dang and Lu, 2007).

Wang *et al.* (2012) developed the dry bulb variety 'Red Flesh Jackfruit' from natural seedlings of Thailand jackfruit. It is an early-fruiting variety, flowering year-round but mainly from March to October. The fruits are oblong with an average weight of 10.3 kg. The flakes are orange-red and crispy with a sweet aroma and taste (TSS 18.87°Bx). The major varieties grown in China are 'Malaixiya No. 1', 'Changyou', 'Siji', 'Xiangmi No. 17' and 'Xiangmi No. 1'.

'Malaixiya No. 1'

'Malaixiya No. 1' was developed and released by Xilian Farm Co., Hainan Agricultural Reclamation Investment Holding Group Co., Hainan Province. It is grown mainly in Hainan, Guangxi and Yunnan. The average yield is 77.0 kg per plant (3-year-old trees) and 150.0 kg per plant (5-year-old trees). The average fruit weight is 12.9 kg, and the fruit has a mild pericarp, hard pulp and mild aroma. The bulbs are yellow in colour and TSS is 17.2°Bx. The edible portion comprises around 46.6% of the fruit (Fig. 4.17).

'Changyou'

This was developed by the Fruit Science Institute of Maoming, Guangdong Province. It is mainly grown in Guangdong and Guangxi. The average yield is 20.0 kg per plant (3-year-old trees) and 80.0 kg per plant (5-year-old trees). The average fruit weight is 4.0 kg, and the fruit has a thin pericarp, hard pulp and strong aroma. The pulp is golden-yellow with TSS 20.5–27.5°Bx. The edible portion comprises 68.0% of the fruit (Fig. 4.18).

'Siji'

'Siji' was released by the Gaozhou Huafeng pollution-free fruit field, Guangdong Province. It is grown mainly in Guangdong and Guangxi. The average yield is 58.0 kg per plant (3-year-old trees) and 135.0 kg per plant (5-year-old trees). The average fruit weight is 15.0 kg, and the fruit has a thin pericarp, hard pulp and strong

Fig. 4.17. 'Malaixiya No. 1' tree (A) and bulbs (B). Photographs courtesy of Dr Lanxi Su.

Fig. 4.18. 'Changyou' tree (A) and bulbs (B). Photographs courtesy of Dr Lanxi Su.

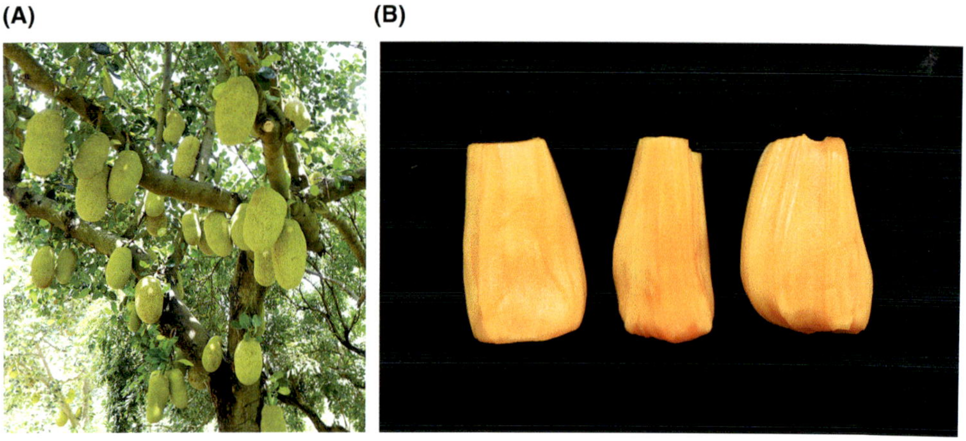

Fig. 4.19. 'Siji' tree (A) and bulbs (B). Photographs courtesy of Dr Lanxi Su.

aroma. The pulp is orange in colour with TSS 20.0–24.0°Bx. The edible portion comprises 48.0% of the fruit (Fig. 4.19).

'Xiangmi No. 17'

This variety was developed by the Spice and Beverage Research Institute, Chinese Academy of Tropical Agricultural Sciences, Wanning, Hainan Province. It is mainly grown in Hainan. The average yield is 40.0 kg per plant (3-year-old trees) and 80.0 kg per plant (5-year-old trees). The average fruit weight is 9.9 kg, and the fruit has a thick pericarp, hard pulp and strong aroma (smells like durian). The pulp is orange-red with TSS 23.5–27.8°Bx. The edible portion comprises 45.5% of the fruit.

'Xiangmi No. 1'

This was also developed by the Spice and Beverage Research Institute and is grown mainly in Hainan. The average yield is 25.0 kg per plant (3-year-old trees) and 100.0 kg per plant (5-year-old trees). The average fruit weight is 4.0 kg, and the fruit has a thin pericarp, soft pulp and strong aroma (with a light smell of durian). The pulp is orange with TSS 28.0–32.0°Bx. The edible portion comprises 42.0% of the fruit.

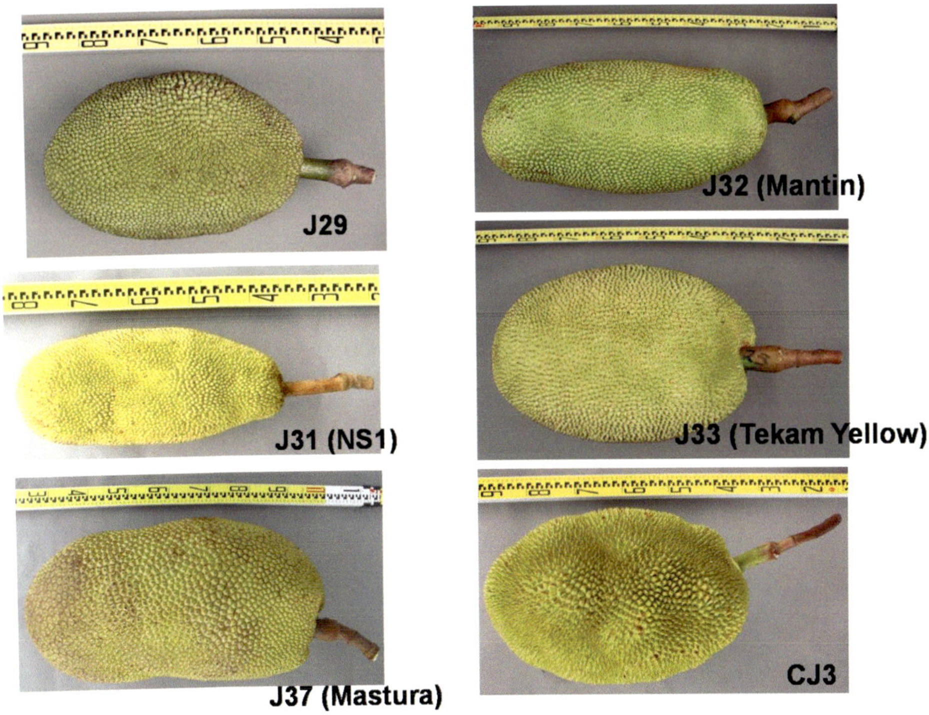

Fig. 4.20. Different varieties of jackfruit from Malaysia. Photograph courtesy of Department of Agriculture, Malaysia.

4.6.8 Malaysia

Jackfruit is considered a major fruit of Malaysia. Two types are reported to grow in Malaysia. The shallow-rooted type produces small trees where fruiting starts as early as 15–20 months after planting. The deep-rooted type bears fruits only at the age of 8–10 years and produces large fruits in much greater numbers (Mitra, 2017).

The recommended clones for cultivation are 'J29', 'J37' ('Mastura'), 'J32' ('Mantin'), 'J31' ('NS1') and 'J33' ('Tekam Yellow') (Fig. 4.20). The clone 'J35' is the most frequently cultivated (Yaccob Ahmed, International Tropical Fruit Network, Malaysia, personal communication); it is a hybrid of 'CJ1' × 'CJ6' and produced fruit of 15–25 kg in weight, with a thick bulb that is golden-yellow coloured, sweet and firm.

'J29'

This variety has a medium to large fruit (10–12 kg) that is globular in shape. The pulp is yellowish-green in colour and thick, with TSS 13°Bx and a strong aroma. It is suitable for fresh consumption.

'J31' ('NS1')

This variety has a medium-sized fruit (7–12 kg) that is oblong/obovate in shape. The pulp is medium thick, firm and yellow-orange, with TSS 13°Bx and a very fragrant aroma.

'J32' ('Mantin')

The fruits are large (15–20 kg) and oblong in shape, with an orange pulp, that is thick and firm (Fig. 4.21). The bulb is moderately sweet.

'J33' ('Tekam Yellow')

The fruits are large (16–20 kg) and oblong in shape, with a yellow pulp that is very sweet (TSS 27–30°Bx) (Fig. 4.22).

(A) (B) (C)

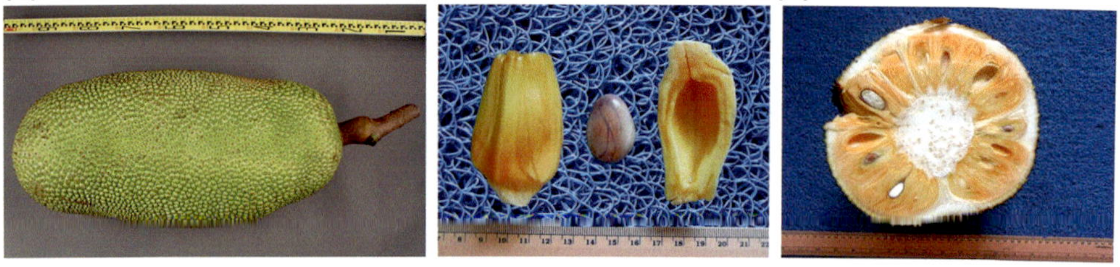

Fig. 4.21. 'J32' ('Mantin') fruit (A) and bulb and seed (B, C). Photographs courtesy of Department of Agriculture, Malaysia.

(A) (B) (C)

Fig. 4.22. 'J33' ('Tekam Yellow') fruit (A, B) and pulp (C). Photographs courtesy of Department of Agriculture, Malaysia.

'J35' ('Crystal Jackfruit 1', 'CJ1')

This variety has very large fruits (28 kg) that are oblong in shape, with a thin, yellow pulp, that is moderately sweet (15°Bx) and very fragrant.

'J37' ('Mastura')

This is a hybrid of 'J35' and 'J36' (Fig. 4.23). The fruits are large (15–25 kg, sometimes up to more than 40 kg) and oblong in shape, with a thick, firm pulp that is golden-yellow and sweet.

4.6.9 The USA

The jackfruit has been grown in south Florida for more than 100 years but has never had commercial importance. Many superior varieties have been imported by Fairchild Tropical Garden (Campbell and McNaughton, 1994), including 'Black Gold' (high yielder, sweet and aromatic) (Fig. 4.24), 'Cheena' (small fruit of 2.5–5 kg, soft and melting flesh, excellent quality), 'Cochin' (small fruit of 1–2.5 kg, firm to crunchy flesh, good quality), 'Golden Nugget' (fruit weight 3–6 kg, high yielder, soft to medium flesh, excellent flavour) (Fig. 4.25), 'Honey Gold' (fruit weight 5–6 kg, firm flesh, sweet, rich flavour and aroma) and 'Lemon Gold' (fruit weight 6–7 kg, sweet and aromatic flavour, firm flesh) from Australia; 'Dang Rasimi' (fruit weight 8–10 kg, firm to soft flesh, sweet flavour) and 'Kun Wi Chan' (fruit weight 15–20 kg, firm to soft flesh, pleasant flavour) (Fig. 4.26) from Thailand; 'J-31' (fruit weight 10–12 kg, firm flesh, sweet rich flavour, earthly aroma) and 'NS1' (fruit weight 4–5 kg, firm flesh, excellent texture, sweet rich flavour) from Malaysia; and 'Tabouey' (fruit weight 9–11 kg, firm flesh, pleasant flavour, very little aroma) from Indonesia. Breeding to develop hybrids was initiated in 1998 using all the available varieties both as male and female parents (Wasielewski and Campbell, 2000), and some of the hybrids were planted in 2003 for evaluation.

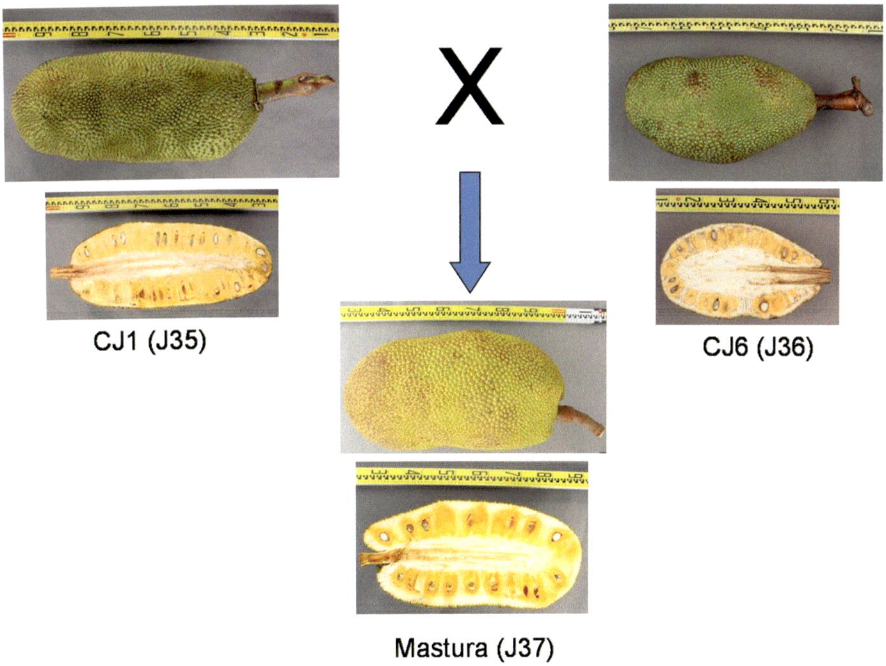

Fig. 4.23. 'J37' ('Mastura') is a hybrid of 'J35' and 'J36'. Photographs courtesy of Department of Agriculture, Malaysia.

Fig. 4.24. 'Black Gold' fruit. Photograph courtesy of Dr I Maguire.

4.6.10 Brazil

There are three varieties of jackfruit (jaca in Brazil) found in Brazil: 'jaca mole', which has softer sweet flesh, 'jaca-dura', which is heavy, and 'jaca-manteiga', which consists of buttery flesh. Interestingly, jackfruit is now one of the most conspicuous invasive plants in Brazil. The favourable climatic (rainfall and mean temperature) and relief (altitudinal range) conditions in the coastal Atlantic rainforest in Brazil may have facilitated the spread of jackfruit throughout the country. Bergallo *et al.* (2016) studied jackfruit invasion in the Atlantic Forest in Ilha Grande State Park, located on an island (Ilha Grande). They observed that jackfruit trees are spread all over Ilha Grande (Fig. 4.27) and that plant density varied significantly along and between trails. They suggested that control of jackfruit trees on the island is urgent as the prevalence of seedlings in jackfruit plots is high, and some native mammal species may be playing a dispersal role (Fig. 4.28).

Fig. 4.25. 'Golden Nugget' fruit. Photograph courtesy of Dr I. Maguire.

4.6.11 Uganda

The history of Uganda indicates that jackfruit was introduced by Asians in the 1890s (Dutton, 1976). Uganda harbours a high genetic diversity of jackfruit (Nakintu *et al.*, 2020). A cross-sectional survey involving 349 participants comparing 205 jackfruit traders and 144 jackfruit farmers was conducted by Nakintu *et al.* (2019). They observed that four different varieties were grown, which were classified based on flesh texture and flesh colour. The variety 'Serebera' has soft, yellow flesh, 'Kanaanansi' has firm, yellow flesh, 'Namusaayi' has firm, red flesh and 'Namata' has firm, white flesh. Farmers harvest the fruits throughout the year with a peak season in December and January.

4.7 Conclusion

Jackfruit is a cross-pollinated and predominately seed-propagated tree. Wide variations are observed in tree structure, flowering and fruiting habits, and fruit quality characteristics. The heterogeneous nature of the seedling population offers tremendous potential for selection of superior types for commercial cultivation. A good number of selections have been made in different jackfruit-growing countries, and some have been recommended for commercial cultivation. As wide genetic variability exists in this crop, there is an apparent need to exploit heterosis for yield gain and improvement in fruit quality. Interspecific hybridization also

Fig. 4.26. 'Kun Wi Chan' fruit. Photograph courtesy of Dr I. Maguire.

deserves attention in order to determine the species that are compatible with jackfruit and their contribution in a particular direction such as tree size (dwarfness), precocity, salt hardiness and low water stress. Compared with major tropical fruits, very little information is available on the inheritance pattern of characteristics and genomic studies in jackfruit. Marker-assisted selection could be an extremely useful tool for genetic improvement of jackfruit, as it allows early selection of elite clones with desirable traits on the basis of markers that are associated with commercially important traits. In jackfruit, concerted efforts are needed to generate molecular tools such as markers, maps, DNA sequences and quantitative trait loci and to make these available to researchers and breeding programmes.

Fig. 4.27. Jackfruit trees in some areas can dominate the landscape, such as in Tijuca National Park and Ilha Grande State Park (Rio de Janeiro, Brazil), where control projects have been developed. Photograph courtesy of Dr H.G Bergallo.

Fig. 4.28. The tasty fruits of the jackfruit attract fauna, and some species of mammals disperse their seeds. Photograph courtesy of Dr H.G. Bergallo.

References

Amzad Hossain, A.K.M. (1996) *Status Report on Genetic Resources of Jackfruit in Bangladesh*. International Plant Genetic Resources Institute (IPGRI), IPGRI, Selangor, Malaysia.

APAARI (Asia-Pacific Association of Agricultural Research Institutions) (2012) *Jackfruit Improvement in the Asia-Pacific Region – A Status Report*. APAARI, Bangkok.

Azad, A.K., Jones, J.G. and Haq, N. (2007) Assessing morphological and isozyme variation of jackfruit (*Artocarpus heterophyllus* Lam.) in Bangladesh. *Agroforestry Systems* 71(2), 109–125. DOI: 10.1007/s10457-007-9039-8.

Bergallo, H.G., Bergallo, A.C., Rocha, H.B. and Rocha, C.F.D. (2016) Invasion by *Artocarpus heterophyllus* (Moraceae) in an island in the Atlantic Forest Biome, Brazil: distribution at the landscape level, density and need for control. *Journal of Coastal Conservation* 20(3), 191–198. DOI: 10.1007/s11852-016-0429-9.

Campbell, R.J. and McNaughton, R. (1994) New jackfruit cultivars under trial in South Florida. *Proceedings of the Interamerican Society of Tropical Horticulture* 38, 201–205.

Campbell, R.J., El-Sawa, S., Wasielewski, J., Ledesma, N. and Ayala-Silva, T. (2004) Breeding and selection of jackfruit for South Florida. In: *Proceedings of the Florida State Horticulture Society.* Krome Memorial Lecture. Available at: www.researchgate.net/profile/Noris-Ledesma/publication/237467885_Breeding_and_Selection_of_Jackfruit_for_South_Florida/links/55d5253908ae43dd17de4fe6/Breeding-and-Selection-of-Jackfruit-for-South-Florida.pdf

Chandel, K.P.S., Rekha, C., Radhamani, J., Malik, S.K. and Chaudhury, R. (1995) Desiccation and freezing sensitivity in recalcitrant seeds of tea, cocoa and jackfruit. *Annals of Botany* 76(5), 443–450. DOI: 10.1006/anbo.1995.1118.

Chellappan, K. and Roche, J.L. (1982) A hybrid Jack. *South Indian Horticulture* 30, 76–81.

Dang, Z.Q. and Lu, G. (2007) Cultural techniques for Hongbao and Sijimitain *Artocarpus* cultivars. *South China Fruits 2007* 37–38.

Diczabalis, Y. (2012) *Tropical Exotic Fruit Industry – Strategic Direction Setting 2012–2015*. Rural Industries Research and Development Corporation (RIRDC) Publication no.12/050, RIRDC Wagga Wagga, NSW, Australia.

Dutton, P.R. (1976) Jackfruit. In: *The Propagation of Tropical Fruit Trees*. Commonwealth Agricultural Bureau, Farnam Royal, Slough, UK, pp. 269–290.

Fu, J.R., Xia, H.Q. and Tang, L.F. (1993) Effects of desiccation on excised embryonic axes of three recalcitrant seeds and studies on cryopreservation. *Seed Science and Technology* 21, 85–95.

Ganeshan, S., Rajasekharan, P.E. and Bhaskaran, S. (2012) *In vitro* conservation of *Artocarpus heterophyllus* Lam. *Tree and Forestry Science and Biotechnology* 6, 126–129.

Gunasena, H.P.M., Ariyadasa, K.P., Wikramasinghe, A., Herath, H.M.W., Wikramasinghe, P. *et al.* (1996) *Manual of Jack Cultivation in Sri Lanka*, Vol. 48. Forestry Information Service, Forest Department, Sri Lanka.

Gupta, S., Singh, K. and Gupta, V. (2019) Genetic resource conservation of horticultural crops in India-Achievements and issues. *Progressive Horticulture* 51(1), 16–29. DOI: 10.5958/2249-5258.2019.00002.2.

Guruprasad, T.R. and Thimmaraju, K.R. (1989) Seed and spine characters in three jackfruit (*Artocarpus heterophyllus* L.) types. *Environment and Ecology* 7, 511–512.

Haq, N. (2006) *Jackfruit* (Artocarpus heterophyllus). Southampton Centre for Underutilized Crops, University of Southampton, Southampton, UK.

Hocking, D., Hocking, A. and Islam, K. (1996) Trees on farms in Bangladesh. *Agroforestry Systems* 33(3), 231–247. DOI: 10.1007/BF00055425.

Ibrahim, M., Islam, M.S., Helali, M.O.H., Alam, A.K.M.S. and Shafique, M.Z. (2013) Morphological fruit characters and nutritional food value of different jackfruit (*Artocarpus heterophyllus* Lam.) cultivars in Rajshahi region of Bangladesh. *Bangladesh Journal of Scientific and Industrial Research* 48(4), 287–292. DOI: 10.3329/bjsir.v48i4.18280.

Karunakaran, G. and Singh, P. (2019) Jackfruit varieties serves niche for growers. *Indian Horticulture* 62, 18–19.

Karunakaran, G., Singh, P. and Ravishankar, H. (2017) New jackfruit for homesteads. *Indian Horticulture* 62, 46–48.

Khan, R., Zerega, N., Hossain, S. and Zuberi, M.I. (2010) Jackfruit (*Artocarpus heterophyllus* Lam.) diversity in Bangladesh: land use and artificial selection 1. *Economic Botany* 64(2), 124–136. DOI: 10.1007/s12231-010-9116-1.

Li, Y., Mao, Q., Feng, F. and Ye, C. (2010) Genetic diversity within a jackfruit (*Artocarpus heterophyllus* Lam.) germplasm collection in China using AFLP markers. *Agricultural Sciences in China* 9(9), 1263–1270. DOI: 10.1016/S1671-2927(09)60215-7.

Li, Y.-Z., Ye, C.-H., Feng, F., Fang, L.-Q. and Tang, L.-R. (2007) Study on jackfruit germplasm from LeiZhou peninsula. *Journal of Hunan Agriculture University* 33, 183–186.

Magdalita, P.M., Dayap, F.T. and Valencia, L.D.C. (2011) Farmer participatory breeding and selection. In: Valavi, S.G., Peter, K.V. and Philly, T. (eds) *The Jackfruit*. Studium Press LLC, Houston, Texas, pp. 89–113.

Maity, C.S., Wangchu, L. and Mitra, S.K. (2002) Genetic variability for physico-chemical attributes in jackfruit (*Artocarpus heterophyllus* Lam.) genotypes of West Bengal. *Indian Agriculturist* 47, 193–199.

Majhi, D. (2012) Characterization of selected clones of mango and jackfruit. PhD thesis, Bidhan Chandra Krishi Viswavidyalaya, West Bengal, India.

Medagoda, I. and Jayawardena, S.D.C. (1997) Status report on genetic resources of jackfruit in Sri Lanka. In: *International Plant Genetic Resources Institute (IPGRI) Project No.BO6*. IPGRI, Selangor, Malaysia.

Mitra, S.K. (2000) *Background Paper on the Genetic Resources of Jackfruit*. FYNSYS Doc. Id. Cmt no.88090. Food and Agricultural Organization of the United Nations, Rome.

Mitra, S.K. (2017) Genetic resources of jackfruit and rambutan. *Acta Horticulturae* 1186, 15–22. DOI: 10.17660/ActaHortic.2017.1186.2.

Mitra, S.K. and Maity, C.S. (2002) A summary of the genetic resources of jackfruit (*Artocarpus heterophyllus* Lam.) in West Bengal, India. *Acta Horticulturae* 575, 269–271. DOI: 10.17660/ActaHortic.2002.575.29.

Mitra, S.K. and Mani, D. (2000) Conservation and utilisation of genetic resources in jackfruit (*Artocarpus heterophyllus* Lamk.) – a potential underutilised fruit. *Acta Horticulturae* 523, 229–232. DOI: 10.17660/ActaHortic.2000.523.29.

Morton, J.F. (1987) Jackfruit (*Artocarpus heterophyllus*). In: *Fruits of Warm Climates*. J.F. Morton, Miami, Florida, pp. 58–64.

Nakintu, J., Olet, E., Andama, M.A. and Lejju, J. (2019) Ethno-varieties and distribution of jackfruit tree (*Artocarpus heterophyllus* Lam.) in Uganda: implications for trade, food security and germplasm conservation. *East African Journal of Science, Technology and Innovation* 1, 27–51. DOI: 10.37425/eajsti.v1i1.66.

Nakintu, J., Albrecht, C., Müller, C.M., Kagoro-Rugunda, G., Andama, M. *et al.* (2020) Exploring the genetic diversity of jackfruit (*Artocarpus heterophyllus* Lam.) grown in Uganda based on SSR markers. *Genetic Resources and Crop Evolution* 67(3), 605–619. DOI: 10.1007/s10722-019-00830-5.

Palupi, D. and Daryono, B.S. (2021) Genetic diversity in jackfruit (*Artocarpus heterophyllus* Lam.) provenances based on morphological parameters. *SABRAO Journal of Breeding and Genetics* 53, 479–498.

Palupi, D., Rahayu, S.S.B. and Daryono, B.S. (2019) Genetic diversity in jackfruit (*Artocarpus heterophyllus* Lam.) based on molecular characters in Indonesia. *SABRAO Journal of Breeding and Genetics* 51, 57–67.

Rai, M., Nath, V, Das, B., Rai, A. and Kumar, M. (2003) Evaluation of jackfruit genotypes for yield and quality attributes under eastern Indian condition. *Orissa Journal of Horticulture* 31, 1–6.

Rajasekharan, P.E. and Ganeshan, S. (2019) Current perspectives on pollen cryopreservation in horticultural species. *Acta Horticulturae* 1234, 47–56. DOI: 10.17660/ActaHortic.2019.1234.6.

Ray, P.K. (2002) Jackfruit. In: Ray, P.K. (ed.) *Breeding Tropical and Subtropical Fruits*. Arora Publishing House, New Delhi, pp. 290–299.

Reddy, B.M.C., Prakash, P., Shashikumar, S., Sampathkumar, P. and Govindaraju, R.L. (2001) Exploration, collection of potential germplasm, elite clones and establishment of field genebank of *Artocarpus heterophyllus* Lam. In: *Proceedings of the National Symposium on Biodiversity vis a vis Resource Exploration: An Introspection*, p. 50.

Richards, A.V. (1950) A note on the cultivation of Singapore jack. *Tropical Agriculturist* 105, 12–13.

Sastrapradja, S. (1975) Tropical fruit germplasm in Southeast Asia. In: Williams, J.T., Lamoureux, C.H. and Soetijipto, N.W. (eds) *Proceedings of Symposium on Southeast Asian Plant Genetic Resources*. Indonesian Institute of Sciences (LIPI), Bogor, Indonesia, pp. 33–46.

Schnell, R.J., Olano, C.T., Campbell, R.J. and Brown, J.S. (2001) AFLP analysis of genetic diversity within a jackfruit germplasm collection. *Scientia Horticulturae* 91, 261–274.

Sharma, S.K., Singh, A.K. and Singh, O.P. (2006) Character association among various quantitative traits in jackfruit (*Artocarpus heterophyllus* Lam). *Advances in Plant Science* 19, 633–638.

Shyamalamma, S., Priyanka, S., Kalpana, B., Munishnamamna, K.B. and Narayanagoda, K. (2017) Genetic variability in selected genotypes from Karnataka. In: *Abstract International Workshop on Production, Value Addition and Marketing of Jackfruit, 9–14 August, 2017*. Regional Agricultural Research Station, Kerala Agricultural University, Wayanad, Kerala, p. 4.

Thammasiri, K. (1998) Cryopreservation of embryonic axes of jackfruit by vitrification. In: *IUFRO Seed Symposium 1998 'Recalcitrant seeds'. Proceedings of the Conference*, Forest Research Institute, Malaysia, Kuala Lumpur, October 12–15, 1998, pp. 153–158.

Tri, M.V. and Vu, N.T. (2020) *Survey of Production Status of Jackfruit in Vietnam*. Project Report. Southern Fruit Research Institute, Tien Giang, Vietnam.

Valencia, L.D.C. and Magdalita, P.M. (2010) Development of 'Mabini' jackfruit (*Artocarpus heterophyllus* Lam.), a dual-purpose variety. *Philippines Journal of Crop Science* 35, 74–79.

Wang, Z., Pan, D., Deng, Z., Li, J., Ye, Y. and Zhu, J. (2012) Breeding and selection of a new jackfruit cultivar 'Red Flesh Jackfruit'. *Journal of Fruit Science* 29, 518–519.

Wangchu, L. (2005) Selection of superior jackfruit genotypes and their identification through electrophoretic banding patterns of isozymes. PhD thesis, Bidhan Chandra Krishi Viswavidyalaya, West Bengal, India.

Wangchu, L., Singh, D. and Mitra, S.K. (2011) Biochemical makers: a useful tool for assessing genetic diversity in jackfruit. *Acta Horticulturae* 890, 91–101. DOI: 10.17660/ActaHortic.2011.890.10.

Wangchu, L., Singh, D. and Mitra, S.K. (2013) Studies on the diversity and selection of superior types in jackfruit (*Artocarpus heterophyllus* Lam.). *Genetic Resources and Crop Evolution* 60(5), 1749–1762. DOI: 10.1007/s10722-013-9985-0.

Wasielewski, J. and Campbell, R.J. (2000) Jackfruit breeding at Fairchild Tropical Garden. In: *Proceedings of the Florida State Horticulture Society*, Miami, FL. USA, pp. 218–219 (Vol. 112).

Witherup, C., Ragone, D., Wiesner-Hanks, T., Irish, B., Scheffler, B. *et al.* (2013) Development of microsatellite loci in *Artocarpus altilis* (Moraceae) and cross-amplification in congeneric species. *Applications in Plant Sciences* 1(7), apps.1200423. DOI: 10.3732/apps.1200423.

Witherup, C., Zuberi, M.I., Hossain, S. and Zerega, N.J.C. (2019) Genetic diversity of Bangladeshi jackfruit (*Artocarpus heterophyllus*) over time and across seedling sources. *Economic Botany* 73(2), 233–248. DOI: 10.1007/s12231-019-09452-5.

Yaacob, O. and Subhadrabandhu, S. (1995) Jackfruit, cempedak, breadfruit, and marang/Tarap (*Artocarpus* spp.). In: *The Production of Economic Fruits in South East Asia*. Oxford University Press, Kuala Lumpur, pp. 122–128.

Ye, C.H., Wang, Y.H., Li, Y.Z. and Feng, F. (2009) Analysis of genetic diversity of jackfruit germplasm using ISSR marking method. *Journal of Fruit Science* 28, 659–665.

5 Biotechnology

Subbareddy Shyamalamma[1]*, Parlakoti M. Sampath[2], Huliyar V. Vidyashree[1], Vasudevamurthy Dileep[1], Marina Dsouza Valeeta[1] and Chikkavalli K. Harshitha[1]

[1]Department of Biotechnology, University of Agricultural Sciences, GKVK, Bengaluru, India; [2]Department of Fruit Science, University of Horticultural Sciences, Bengaluru, India

5.1 Introduction

Jackfruit (*Artocarpus heterophyllus*) belongs to the family Moraceae. It is believed to have originated in the south-western rainforests of India (Naik, 1949) and has been introduced and cultivated in many tropical countries. Besides India, jackfruit is commonly grown in many parts of South-east Asia (Rahman *et al.*, 1999), in the evergreen forest zone of West Africa, in northern Australia (Azad *et al.*, 2007) and in south Florida (Schnell *et al.*, 2001). As it is protoandrous and cross-pollinated, and is propagated mainly by seed, jackfruit shows wide variability in many characteristics including age of first bearing, and the texture, flavour and colour of the fruit. Jackfruits are generally classified into two main types (Odoemelam, 2005): (i) those having fruits with small, fibrous, soft and spongy flakes with very sweet carpels, called 'Koozhachakka' in Malayalam and 'Biluva' or 'Ambli' in Kannada; and (ii) those with fruits that are crunchy but not as sweet, with crisp carpels and of high quality, called as 'Varikka' in Malayalam and 'Bakke' in Kannada. So far, there are few developed varieties, namely 'PLR-1', 'PLR-2' and 'PPI-1' from Tamil Nadu Agricultural University; 'Muttam Varikka' and 'Sindhura' from Kerala Agricultural University, India; 'Swarna', 'Lalbagh Madhura' and 'Byrachandra' from the University of Agricultural Sciences, Bangalore, India; and 'Konkan Prolific' from Dr Balasaheb Sawant Konkan Krishi Vidyapeeth University, Dapoli, India. Some exotic varieties such as 'J33', 'Dang Surya' and 'Nang Dak' from Malaysia and 'Vietnam Super Early' from Vietnam have been introduced. However, in different localities, local varieties have different names based on their variability in yield, fruit shape and size, rind colour, pulp colour, number of flakes, texture of flakes, total sugars and many other characteristics.

5.2 Taxonomy and Cytology

Included in the family Moraceae along with jackfruit are *Ficus* spp. (fig) and *Morus* spp. (mulberry) (Chandler, 1958; Popenoe, 1974). Jackfruit crosses freely with *Artocarpus integer*, *Artocarpus lanceifolius* and *Artocarpus rigidus*, which are closely related species of *A. heterophyllus* (Kanzaki *et al.*, 1997). The family Moraceae encompasses about 1000 species in 67 genera, mostly tropical shrubs and trees but also a few vines and herbs (Merrill, 1912;

*Corresponding author: shyamala_reddyrs@yahoo.co.in

DOI: 10.1079/9781800622319.0005

Bailey, 1949). The genus *Artocarpus* comprises around 50 species, 11 of which are known to produce edible fruits (Barrau, 1976; Campbell, 1984; Corner, 1988). Very little information is available on the cytology of the *A. heterophyllus* gene pool. Darlington and Wylie (1956) reported that it is tetraploid with a somatic chromosome number of 56 ($2n = 4x = 56$); thus, the basic chromosome number (x) is 14. In contrast, in a study carried out by Yang *et al.* (2015) pertaining to ploidy and genomic size determination of jackfruit lines, it was reported that the ten jackfruit lines under study were diploid ($2n = 2x = 56$). The genome size of the soft-fleshed type (1902 ± 8.08 Mb) was larger than that of the firm-fleshed types (1243.75 ± 9.21 to 1540 ± 3.55 Mb) indicating that the firm-fleshed jackfruit genotypes have evolved from the soft-fleshed types. These cytological complexities related to the ploidy level, and the genome sizes have made the application of the molecular markers in jackfruit a cumbersome process.

5.3 Applications of Markers in Jackfruit

5.3.1 Morphological markers

Morphological markers are detected by eye and their inheritance can be monitored visually without specialized biochemical or molecular techniques. In jackfruit, several morphological traits such as trunk character, leaf shape and size, fruit shape and size, latex content in ripe fruits and pulp colour can act as morphological markers. For example, 'Rudrakshi' jackfruit has a typical round fruit shape with flat spines that are hexagonally arranged on the rind of the fruit. The pulp colour varies from creamish-white to yellow to orange and red. The red types are preferred in the market for their attractive colour and are popular as 'Chandra Halasu' in Karnataka. Morphological markers are affected by the environment and their expression will be altered by epistatic and pleiotropic gene interactions. Morphological markers are limited in number; their alleles interact in a dominant-recessive manner, making it impossible to distinguish heterozygous from homozygous individuals. Many of these traits are of polygenic inheritance and are expressed only after a long juvenile phase (Hamrick *et al.*, 1992). However, characterization of genotypes at the genetic level, supplemented by phenotypic characters, provides the first step towards more efficient conservation, maintenance and utilization of existing genetic diversity (Prakash *et al.*, 2002).

5.3.2 Biochemical markers

The phenotypic approach to characterize plant populations has been criticized because it does not provide genetic information (Simpson and Withers, 1986). Biochemical markers help in assisting selection of superior types and determining the extent of genetic diversity within the selected clones. Pushpakumara *et al.* (2005) tried isozyme analysis in jackfruit, but it showed low levels of polymorphism and only two out of 16 enzymes (12.5%) were polymorphic. There was no evidence to distinguish soft- and firm-fleshed genotypes. The success of biochemical markers may also depend on the stage at which the plant material is used for extraction of enzymes, as enzyme expression at early stages (e.g. embryo, leaf tissues in seedling) may not be the same as at later stages (e.g. tissue from ripening syncarps). Fifty accessions from Bangladesh were evaluated by Azad *et al.* (2007) for four enzyme systems. The isozyme patterns were determined on the basis of the number and position of loci. They discovered that morphological traits such as weight, length, girth of the fruits and percentage of pulp correlated poorly with environmental factors, suggesting that these characteristics are more likely to be genetically controlled.

Peroxidase and esterase isozymes were investigated from leaf tissue samples of 44 selected (superior) jackfruit genotypes from West Bengal (Wangchu et al., 2011). These selected genotypes were collected following jackfruit International Plant Genetic Resources Institute descriptors (IPGRI, 2000) after surveying 1500 trees in the districts of Nadia, North 24 Parganas and Cooch Behar in West Bengal. Polymorphism was observed in both of the enzyme systems studied. In the peroxidase assay, seven loci were identified among 44 genotypes and the relative mobility values of the loci ranged between 0.29 and 0.50. In the esterase assay, six loci were identified and the relative mobility values of the loci ranged between 0.22 and 0.65. It was also observed that genotypes collected from nearby locations did not show similar banding patterns, indicating several causes for genetic diversity such as geographical origin, exchange of genetic stocks, genetic drift, spontaneous variation, natural and artificial selection, cross-pollination and seedling origin of the plant. In this study, both the young and mature leaves were used for biochemical extraction, and it was found that the mature leaves gave better and clearer banding patterns compared with young leaves. However, isozyme markers are also known to be affected by both environment and post-translation modification, and their practical uses are limited (Akashi *et al.*, 2002).

5.3.3 Molecular markers

A molecular marker is a DNA sequence that is readily detected and whose inheritance can easily be monitored. They rely on DNA assays, in contrast to morphological markers based on visible traits and biochemical molecular markers based on protein products. Hence, DNA is an ideal molecule for studying polymorphisms. These molecular markers can be linked to important traits and used for early selection of potentially desirable genotypes. Molecular markers are capable of producing patterns that are unique for each individual genotype.

The molecular markers that have been used for analysis of jackfruit are shown in Table 5.1 and discussed below.

Random amplified polymorphic DNA

Random amplified polymorphic DNA (RAPD) variation has been assessed in a number of studies from 2000 onwards. In one such study, 65 accessions of jackfruit from Leizhou Peninsula, southern China, were analysed. 78 loci were detected using 16 RAPD primers, and 69 of these were polymorphic (Ye *et al.*, 2005). The cluster analysis categorized the genotypes into three groups. However, this could not distinguish the genotypes based on their bearing behaviour and latex content.

Genetic diversity in 12 high-yielding jackfruit accessions obtained from different locations in southern India was estimated with RAPD markers by Simon *et al.* (2007). Overall, 171 amplified fragments were obtained using 23 random primers, of which 115 (67.3%) were polymorphic and shared between at least two individuals, while 46 (26.9%) were monomorphic and common to all individuals. Only ten (5.8%) were polymorphic and unique. The genetic dissimilarity matrix revealed a maximum genetic distance of 7.9% between clones of 'Muttam Varikka' and 'Chandra Halasu' from distant locations, while the minimum genetic distance (5%) was between 'Muttam Varikka' and 'Kerala', indicating their similar geographical origin. Ward's method of cluster analysis grouped all individuals on the dendrogram into two major clusters according to their geographical location.

Table 5.1. Molecular markers used in jackfruit.

Type of molecular marker	Reference
RAPD	Ye *et al.* (2005); Simon *et al.* (2007); Pushpakumara and Harris (2007); Prasad *et al.* (2014); Krishnan *et al.* (2015)
SCAR	Singh *et al.* (2011)
AFLP	Schnell *et al.* (2001); Shyamalamma *et al.* (2008); Li *et al.* (2010)
SSR	Thumilan *et al.* (2016); Kavya *et al.* (2019)
ISSR	Sane *et al.* (2009); Wann *et al.* (2013)
matK gene	Prem *et al.* (2017)

RAPD, random amplified polymorphic DNA; SCAR, sequence-characterized amplified region; AFLP, amplified restriction fragment length polymorphism; SSR, simple sequence repeat; ISSR, inter-simple sequence repeat.

Prasad *et al.* (2014) screened two RAPD primers (OP A9 and OP A8) in five different jackfruit genotypes. The marker assay yielded 57 amplified fragments in which primer OP A9 had a higher polymorphism (24%) than OP A8 (22%). Krishnan *et al.* (2015) screened 30 RAPD primers to determine the genetic relationship among six superior jackfruit selections. Among these, ten primers gave good amplification that could be reproduced. Primer OPA-4 (AATCGGGCTG) gave the maximum number of polymorphic loci (14), while OPD-19 (CTGGGGACTT) produced the least (nine). Four major clusters were obtained for genotypes, which followed geographical separation.

Sequence-characterized amplified region

An attempt to develop trait-specific markers to differentiate between soft- and firm-fleshed fruit types of jackfruit was carried out by Pushpakumara and Harris (2007) using RAPD markers in Sri Lanka. These two jackfruit types of fruit vary in the firmness of the edible flesh of the ripened fruit (Acedo, 1992; Yaacob and Subhadrabandhu, 1995; Pushpakumara *et al.*, 1997). Variation in flesh texture between the two fruit types has not been consistent through generations, and seeds planted from either fruit type yields a mixture of both. However, little is known about their pattern of segregation. The seedlings of these two types cannot be morphologically differentiated and it is possible only when ripe syncarps are produced, which generally takes 8–10 years from planting. Identification of the fruit types at the seedling stage would be useful in the management of genetic resources for conservation and genetic improvement programmes (Pushpakumara *et al.*, 1997). The two fruit types of this species are not consistently distinguishable on the basis of morphological traits of mature trees (Soepadmo, 1991; Pushpakumara *et al.*, 1997). Furthermore, apart from firmness of the ripened flesh, no morphological or phenological differences were observed between two fruit types in a study carried out on flowering and fruiting morphology and phenology (Pushpakumara *et al.*, 1997). However, the authors found that the two fruit types could be distinguished using one RAPD product, namely OPB-01, with a fragment size of 123 kb for firm-fleshed genotypes, but which was absent in soft-fleshed genotypes. About 100 primers were screened and only five primers showed polymorphism for fruit type and were considered to be useful markers to differentiate fruit types of jackfruit. The reproducibility test was carried out for these five primers with 30 trees of each type.

The occurrence of two fruit types in jackfruit may also be a result of different enzyme activities, or the inhibition of an enzyme/set of enzymes at fruit ripening. In many plant species, the process of fruit ripening and softening is largely the result of enzyme-mediated physico-chemical changes, which are genetically and developmentally regulated (Brady, 1987). Therefore, differential gene expression patterns that are correlated with changes in growth and physiology have been a source of concern in many studies (Christoffersen *et al.*, 1982; Oh *et al.*, 1995). The results of such studies have revealed that fruit ripening is under the control of genes expressed during the ripening process. Studies on biochemical changes during ripening of the firm- and soft-fleshed fruit types of jackfruit revealed a common process of ripening in both fruit types (Selvaraj and Pal, 1989; Rahman *et al.*, 1995). It was also suggested that the texture of the flesh of the ripened syncarps was related to the degree of change in activities of enzymes such as pectin esterase and polygalacturonase towards the early ripening stage. These changes were more marked in the soft-fleshed type, and were either arrested or delayed in the hard-fleshed fruit type. Although such evidence supported the developmental expression of enzymes (and hence genes), little is known about gene expression during fruit ripening in jackfruit (Pushpakumara *et al.*, 2005). Cloning, sequencing and conversion of the RAPD marker to sequence-characterized amplified region (SCAR) marker for verification of the result and development of highly reproducible and easy markers would assist the effective identification of fruit types at the early stage of growth.

Another such trial to develop trait-specific markers was undertaken by Singh *et al.* (2011) to identify a SCAR marker related to fruit cracking in some jackfruit genotypes. The RAPD primer

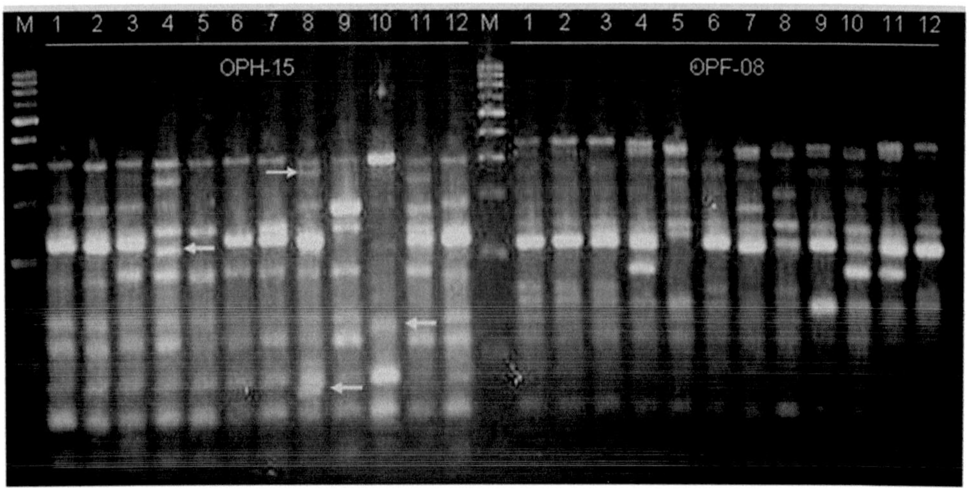

Fig. 5.1. Random amplified polymorphic DNA gel profile of jackfruit genotypes showing specific bands for fruit cracking in lanes 4, 8 and 10 (arrows).

OPH-15 produced polymorphic loci of 810 bp in fruit cracking genotypes (Fig. 5.1). The result was confirmed for reproducibility. Furthermore, the marker was sequenced and a homology search was done using BLASTX. The gene sequence has been deposited in Genbank (National Center for Biotechnology Information). This can be used as a marker to identify jackfruit genotypes with the fruit cracking characteristic at the seedling stage.

Amplified restriction fragment length polymorphism

Amplified restriction fragment length polymorphism (AFLP) markers are reported to be the best markers for genetic diversity studies due to their ability to scan the whole genome with two distinct types of primer sequences. The AFLP technique was adapted by Schnell *et al.* (2001) for genetic diversity analysis in 26 accessions of jackfruit collected from eight countries. They screened 30 primers of which 12 produced good-quality polymorphic fragments with a moderate level of polymorphism (49.2%) among the genotypes. Cluster analysis and principal component analysis grouped all of the jackfruit accessions of South-east Asian origin into one major cluster with little bootstrap support for groupings within the cluster. The Indian accessions were grouped into a different cluster, as did the hybrid and breadfruit accessions.

AFLP markers were used to investigate the genetic diversity of 50 jackfruit accessions collected from different parts of South India by Shyamalamma *et al.* (2008). They selected eight primers from the 16 primer pairs for screening of jackfruit genotypes. These primer combinations amplified 5976 loci, of which 1267 (21%) were polymorphic. Among the jackfruit accessions, the similarity coefficient ranged from 0.137 to 0.978; the accessions also shared a large number of monomorphic fragments (78%). Three major clusters were obtained by cluster analysis and principal component analysis. The genotypes growing in regions with very dry conditions and in locations having medium to heavy rainfall grouped in different clusters, and a good correlation between genetic and geographical data was found.

Similarly, Li *et al.* (2010) analysed the genetic diversity of 50 jackfruit accessions from three provinces in China using AFLP markers. A total of 320 unambiguous loci were produced by eight primer combinations, and 65 (20.3%) of them were polymorphic. The dendrogram revealed five groups, which could not distinguish the genotypes based on geographical origin, fruit quality and bearing behaviour.

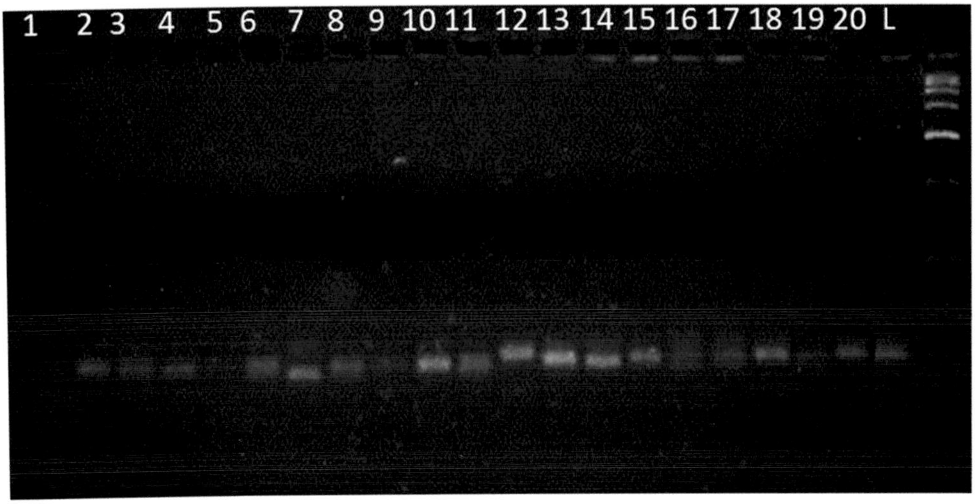

Fig. 5.2. Simple sequence repeat marker profile of jackfruit genotypes for different pulp colours. Lanes 1–4, red; lanes 6–10, orange; lanes 11–15, yellow; lanes 16–20, cream.

Simple sequence repeats

Simple sequence repeat (SSR) markers for jackfruit diversity estimation were developed by partial genome sequencing of jackfruit (Kavya *et al.*, 2019). The primers thus obtained were used to estimate genetic diversity among 20 jackfruit genotypes collected from different parts of Karnataka to differentiate the genotypes with distinct pulp colours. Based on the colour of the jackfruit pulp, the 20 genotypes were classified into four types: red, orange, yellow and cream. A total of 22 SSR primers gave good amplification in 20 jackfruit genotypes and nine primers were found to be polymorphic (Fig. 5.2). The similarity matrix generated from the marker data showed correlation between pulp colour and genotypes in such a way that the genotypes with similar pulp colour had a high similarity index. All the genotypes with red colour pulp were closely related among themselves and also closely related to the orange genotypes. The cream genotypes had similarity with the yellow genotypes rather than the orange and red genotypes. Three major clusters were obtained on the dendrogram where cluster I grouped three genotypes with cream-coloured pulp. Cluster II was further classified into two subclusters, IIa and IIb, which comprised two cream and two yellow pulp genotypes. Cluster III comprised a large number (13) of the genotypes with yellow, orange and red pulp. The genotypes for red pulp were closely linked to those for orange pulp irrespective of the place of collection, indicating genetic relationships for pulp colour rather than place of cultivation. Further studies are recommended to screen for and determine the sequences linked to pulp colour.

Thumilan *et al.* (2016) developed and validated 206 expressed sequence tag-derived SSR markers using transcriptome data generated from leaf tissue from a drought-tolerant mulberry (family Moraceae) genotype, Dudia white. A total of 264 primers to the most appropriate repeat regions were designed, of which 206 were locus specific; of these, 60 (29%) were transferable to jackfruit.

Inter-simple sequence repeats

Sane *et al.* (2009) distinguished 18 accessions of jackfruit maintained in the field gene bank of Indian Institute of Horticultural Research Bangalore by inter-simple sequence repeat (ISSR) analysis. The cluster diagram depicting genetic relationships between accessions revealed that the accessions 'Gumless', 'Dorichandra' and 'Thenvarikka' were the most divergent. The morphological characters when collated with

DNA profiles of these accessions showed that 'Tenvarike', which is a divergent accession, had desirable fruit characters such as low latex exudation and low flake fibre content.

A study was undertaken to assess the molecular differences between clonal and seedling progenies collected from two distinct mother plants known to produce good-quality fruits. The clonal progenies were propagated by grafting from a single mother plant, and the seedling progenies were raised from seeds of an elite mother plant known for quality fruits. The major objective was to elucidate molecular variations between the clonal and seedling progenies using ISSR markers among 25 progenies of jackfruit (Wann *et al.*, 2013). Out of 25 ISSR primers used, only six gave good amplification. Higher polymorphism (34.78%) was exhibited by primer 823 and lower polymorphism (10.00%) by primer 815. The cluster analysis grouped the progenies of seedling and clonal origin into four groups where clusters I, II, III and IVb grouped all the clonal progenies and cluster IVa contained all the seedling progenies, clearly depicting the variations at the genome level.

matK *marker gene*

Maturase K (*matK*) is a plant plastidial gene (Zoschke *et al.*, 2010) and the protein it encodes is an intron maturase, which splices introns. Among other maturases, this protein retains only a well-conserved domain and remnants of a reverse transcriptase domain (Mohr *et al.*, 1993). Universal *matK* can be used for DNA barcoding of angiosperms (Yu *et al.*, 2011). The level of genetic diversity and genetic relationships among six varieties of jackfruit were assessed using the *matK* gene based on a polymerase chain reaction technique and restriction fragment length polymorphism markers (Prem *et al.*, 2017). The partial sequence of the *matK* gene of six different jackfruit varieties was used to analyse their level of genetic diversity and genetic relationships. The size of the amplified products was approximately 700 bp. After sequencing and sequence editing, sequence information on a 674 bp region was obtained for analysis. Alignment of sequences revealed two haplotypes out of 674 sites. The nucleotide frequencies were 30.00% (A), 37.69% (T/U), 17.93% (C) and 14.39% (G). The sequence of the 'Singapore Varikka' variety had a change in a single nucleotide in the 654th bp locus to adenine (A),whereas the other five genotypes had cytosine (C).

5.4 Jackfruit Genome Sequence

Draft genome sequence information for jackfruit and breadfruit has been reported by Sahu *et al.* (2020). They have worked on jackfruit in comparison with breadfruit and published genomic information for these two economically important species, reporting 273 Gb and 227 Gb of raw data from jackfruit and breadfruit, respectively. The high-quality reads from jackfruit were assembled into 162,440 scaffolds totalling 982 Mb with 35,858 genes. Similarly, the breadfruit reads were assembled into 180,971 scaffolds totalling 833 Mb with 34,010 genes. A total of 2822 and 2034 expanded gene families were found in jackfruit and breadfruit, respectively, enriched in pathways including starch and sucrose metabolism, photosynthesis and others. The copy number of several starch synthesis-related genes was found to be increased in jackfruit and breadfruit compared with closely related species, and the tissue-specific expression might imply their sugar-rich and starch-rich characteristics. Overall, the publication of high-quality genomes for jackfruit and breadfruit provides information about their specific composition and the underlying genes involved in sugar and starch metabolism. The availability of these genomes on the African Orphan Crops Consortium (AOCC)/Online Resource for Community Annotation of Eukaryotes (ORCAE) platform (https://bioinformatics.psb.ugent.be/orcae/aocc, accessed 7 July 2022) will enable various stakeholders to access and improve the annotations of these genomes.

5.5 Advances in Plant Tissue Culture

Propagation of jackfruit plant from seeds is not widely accepted, because of the highly heterozygous nature of seedlings. To get true-to-type quality fruits and also to carry out vegetative propagation of jackfruit throughout the year with mass production of plants, tissue culture techniques could be used conveniently with standardization of a vegetative propagation protocol. The work carried out in this direction by various scientists is summarized below.

Nodal explants of jackfruit cultured on Murashige and Skoog (MS) medium were induced to form multiple shoots, with supplementation with 1.0 mg 6-benzylaminopurine (BAP) l^{-1} and 0.5 mg kinetin l^{-1} (Roy *et al.*, 1989). Shoots formed *in vitro* were also subcultured in similar fresh medium, and five to seven new shoots developed. Rooting was induced on *in vitro*-proliferated shoots by culturing in half-strength MS salts supplemented with 1.0 mg l^{-1} each of 1-naphthaleneacetic acid (NAA) and indole-3-butyric acid (IBA). The regenerants were successfully transferred to pots in a greenhouse and later to the field.

To maintain the true type quality fruit, this tissue culture technique could also be used for the propagation of jackfruit throughout the year. Ashrafuzzaman *et al.* (2012) carried out micropropagation of jackfruit using healthy and juvenile shoot tips as explants and cultured them in MS medium complemented with different concentrations (1, 2, 3 and 4 mg l^{-1}) of plant growth hormone (BAP) for multiple shoot development. Regeneration of shoots increased comparatively better when the MS medium was enriched with 2 mg l^{-1} BAP (Fig. 5.3). With increased subculture (up to a maximum of ten), the frequency of shoot proliferation was enhanced. These shoots were then cultured on half-strength MS medium supplemented with different concentrations of IBA. IBA at 2 mg l^{-1} induced a higher number of roots/explants, increased root length and early root induction. Amany *et al.* (2007) tried to develop a jackfruit propagation protocol through tissue culture. They cultured explants on MS medium supplemented with 3.0 mg BAP l^{-1} and 0.1 mg NAA l^{-1}. MS medium surpassed other nutrient media (B5 and WPM) in terms of production of average number of shoots, leaves/explant and shoot length. Sorbitol at 100 mM scored the highest growth parameters compared with sucrose. IBA at 1.50 mg l^{-1} resulted in the highest number of roots per explant. Moreover, the greatest number of roots was obtained on half-strength MS medium supplemented with 1.50 mg IBA l^{-1}. Jackfruit

Fig. 5.3. Induction of jackfruit shoots by *in vitro* culture in 2 mg l^{-1} 6-benzylaminopurine. (A) Single shoot. (B) Multiple shoots.

plantlets were successfully acclimatized in the greenhouse with 90% survival using a mixture of peat and sand (2:1, w/w).

In order to check the seasonal influence on *in vitro* propagation in jackfruit, Eglal *et al.* (2015), collected healthy and juvenile shoot tips and nodal segments during the four different seasons (winter, spring, summer and autumn) and cultured them on MS medium. Explants that were collected in winter gave the highest survival percentage (100%) without browning of explants. They also gave the highest growth to survival percentage (90% and 50% for shoot tips and nodal segments, respectively). Shoots could be multiplied on MS medium supplemented with BAP (0.0, 1.0, 2.0 and 5.0 mg l^{-1}) and kinetin (0.0, 0.5, 1.0, 1.5 and 2.0 mg l^{-1}). The maximum number of proliferated shoots (6.6 shoots per explant) was obtained on MS medium supplemented with 2.0 mg BAP l^{-1} and 0.5 mg kinetin l^{-1}. The highest increase in shoot length (5.14 cm) was recorded on medium containing 3.0 mg gibberellic acid l^{-1}. These shoots were then cultured on half-strength MS medium supplemented with IBA and NAA at concentrations of 0.5 and 1.0 mg l^{-1}, respectively; these auxins were added to the medium either individually or in combination. It was observed that medium containing 1.0 mg l^{-1} of both IBA and NAA gave the highest rooting percentage (80%), the highest average number of roots per explant (4.3) and the highest average root length (4.0 cm).

Shoot tips collected from mature trees and sterilized with 70% ethanol and mercuric chloride before culturing on MS medium supplemented with different combinations of BAP and NAA gave successful results. The combination of BAP and NAA resulted in significant differences for shoot number, shoot length and leaf number ($p < 0.01$), whereas 2 mg BAP l^{-1} alone was found to be the best with a mean shoot number of 5.12 and length of 0.89 cm. This optimized protocol could be used as a baseline for further studies on *in vitro* propagation of jackfruit (Ali *et al.*, 2017).

5.6 Conclusion

Jackfruit is a perennial fruit tree that is easy to cultivate and maintain for years but requires crop improvement using biotechnology-based tools such as marker-assisted selection, genetic engineering, and plant tissue culture for developing and deploying the best varieties for various applications. Although markers play a significant role in the selection of fruit traits of commercial importance, their application in jackfruit is limited due to the tetraploid nature of the crop. The genome has been sequenced and some information has been made available, but information on the genes and their validation is a challenging task ahead. However, efforts have been made by several workers to identify the markers linked to fruit quality traits (soft- and firm-fleshed types, fruit cracking and pulp colour). Thus, there is a need for whole-genome sequencing in jackfruit to identify the marker sequences linked to traits of commercial importance such as fruit size and shape, pulp colour and age of fruit bearing. Gene annotation for important fruit quality traits would hasten crop improvement in jackfruit. Furthermore, plant tissue culture techniques have huge applications in mass production and will aid in crop improvement.

References

Acedo, A.L. (1992) *Jackfruit Biology, Production, Use, and Philippine Research*. Monograph no.1, Forestry/ Fuelwood Research and Development (F/FRED) Project, Winrock International Institute for Agricultural Development, North Little Rock, Arkansas.

Akashi, Y., Fukuda, N., Wako, T., Masuda, M. and Kato, K. (2002) Genetic variation and phylogenetic relationships in East and South Asian melons, *Cucumis melo* L., based on the analysis of five isozymes. *Euphytica* 125(3), 385–396. DOI: 10.1023/A:1016086206423.

Ali, J., Bantte, K. and Feyissa, T. (2017) Protocol optimization for *in vitro* shoot multiplication of jackfruit (*Artocarpus heterophyllus* L.). *African Journal of Biotechnology* 16, 87–90.

Amany, M.H., Enas, A.M.A. and Ehab, S.B. (2007) In *vitro* propagation of jackfruit (*Artocarpus heterophyllus* L. *Journal of Applied Sciences of Research* 3, 218–226.

Ashrafuzzaman, M., Kar, S., Khanam, D. and Prodhan, S.H. (2012) In *vitro* regeneration and multiplication of jackfruit (*Artocarpus heterophyllus* L. *Research Journal of Biology* 2, 59–65.

Azad, A.K., Jones, J.G. and Haq, N. (2007) Assessing morphological and isozyme variation of jackfruit (*Artocarpus heterophyllus* Lam.) in Bangladesh. *Agroforestry Systems* 71(2), 109–125. DOI: 10.1007/s10457-007-9039-8.

Bailey, L.H. (1949) *Manual of Cultivated Plants*. Macmillan Co., New York, New York.

Barrau, J. (1976) Breadfruit and relatives. In: Simmonds, N.W. (ed.) *Evolution of Crop Plants*. Longman, London, pp. 201–202.

Brady, C.J. (1987) Fruit ripening. *Annual Review of Plant Physiology* 38(1), 155–178. DOI: 10.1146/annurev.pp.38.060187.001103.

Campbell, C.W. (1984) Tropical fruits and nuts. In: Martin, F.W. (ed.) *CRC Handbook of Tropical Food Crops*. CRC Press, Boca Raton, Florida.

Chandler, W.H. (1958) *Evergreen Orchards*, 2nd edn. Henry Kimpton, London.

Christoffersen, R.E., Warm, E. and Laties, G.G. (1982) Gene expression during fruit ripening in avocado. *Planta* 155(1), 52–57. DOI: 10.1007/BF00402931.

Corner, E.J.H. (1988) Notes on the systematics and distribution of Malayan phanerogams II. The jack and the chempedak. *Garden Bulletin* 10, 56–81.

Darlington, C.D. and Wylie, A.P. (1956) *Chromosome Atlas of Flowering Plants*. George Allen and Unwin Ltd, London.

Eglal, M.H., Reda, A.M. and Naglaa, A.A.E. (2015) In *vitro* rapid propagation of jackfruit (*Artocarpus heterophyllus* Lam.). *American–Eurasian Journal of Agricultural & Environment Sciences* 15, 147–153.

Hamrick, J.L., Godt, M.J.W. and Sherman-Broyles, S.L. (1992) Factors influencing levels of genetic diversity in woody plant species. *New Forests* 6(1–4), 95–124. DOI: 10.1007/BF00120641.

International Plant Genetic Resources Institute (2000) Plant descriptors. In: *Descriptors for Jackfruit* (Artocarpus heterophyllus *L.)*. IPGRI, Rome, pp. 30–48.

Kanzaki, S., Yonemori, K., Sugiura, A. and Subhadrabandhu, S. (1997) Phylogenetic relationships between the jackfruit, the breadfruit and nine other *Artocarpus* spp. from RFLP analysis of an amplified region of chloroplast DNA. *Scientia Horticulturae* 70(1), 57–66. DOI: 10.1016/S0304-4238(97)00045-9.

Kavya, K., Shyamalamma, S. and Gayatri, S. (2019) Morphological and molecular genetic diversity analysis using SSR markers in Jackfruit (*Artocarpus heterophyllus* Lam.) genotypes for pulp colour. *Indian Journal Of Agricultural Research* 53, 8–16. DOI: 10.18805/IJARe.A-5066.

Krishnan, A.G., Sabu, T.S., Sible, G.V. and Lindamol, X. (2015) Genetic diversity analysis in jackfruit selections of kuttanad region using RAPD technique. *International Journal of Scientific and Research Publications* 5, 1–6.

Li, Y., Mao, Q., Feng, F. and Ye, C. (2010) Genetic diversity within a jackfruit (*Artocarpus heterophyllus* Lam.) germplasm collection in China using AFLP markers. *Agricultural Sciences in China* 9, 1263–1270. DOI: 10.1016/S1671-2927(09)60215-7.

Merrill, E.D. (1912) *A Flora of Manila (1976 Reprint)*. Bureau of Printing, Manila. DOI: 10.5962/bhl.title.54449.

Mohr, G., Perlman, P.S. and Lambowitz, A.M. (1993) Evolutionary relationships among group II intron-encoded proteins and identification of a conserved domain that may be related to maturase function. *Nucleic Acids Research* 21(22), 4991–4997. DOI: 10.1093/nar/21.22.4991.

Naik, K.C. (1949) Fruits common to the plains and humid zones. In: *South Indian Fruits and Their Culture*. P. Varadachery & Co, Madras, India, pp. 300–302.

Odoemelam, S.A. (2005) Functional properties of raw and heat processed jackfruit (*Artocarpus heterophyllus*) flour. *Pakistan Journal of Nutrition* 4(6), 366–370. DOI: 10.3923/pjn.2005.366.370.

Oh, B.J., Balint, D.E. and Giovannoni, J.J. (1995) A modified procedure for PCR-based differential display and demonstration of use in plants for isolation of genes related to fruit ripening. *Plant Molecular Biology Reporter* 13, 70–81.

Popenoe, W. (1974) The breadfruit and its relatives. In: *Manual of Tropical and Sub-Tropical Fruits*. Haffner Press, New York, pp. 414–419.

Prakash, D.P., Narayanaswamy, P. and Sondur, S.N. (2002) Analysis of molecular diversity in guava using RAPD markers. *The Journal of Horticultural Science and Biotechnology* 77(3), 287–293. DOI: 10.1080/14620316.2002.11511494.

Prasad, M.P., Kirti, P. and Ceera, M. (2014) Phytochemical, antioxidant activity and determination of genetic diversity in *Artocarpus heterophyllus* using RAPD molecular markers. *International Journal of Sciences & Research* 10, 44–49.

Prem, J.V., Sajeshkumar, N.K., Jiby, J.M. and Sophyiamol, J. (2017) *Studies on Genetic Relationships among Six Varieties of Jackfruit* (Artocarpus Heterophyllus *Lam.*) *in Kerala Employing* "matK" *Gene Using PCR Technique and RFLP Markers*. Grin Publisher, Norderstedt, Germany.

Pushpakumara, D.K.N.G. and Harris, S.A. (2007) Potential of RAPD markers for identification of fruit types of *Artocarpus*. *Journal of the National Science Foundation of Sri Lanka* 35(3), 175. DOI: 10.4038/jnsfsr.v35i3.2016.

Pushpakumara, D.K.N.G., Boshier, D.H. and Harris, S.A. (1997) Mating system in *Artocarpus heterophyllus* Lam. *Tropical Agricultural Research* 9, 1–14.

Pushpakumara, D.K.N.G., Silva, G.L.L.P., Harris, S.A. and Jianlin, H. (2005) *DNA Extraction, PCR Optimization and Southern Hybridization Laboratory Protocols for Woody Perennials: Jackfruit* (Artocarpus Heterophyllus *Lam.*) *and Farm Animals: Cattle* (Bos Taunrus, Bos Indicus) *and Sheep* (Ovis Aries). CARP-ICRAF Tree Domestication Project of Sri Lanka, 114/9, Wijerama Mawatha, Colombo 7, Sri Lanka.

Rahman, A.K.M.M., Huq, E., Mian, A.J. and Chesson, A. (1995) Microscopic and chemical changes occurring during the ripening of two forms of jackfruit (*Artocarpus heterophyllus* L.). *Food Chemistry* 52(4), 405–410. DOI: 10.1016/0308-8146(95)93290-8.

Rahman, M.A., Nahar, N., Jabbar Mian, A. and Mosihuzzaman, M. (1999) Variation of carbohydrate composition of two forms of fruit from jack tree (*Artocarpus heterophyllus* L.) with maturity and climatic conditions. *Food Chemistry* 65(1), 91–97. DOI: 10.1016/S0308-8146(98)00175-7.

Roy, S.K., Rahman, S.L. and Majuar, R. (1989) In *vitro* propagation of jackfruit (*Artocarpus heterophyllus* Lam.). *Indian Journal of Horticulture* 65, 355–358.

Sahu, S.K., Liu, M., Yssel, A., Kariba, R., Muthemba, S. *et al.* (2020) Draft genomes of two *Artocarpus* plants, jackfruit (A. *heterophyllus*) and breadfruit (A. *altilis*). *Genes* 11, 27.

Sane, A., Kavya, G., Ravishankar, H. and Patil, P. (2009) Morphological and molecular (ISSR) characterization of jackfruit (*Artocarpus heterophyllus* Lam.) accessions. In: *International Conference on Horticulture (ICH-2009)*, November 9–12, 2009.

Schnell, R.J., Olano, C.T., Campbell, R.J. and Brown, J.S. (2001) AFLP analysis of genetic diversity within a jackfruit germplasm collection. *Scientia Horticulturae* 91(3–4), 261–274. DOI: 10.1016/S0304-4238(01)00270-9.

Selvaraj, Y. and Pal, D.K. (1989) Biochemical changes during the ripening of jackfruit (*Artocarpus heterophyllus* L.). *Journal of Food Science & Technology* 26, 304–307.

Shyamalamma, S., Chandra, S.B.C., Hegde, M. and Naryanswamy, P. (2008) Evaluation of genetic diversity in jackfruit (*Artocarpus heterophyllus* Lam.) based on amplified fragment length polymorphism markers. *Genetics and Molecular Research* 7(3), 645–656. DOI: 10.4238/vol7-3gmr457.

Simon, L., Shyamalamma, S. and Narayanaswamy, P. (2007) Morphological and molecular analysis of genetic diversity in jackfruit. *The Journal of Horticultural Science and Biotechnology* 82(5), 764–768. DOI: 10.1080/14620316.2007.11512302.

Simpson, M.J.A. and Withers, L.A. (1986) *Characterization of Plant Genetic Resources using Isozyme Electrophoresis: A Guide to the Literature*. International Board for Plant Genetic Resources, Rome.

Singh, S.R., Narayanaswamy, P., Yumnam, S.S. and Banik, B.C. (2011) Development of RAPD marker related to fruit cracking in jackfruit (*Artocarpus heterophyllus* Lam). *Asian Journal of Horticulture* 6, 199–202.

Soepadmo, E. (1991) *Artocarpus heterophyllus* Lam. In: Verheij, E.W.M. and Coronel, R.E. (eds) *Plant Resources of Southeast Asia, No.2: Edible Fruits and Nuts*. PUDOC, Wageningen, Netherlands, pp. 86–91.

Thumilan, B.M., Sajeevan, R.S., Biradar, J., Madhuri, T., N Nataraja, K. *et al.* (2016) Development and characterization of genic SSR markers from Indian mulberry transcriptome and their transferability to related species of Moraceae. *PloS One* 11(9), e0162909. DOI: 10.1371/journal.pone.0162909.

Wangchu, L., Singh, D. and Mitra, S.K. (2011) Biochemical markers: a useful tool for assessing genetic diversity in jackfruit *(Artocarpus heterophyllus* Lam.). *Acta Horticulturae* 890, 91–101. DOI: 10.17660/ActaHortic.2011.890.10.

Wann, T., Shyamalamma, S., Priyanka, S., Vanitha, P.A. and Bhaskar, R.D.V. (2013) Assessment of genetic variation and field adaptability in clonal and seedling progenies of jackfruit through morphological and molecular evaluation. In: *Proceedings of the National Seminar on Tree Biotechnology 2013:*

Emerging Opportunities in Forestry and Tree Science, Institute of Forest Genetic and Tree Breeding, September 23–24, 2013.

Yaacob, O. and Subhadrabandhu, S. (1995) *The Production of Economic Fruit in Southeast Asia*. Oxford University Press, Oxford, UK.

Yang, Z.-Y., Wu, C.-L., Feng, F., Lu, Q.-. F., Wang, J.-N. *et al.* (2015) Determination and analysis of ploidy and genome size of different jackfruit lines. *Journal of Fruit Science* 32, 567–571.

Ye, C., Li, Y. and Feng, F. (2005) Analysis of genetic diversity of jackfruit germplasm grown in Leizhou Peninsula, China, using RAPD marking method. *Acta Horticulturae Sinica* 32, 1088–1091.

Yu, J., Xue, J.-H. and Zhou, S.-L. (2011) New universal matK primers for DNA barcoding angiosperms. *Journal of Systematics and Evolution* 49(3), 176–181. DOI: 10.1111/j.1759-6831.2011.00134.x.

Zoschke, R., Nakamura, M., Liere, K., Sugiura, M., Börner, T. *et al.* (2010) An organellar maturase associates with multiple group II introns. *Proceedings of the National Academy of Sciences of the United States of America* 107(7), 3245–3250. DOI: 10.1073/pnas.0909400107.

6 Propagation

R.M.S. Ruwan Chamara[1], Asanga D. Nagalla[2], H.M. Prathibhani C. Kumarihami[1], D.M. Samantha B. Dissanayaka[1], D.K.N. Gamini Pushpakumara[1] and Chalinda K. Beneragama[1]*

[1]Department of Crop Science, Faculty of Agriculture, University of Peradeniya, Peradeniya, Sri Lanka; [2]Department of Agricultural and Environmental Biology, Graduate School of Agricultural and Life Science, University of Tokyo, Tokyo, Japan

6.1 Introduction

Jackfruit can be propagated by both sexual and asexual means. Asexual or vegetative propagation is suggested for propagation of jackfruit because plants raised from seeds never bear true-to-type fruits the same as the mother plant. Jackfruit is a monoecious species that produces separate female and male inflorescences on the same tree. A detailed explanation of the flowering and fruiting phenology of individual jackfruit trees is described in Chapter 9 (this volume).

6.2 Sexual Propagation

Jackfruit has very low seed set, with a mean of 3% of the flowers in an inflorescence producing seed as a result of cross-pollination (Pushpakumara, 2011). The seeds of jackfruit are light brown, 2–3 cm long and 1–1.5 cm wide (Prakash *et al.*, 2009). The seeds are covered by the flesh and enclosed in a white aril surrounding a thin brown spermoderm. The seeds are recalcitrant and sensitive to both drying and freezing, and therefore cannot be stored for more than a few weeks (Witherup *et al.*, 2019). Depending on the variety, as many as 500 seeds may be present in each fruit (Prakash *et al.*, 2009; Ranasinghe *et al.*, 2019). Jackfruit is commonly propagated from seed, although vegetative propagation methods have been used to obtain uniform varieties in countries such as Thailand, Vietnam, Australia, USA (Florida) and Indonesia where jackfruit has been introduced (Campbell and Ledesma, 2003). In countries where jackfruit originated or has been domesticated, seed propagation is still the most common form of propagation as it is the cheapest, easiest and most convenient for growers. Moreover, jackfruit has remained a predominantly seed-propagated crop in India and Bangladesh with few varieties (Haq, 2006).

Collection of seeds for propagation should be done from a mother plant with good-quality plant characteristics. It is advised to collect the seeds from trees that are older than 15 years and show good growth and a strong trunk (Rajmohan, 2011). The mother plant should also be free from pests and diseases. Other good-quality characteristics are a high yielding ability with regular bearing, a uniform and attractive fruit shape, fruit weight of 2–5 kg and the ability to produce 50–300 fruits year^{-1}

*Corresponding author: chalindab@agri.pdn.ac.lk

©CAB International 2023. *Jackfruit : Botany, Production and Uses* (ed. S. Mitra)
DOI: 10.1079/9781800622319.0006

(Rajmohan, 2011). It is also desirable to take seeds from fruits with juicy, sweet, medium-soft flesh having a Brix value of more than 20%. Seed size and germination ability are positively correlated. Larger seeds obtained from a quality mother plant have greater germination potential and can produce vigorous seedlings (Sonwalkar, 1951; Hayes, 1953; Khan, 2004). Starch is found as an abundant reserve stored in cotyledons. Both α-and β-amylases are very active in ungerminated seeds and during germination, there is continuous depletion of starch content as a result of amylase activity. Metabolism of soluble carbohydrates during germination is unique in jackfruit seeds because loss of viability is under the control of soluble and insoluble carbohydrate status prevailing during germination-associated metabolism of seeds, which have been categorized as recalcitrant (Sheela and Salim, 2010).

It is best to plant the seeds soon after extraction from the fruit as the germination ability is maximal at this time and seeds do not retain their viability for long periods after extraction from the fruit. If the seeds are stored for some time in open conditions after extraction from the fruit, there is a high possibility that the germination percentage will be significantly reduced or that they might not germinate at all (Rajmohan, 2011). The germination percentage of fresh jackfruit seeds is more than 93% (Pushpakumara, 2011). It has been shown that the germination percentage reduced rapidly to as low as below 40% after 1 month if the seeds were stored in open conditions. Krishnasamy (1990) working with seeds of the variety 'Palur Jack' observed 62% moisture content in freshly extracted seeds (day 0) and 100% germination. By day 10, the moisture content had declined to 42% but the seeds still showed 98% germination, on day 14 the moisture content was 36.3% with 10% germination, and on day 15 with 34.1% seed moisture content, germination was 0%. It has been suggested that the critical seed moisture content in this variety is 39% at which approximately 93% of seeds will germinate. A decline in viability with moisture level was reported to be associated with increased leachate conductivity, lipid peroxidation products and/or soluble carbohydrates (Chandel *et al.*, 1995). However, seed viability could be preserved for a certain period of time when seeds were buried in coir dust and dry sand or stored in air-tight polythene containers (Soepadmo, 1992). Seeds can be kept viable for 3 months if they are stored in air-tight polythene containers under conditions of 40% moisture and 20°C (Singh, 1969). Tang and Fu (1993) suggested washing freshly extracted seeds, partially drying them and then placing them in perforated polyethylene bags, each inside another bag containing perlite with a 50% moisture content. They observed that in bags stored at 15°C, seed viability was retained for over 2 years.

After extraction of seeds from a ripe fruit, the mucilage surrounding the seeds should be removed before planting by thoroughly washing the seeds (Quasem and Sharkur, 1984). For direct sowing, two to three seeds should be planted at 2–3 cm depth in pits of 50 × 50 × 50 cm filled with well-decomposed cow dung manure or compost. The pits should be covered with organic mulch and watered regularly. Seed germination begins within 7–10 days and takes around 3–5 weeks to complete if the seeds are not exposed to any pre-sowing seed treatment. This time period can be shortened by soaking the seeds in water for 24 h before planting (Medagoda and Tennakoon, 2001; Gawankar *et al.*, 2019, 2020). Seed germination starts within a week after sowing and is completed in approximately 35 days; seed treatments do not improve the germination percentage or the rate of germination (Pushpakumara, 2011). However, pre-sowing seed treatments with growth regulators have been shown to improve not only the seed germination ability but also subsequent growth of seedlings and their vigour (Shanmugavelu, 1971; Haq, 2006; Harshavardhan and Rajasekhar, 2012). Soaking seeds in gibberellic acid (GA_3) at a concentration of 500 mg l^{-1} for 48 h increased germination of seeds and shoot growth (Shanmugavelu, 1971; Maiti *et al.*, 2003). Singh *et al.* (2002) recorded the highest germination (95.3%) and coefficient of velocity of germination (27.67), the shortest span of germination (13 days) and the fastest germination (3.6 days) by treating the seeds with GA_3 at 100 mg l^{-1} for 12 h before sowing. Similarly, seeds treated with naphthaleneacetic acid (NAA) prior to sowing enhanced the germination percentage (Haq, 2006). Soaking seeds for 8 h in kinetin (50 mg $^{l-}$1) caused a higher vigour index of seedlings (Prakash, 1998).

Soaking seeds in GA_3 at 200 mg l^{-1} for 24 h resulted in taller seedlings with a better absolute growth rate and fewer days required to attain graftable-size seedlings, while soaking seeds in potassium nitrate at 0.5% for 24 h produced seedlings with maximum leaf area per seedling (Harshavardhan and Rajasekhar, 2012).

As well as direct sowing in the field, seedlings can also be raised in polythene bags and then transplanted in the field after 1–3 months. Where seeds are planted in a seedbed instead of polythene bags, there should be special attention to ensure that the seedlings are transplanted in the field before the food reserves of the cotyledon are depleted. This is particularly important for seedlings to recover from the transplantation shock in the field (Khan, 2004). Pieces of coconut husk with enough volume of soil for the root growth of seedlings can also be used to germinate seeds and raise seedlings during the early stages of growth. Once germination has taken place and the seedlings have reached a desirable height, the seedlings together with the husk can be planted in the field. This method is common in some countries and allows the seedlings to be kept for as long as 1 or 2 years (Gunasena *et al.*, 1996). Seed propagation has also become an important step to produce rootstocks required for vegetative propagation. To produce seedlings required for rootstocks in vegetative propagation, seeds are sown in seed beds, containers or pots. At the four-leaf stage, the seedlings are transplanted to larger polythene bags. Compared with vegetative propagated trees, seedling trees grow larger in size and have a better timber value. However, the sexual propagation method has some disadvantages such as bearing late and variations in yield and fruit quality.

6.3 Vegetative Propagation

Good-quality planting materials are essential for the successful cultivation of fruit crops. The vast majority of angiosperms propagate sexually through seeds. However, for large-scale propagation and preservation of selected genetic traits, sexual reproduction risks losing these characteristics (Henrique *et al.*, 2006). Similarly, because it is a cross-pollinated crop, the propagation of jackfruit plants from seeds is not generally recognized due to high heterozygosis. However, sexual propagation is the key to breeding and developing new varieties with desirable characteristics such as high yield and pest and disease resistance (Jagadeesh *et al.*, 2007; Harb *et al.*, 2015).

In contrast, the importance of vegetative propagation in maintaining genetic consistency and keeping the identity of a superior clone or variety is well known (Aralikatti et al., 2011). Accordingly, vegetative propagation is a suitable approach as it maintains desirable individual traits, resulting in a genetically identical plant to the original mother plant. Vegetative propagation is, therefore, an essential aspect of tree improvement, as it is required to establish clonal seed orchards and might imply significant gains in productivity and higher orchard uniformity (Junqueira *et al.*, 2006; Basuchaudhuri, 2021).

Hence, there is a critical need to identify appropriate vegetative propagation techniques for the rapid multiplication of jackfruit plants. Jackfruit is considered a rather more challenging species to propagate by vegetative means than other fruit plants, probably due to its latex, which obstructs the usual callus development process (Soepadmo, 1991; Khatun *et al.*, 2008). Jackfruit may be vegetatively multiplied in several different ways, which are described below.

6.3.1 Propagation by cuttings

Vegetative propagation of jackfruit using stem cuttings has been one strategic approach for many years. The absence of high-quality planting materials continues to be a significant impediment to large-scale plantation operations. Therefore, considering the lack of rootstocks, scions, labour and other resources, vegetative propagation through cuttings can be a more efficient and cost-effective multiplication technique.

A cutting is a vegetative plant part detached from its parent plant to regenerate and develop into a new plant (Relf and Ball, 2009). Cuttings come in various forms, including stem, leaf and root cuttings. They are commonly used to propagate many types of plants, including both woody and herbaceous. Stem cuttings are more easily propagated than root cuttings in general, and hardwood, semi-hardwood and softwood stem cuttings are the three main types of stem cuttings available (Relf and Ball, 2009; Costa *et al.*, 2017). The standard method of propagating jackfruit is by asexual propagation from stem cuttings because it is the most convenient and low-cost method of attaining a fully established healthy tree in the least time (Costa *et al.*, 2017).

Cuttings from the mother plant are taken with a sharp blade to avoid damaging the mother plant. If flower buds are present, they should be removed to enable the cutting to use its energy and reserve for root and shoot formation instead of fruit and seed production. Transpiration water loss, light availability and air movement of the cuttings can be controlled by trimming up to half the leaf length on large-leaved cuttings. Rooting hormones are commonly used to hasten the rooting of cuttings and are applied to the base of the cutting. The cuttings are placed in pots or trays filled with medium and kept in a moderately shady area. They can be covered with clear polythene in a propagator to prevent them from drying out.

Cuttings may not always be successful in rooting, and the reasons for this are largely unknown. Rooting and subsequent seedling production performance of jackfruit cuttings are dependent on many factors such as size, health and the time of the collection, condition of the cutting, age of the mother plant, pruning of the mother plant, hormonal treatments, condition and type of rooting medium, irrigation, moisture, temperature, light and internal factors (carbohydrates and photosynthesis, water status, water uptake, leaf transpiration), as well as sanitation before and during propagation (Tsipouridis *et al.*, 2006; Costa *et al.*, 2017; Faleiro *et al.*, 2019).

Rooting medium

The cutting must be propagated in a high-quality rooting medium because the medium plays a significant role in successful rooting and subsequent establishment and growth of cuttings. It should be porous to allow root aeration and drainage, while also retaining water and nutrients. The cutting must have a ready moisture supply at the cut surface to form a new root system. However, it is clear from observations that it is essential to maintain the correct balance of air and water in the medium and the temperature at the base of the cuttings. Clean and sterile rooting substrate should be used for propagation. Fertilizers should be used carefully in the medium where excessive fertility might weaken or discourage root development (Howard, 1971). The degree of bottom heat controls the rate at which roots develop; hence, the bottom temperature can be regulated to shorten or delay the rooting phase to fit in with other nursery operations (Howard and Garner, 1965; Howard, 1971). At the optimal of 21°C, roots emerge on many clones in 2 weeks, and cuttings are wholly established and ready to transplant in approximately 1 month. Thermostatically driven electrical heating wires can provide heat at a capacity of approximately 15 W ft^{-2} (Howard, 1971).

Several studies have identified suitable media for rooting cuttings, such as sphagnum peat, vermiculite, sand and others. Sphagnum peat moss and vermiculite are the main components of this mix, both of which are generally free of diseases, weed seeds and insects (Howard, 1971). A blend of peat moss, vermiculite and perlite in equal ratios is beneficial for moisture retention in the growing medium (Relf and Ball, 2009). Chatterjee and Mukherjee (1980a) tested the effect of rooting medium on the success of rooting of jackfruit cuttings. Of the three media used for cuttings, vermiculite proved to be the best (86.00% rooting and 58.00% survival after 1 year), while sand was the worst (70.00% rooting, 43.00% survival after 1 year).

Hormone treatments

Cuttings are treated with hormone before planting in medium to help them develop a healthy and well-developed root system that will allow improved field establishment (Sharma *et al.*, 2009). Auxin is essential in forming adventitious roots (Howard, 1968; Hartmann *et al.*, 1997; Henrique *et al.*, 2006), particularly for propagating hardwood cuttings. Usually, natural auxin production in the cutting is inadequate to induce roots in difficult-to-root species so synthetic auxins are applied at the basal end of the cuttings to achieve consistent and uniform roots (Basuchaudhuri, 2021). Exogenous exposure to indole-3-butyric acid (IBA) also enhances rooting in stem cuttings owing to its ability to induce active cambium regeneration, cell division and cell multiplication (Rymbai and Reddy, 2010).

However, root development in cuttings is primarily due to a balance of auxins and other hormones such as cytokinins and gibberellins. This balance affects root initiation, root elongation and axillary shoot development, and, as a result, the shoot-to-root ratio of the newly rooted plant in the future. The physiological status of cuttings in terms of carbohydrates and other nutrients may also influence the efficacy of auxins (Relf and Ball, 2009).

Chatterjee and Mukherjee (1980a) tested the effect of invigoration and etiolation on rooting of cuttings. The best result was obtained with invigorated shoots under etiolation and 5000 p.p.m. IBA treatment. Kamaluddin *et al.* (1997) observed that leafy tip cuttings of jackfruit seedlings rooted following treatment with 0.8% IBA solution. Rahman and Blake (1988) reported that the rooting success of jackfruit cuttings was increased by ringing the shoots used for cuttings and etiolating for 20 days before detachment of the cuttings and dipping the cuttings for 2 min in a solution of IBA at 750 p.p.m. before planting in a rooting medium consisting of compost:vermiculite mixture (1:1, v/v).

Light

Hossain and Kamaluddin (2004) studied the impact of different shade levels on shoot growth and morphology and the rooting ability of cuttings. Cuttings were grown in three different types of shades: vegetational shade (75% full sun), lateral shade (40% full sun) and an overhead shade (10% full sun). Cuttings taken from vegetational, lateral and overhead shade shoots achieved rooting percentages of 80%, 87% and 90%, respectively. Cuttings obtained from the lateral shadow had the most roots (6.2 per cutting) and root dry mass (20.7 mg), followed by cuttings obtained from the overhead shade.

Advantages and disadvantages of using cuttings in propagation

Propagation by cuttings is a less expensive method for jackfruit than grafted or budded plants. Propagation of rootstock from rooted cuttings can overcome seed viability, germination, storage and dormancy problems associated with seeds, can shorten the time to flowering or encourage consistent flowering, can maintain superior genotypes and can contribute to genetic uniformity (Aralikatti et al., 2011; Costa *et al.*, 2017; Basuchaudhuri, 2021). However, plants propagated by cuttings take longer to reach a saleable size, and not all varieties can thrive without a rootstock. When the tree is less vigorous, it produces fewer and smaller fruits than grafted plants, although this greatly depends on the characteristics of the variety (Costa *et al.*, 2017).

6.3.2 Propagation by grafting

Grafting is an asexual, vegetative plant propagation technique that dates back many years. It is commonly done by joining two plant segments, the shoot section known as 'scion' and the root piece called the 'rootstock' (Goldschmidt, 2014). As it is a perennial fruit crop, jackfruit is also propagated vegetatively by grafting. However, the success of grafting and the performance of the grafted plant are influenced by a number of factors, including time and method of grafting, age and the quality of rootstock and scion, and other nursery management practices. Additionally, other aspects such as graft compatibility and pathogenic factors play crucial roles in grafting success. A favourable interaction among all these components results in a successful graft (Islam *et al.*, 2004).

Rootstocks are grown in polythene bags by planting seeds treated with fungicides (Fig. 6.1A). Growing seedlings are maintained under shade in a nursery by adopting general nursery operation practices. Scions are collected from healthy, superior trees of jackfruit varieties in the morning hours on the grafting day and defoliated with sharp secateurs (Fig. 6.1B). The terminally pruned rootstocks are then given a longitudinal cut about 5 cm long. A scion shoot of about the same thickness and 8–10 cm in length is given two gentle sloping cuts of about 5 cm on opposite sides by removing the bark and a little wood, producing a wedge shape (ensuring that some bark is retained on the remaining two sides) (Fig. 6.1C). This wedge-shaped scion is

Fig. 6.1. Softwood grafting. (A) Three-month-old root stock. (B) Two-month-old scion wood. (C) Wedge-shaped cut of scion wood. (D) Grafted plant firmly taped with polythene ribbon. (E) Grafted scion covered with a transparent polythene cap. (F) Successful grafted plant ready for planting. Photographs courtesy of Dr G. Karunakaran.

inserted into the 'V'-shaped slit of the stock and firmly secured with a 150-gauge thickness white transparent polythene strip of 1.5 cm width and 30–45 cm length to keep the stock and scion in firm contact (Fig. 6.1D). The scions are enclosed with a transparent cylindrical bag to prevent water entry into the grafted part and avoid excessive transpiration of the scion (Fig. 6.1E). The grafted plants are then transferred directly to a mist chamber (relative humidity 85–95%, temperature 35–40°C), kept here for about 30 days and then shifted to a shade house (25% sunlight). Scions that remain green are taken to be successful grafts after 30 days (Fig. 6.1F).

Recent studies have brought us a step closer to adaptive grafting techniques that could work well in jackfruit propagation. The age of the rootstock is a critical feature in grafting, and use of the correct-age rootstock leads to a high percentage of success. Aralikatti *et al.* (2011) found that 7 month-old rootstocks had the highest graft success (51.9%) and the most leaves (average 2.33) and sprouts, followed by 9-month-old rootstocks. Aseef *et al.* (2018) also investigated the effect of the age of rootstocks on the growth pattern of grafted scions in jackfruit by adopting softwood grafting. Graft success rate, number of leaves and length of new shoots were all considerably greater using rootstocks of 14 months old. Naik and Kumar (2018), [illegible] obtained the highest success (85.40%) for softwood grafting by using 30-day-old rootstocks. Gunjate *et al.* (1980) selected germinated seedlings that were 10–15 days old as rootstocks. Scion sticks were made from mature terminal shoots of 10 cm long, dark green in colour and with plump dormant apical buds from selected mother trees. Up to 90% of the grafts were successful. Softwood grafting (Fig. 6.1) outperformed alternative propagation strategies such as inarching and epicotyl grafting. The study also revealed that softwood grafting on 2-month-old rootstock caused early graft union formation and had the highest sprouting rate (Baskaran *et al.*, 2008).

Swamy and Melanta (1994) obtained success by grafting (using softwood) on 6-month-old rootstock when carried out in February–March. Selvi *et al.* (2008) reported maximum success in softwood grafting under a net house with early graft union (20.9 days), and with the highest graft take (17.96%) and sprouting (23.3%) on day 60. Khatun *et al.* (2008) studied the performance of modified cleft grafting in jackfruit at various times of the year. The findings indicated that grafting in August had the highest success rate (39%) and the shortest time to first flush and first leaf opening (4.66 and 5.00 days, respectively). However, success rates were insufficient in March, April and May. Grafting in August resulted in the most vegetative growth of the grafted plants in terms of height, diameter, number of leaves and length and width of leaves. As a result, it can be concluded that the age of the rootstock has a significant impact on graft success and subsequent growth in jackfruit.

It was observed in a variety of fruit trees that pre-defoliation of the scion, which promotes the bud to grow earlier, has a significant influence on grafting success. The performance of defoliated scions was better than the performance of undefoliated scions (Maiti and Biswas, 1980). Islam *et al.* (2004) investigated the effects of rootstock age and scion defoliation on epicotyl grafting in jackfruit. Rootstock that was 10 days old and had the scion defoliated 10 days before grafting performed best in all parameters: 10-day-old rootstock had the highest success (34.33%) and survivability (64.33%), while scions defoliated 10 days before grafting showed maximum success (49.56%) and survivability (79.44%). However, Desai and Desai (1989) found no significant difference in success between defoliated and undefoliated scions.

Desai and Desai (1989) further revealed that 6-month-old matured scion sticks led to significantly higher sprouting (73.33%) and survival (68.33%) of grafts. With an increase in the length of the scion sticks from 2.5 to 10 cm, the percentage of success also increased from 6.66% to 58.33%. Scion sticks of 7.5 and 10 cm length gave the highest success (58.33%). Tanushree *et al.* (2019) investigated the relationship between graft success and survival percentage over time in jackfruits. The results showed that 90 days after grafting, the graft success and survival rate ranged from 20.04% to 40.67%, while at 120 days after grafting, it ranged from 57.08% to 71.67%.

Advantages and disadvantages of grafting

Softwood grafting in jackfruit is a rapid and straightforward method of vegetative propagation (Desai and Desai, 1989; Tanushree *et al.*, 2019). This epicotyl grafting method reduces the long juvenile period, shortens production time by 1–3 years and preserves genetic consistency. Clonal propagation of jackfruit is required for the establishment of a commercial orchard in order to keep the ideal attributes of the mother plant (Mathew and Joseph, 1982; Aseef *et al.*, 2018).

Grafting requires investment in additional labour, money and other resources for rootstock production and orchard management. Furthermore, skilled propagators and additional time are required for successful grafting. Despite significant advancements in grafting techniques, the long-term survival of grafted plants remains rather uncertain. The complications of rootstock–scion interactions take on a new dimension when pathogenic organisms come into the picture. More enigmatic is the graft hybrid conundrum and the broader evolutionary significance of grafting. These fundamental concerns have rarely been addressed in horticulture grafting research.

6.3.3 Propagation by budding

Budding is also commonly used as a vegetative propagation technique in jackfruit. Budding is an asexual reproduction procedure that processes the development of a new plant from a generative anatomical tissue belonging to the parent plant. The foreign plant material, called the 'insert' is grafted on to a host plant known as the 'donor'. The insert should be a proliferating tissue such as a bud. Several budding techniques have been established based on the shape of the cut on the donor plant or rootstock. Commonly, chip budding and T-budding are the most widely used budding methods in woody plants.

The standard methods for budding in jackfruit are patch, shield (T-budding), flute (or chip), ring and forkert budding. However, several previous reports suggested that the success rate of jackfruit budding depends on the season of the year (Haq, 2006; APAARI, 2012). Based on previous experiments, chip budding with a patch showed 90% success in June–July in India; however, it was not successful in March. The T-budding method has been reported as frequently unsuccessful (Singh *et al.*, 1982). VNR Seeds, India, developed the protocol of patch budding (Fig. 6.2) of jackfruit (D.N. Chawda, personal communication) in which about 95% success was achieved when the budded plants were kept under shade nets inside the greenhouse at 28–30°C, >80% relative humidity and a light intensity of 8000–10,000 Lux.

In contrast, ring budding appears to be more successful around May with 80% success but was reduced to 20% in June in Bangladesh. This result was obtained from an experiment that compared five methods of budding: T-budding, ring budding, patch budding, I-budding and flute budding (Biswas and Hossain, 1984). The scions were budded on to 1-year-old rootstocks. However, Ravi *et al.* (2021) compared different grafting and budding methods (inarching, wedge grafting, veneer grafting, patch budding and ring budding) and obtained the best result with wedge grafting (77.56% saleable grafts) when performed in the month of March. Kelaskar *et al.* (1991) recorded the highest success in patch budding when carried out in February, which also produced more leaves and greater shoot length than when carried out in other months of the year. Collectively, certain budding techniques need to be matched with planting seasons, confirmed by another study on patch budding reported to be highly successful (80%) during July–August in India (Konhar *et al.*, 1990). In addition to the budding method and planting seasons, physical and surrounding environmental factors also affect budding success. Keeping some leaves on the rootstock during the budding and greenhouse shade management (fully or partially open greenhouse conditions) does not affect the success rate. As it is important to maintain high-humidity conditions in the bud union, it is recommended that polythene strips are used for tying the bud union (Fig. 6.2E) (Kelaskar *et al.*, 1991).

6.3.4 Propagation by layering

Propagation by air layering or marcottage produces identical new plants by wounding the target region and exposing the inner tissue, which facilitates growth of adventitious roots from the wound. In commercial horticultural practices, rooting stimulators (e.g. specific plant hormones) are applied to increase the success rate. After successful layering, new plants can be obtained while connected to the mother plant.

Air layering is a widely used layering practice for jackfruit vegetative propagation. Using this method, rooting of the aerial stems is facilitated while attached to the mother plant. Initially, a ring of bark about 5 cm wide is removed from 2-year-old shoots. This wound needs to be covered using rooting medium. Commonly used rooting media include coir husk, coir dust, fine wood shavings and moist sphagnum moss. After adding the rooting medium, it should be properly secured with a polythene sheet. The roots are expected to emerge after 2–3 months during monsoon season. After rooting and hardening, the new plants can be separated from the mother plants.

The success rate of air layering is about 60–100% (Madhav, 1965). Growth regulators are commonly used to stimulate the rooting and improve the emergence of daughter plants. Some of these are auxins are IBA, NAA and indole acetic acid. Several studies have shown that the correct concentration of auxin should

Fig. 6.2. Patch budding. (A) Rootstock. (B) Bud stick. (C) Removal of the bud from the scion. (D) Preparation of the rootstock for budding. (E) Budding operation. (F) Beheading of rootstock after bud swelling. Photographs courtesy of Dr Narayan Chawda, VNR Seeds, India.

be prepared and added to the cut surface of the stem before covering with the rooting medium. A study comparing IBA and NAA application with 5000, 10,000 and 15,000 p.p.m. showed that 10,000 p.p.m. IBA resulted in the best root layer formation and the highest survival rate (Tomar, 2011). In addition, important root properties such as number of primary roots, thickest diameter of the longest root and length of the longest root were enhanced compared with other chemical treatments. Based on this study, the month of July was found to best for performing layering. Application of NAA at 10,000 p.p.m. was also reported as a fair treatment; however, the survival percentage of layered plants and new root-related properties were less than those obtained with 10,000 p.p.m. IBA (Tomar, 2011). Previous research has suggested using two auxin applications simultaneously for better rooting (Dhua *et al.*, 1983; Singh *et al.*, 1995; Sengupta and Thakur, 2000).

Etoilation of shoots for 10–30 days prior to air layering operation followed by auxin treatment was reported to be more effective in rooting of air layers compared with auxin treatment alone. Dhua and Sen (1984) obtained maximum rooting (100%) of air layers by etiolation of shoots for 15 days and treatment with IBA (1000 mg l^{-1}) and ferulic acid (1000 p.p.m.). The highest survival rate (80%) in the field was obtained in etiolated (30 days) air layers treated with IBA at 1000 or 5000 mg l^{-1} (Hatibarua *et al.*, 1997; Alila *et al.*, 2000).

Pretreatment of stock plants with ethrel (250 p.p.m.) or cycocel (1000 p.p.m.) followed by treatment with IBA (5000 mg l^{-1}) and *p*-hydroxybenzoic acid at 200 p.p.m. during preparation of air layers caused 100% rooting, of which 70% of layers survived (Dutta and Mitra, 1992).

Advantages and disadvantages of air layering

Compared with other vegetative propagation methods for jackfruit, air layering or marcottage has some advantages, including ensured rooting success, a relatively simple practice to perform with a small number of mother plants and a relatively faster method to obtain daughter plants. However, this method has some disadvantages such as the comparative expensive because of the use of rooting stimulant applications, and the fact that a limited number of layers can be produced using a single mother plant. Overall, however, air layering is one of the best techniques for jackfruit propagation.

6.3.5 Stool layering

Stool (mound) layering consists of cutting the stem of a plant back to the ground during the non-growing season and covering the basal portion of the newly developing young shoots in the spring with a mound of soil. Covering with soil keeps the shoots etiolated and encourages root formation. Success in stooling of 1-year-old jackfruit plants was reported by Chatterjee and Mukherjee (1980b). Soil was heaped up to 10 cm round the basal shoots for 15–20 days, after which it was removed. A ring of bark was taken from each etiolated shoot and IBA (5000 mg l^{-1}) in lanolin paste was applied. At callus formation, the shoots were again earthed up. About 1 month later, the rooted shoots were separated and planted out. This method gave an overall average of 75% rooted shoots of which 71% survived. IBA was found to be better than NAA in promoting rooting of shoots (Alila *et al.*, 1997). Kumar *et al.* (2009) reported 73.3% rooting and 83.3% survival of rooted shoots with application of IBA at 800 mg l^{-1}.

6.4 Micropropagation

Micropropagation is widely used for rapid multiplication of a number of plants. It has numerous advantages over conventional propagation methods, including mass clonal propagation of valuable genotypes, an extremely rapid rate of multiplication, expeditious release of improved varieties in a short time, the limited quantity of plant tissues required as the initial explants, year-round production, production of disease-free plant materials and germplasm conservation (Mannan *et al.*, 2006; El-Zaher, 2008; Kamal *et al.*, 2018; Okunade *et al.*, 2018). Micropropagation of jackfruit was first reported by Rao *et al.* (1981) and there are still only a limited number of reports on tissue culture propagation of jackfruit (Tables 6.1 and 6.2).

Table 6.1. Sterilization of jackfruit explants.

Type of explant	Sterilization agents and method	Reference
Nodal cuttings	Wash several times with DW, treat with 0.5% $HgCl_2$ for 2–3 min, wash with SDW	Roy *et al.* (1990)
Apical buds	Wash under running tap water, treat with 1% Cetavlon + 0.1% $HgCl_2$ for 5 min, wash with SDW 4–5 times	Amin and Jaiswal (1993)
Shoot tips	Wash under running tap water, treat with 0.7% PVP solution containing 2% sucrose for 3 min, wash with SDW, treat with 0.2% $HgCl_2$ for 5 min, wash with SDW for 3 min 3–5 times	Roy *et al.* (1993)
Shoot tips, nodal cuttings	Wash under running tap water, treat with 0.2% $HgCl_2$ for 5 min, wash with SDW 3–5 times	Roy and Roy (1996)
Young leaves, shoot tips, nodal cuttings	Wash under running tap water for 30 min, treat with 0.1% $HgCl_2$ for 10 min, wash with SDW 4 times	Ara *et al.* (2000)
Shoot tips	Brush using a soft brush with a mixture of liquid detergent and 5.3% NaOCl, add 5.3% NaOCl for 1 min, wash with SDW 3 times, add 95% ethanol for 1 min, treat with 20% of 5.3% NaOCl with two drops of Teepol for 10 min, wash with SDW 3 times, add 10% of 5.3% NaOCl with two drops of Teepol for 10 min, wash with SDW 3 times	Lee and Keng (2005)
Shoot tips, apical buds	Wash under running tap water, treat with 0.7% PVP solution containing 2% sucrose for 15 min, wash with SDW, add 70% ethanol for 30 s, wash with SDW, add 0.2% $HgCl_2$ for 10 min, wash with SDW water 3–5 times	Mannan *et al.* (2006)
Shoot tips, nodal cuttings	Wash in 5–10% liquid soap solution for 2–3 min, wash in tap water, add 3 g $HgCl_2$ l^{-1} for 1 min, wash with DW 3 times, treat with 40% Clorox (5.25% NaOCl) for 20 min (shoot tips), treat with 3 g $HgCl_2$ l^{-1} for 1 min, wash with DW 3 times, treat with 50% or 70% Clorox for 20 min (intermediate and basal nodal cuttings, respectively); one drop of Tween-20 per 100 ml solution is used as surfactant	Amany *et al.* (2007)
Apical buds, young leaves, nodal cutting, cotyledons	Wash with tap water with a solution of soap for 3 min, wash with DW for 5 min, soak in antioxidant solution (100 mg citric acid l^{-1} and 150 mg ascorbic acid l^{-1}) for 2 h, treat with 70% ethanol for 2 min, treat with 0.2% $HgCl_2$ for 5 min, treat with15% Clorox (5.25% NaOCl) for 5 min, wash with SDW 3 times	El-Zaher (2008)
Shoot tips, nodal cuttings	Wash under running tap water for 30 min, treat with Tween-80 plus Savlon for 10 min, wash with DW + 0.1% $HgCl_2$ for 5–15 min, wash with SDW 7 times	Azam *et al.* (2009)
Shoot tips, nodal cuttings	Treat with 0.5%, 0.75%, 1.0% and 1.25% Clorox for 15 min each with two drops of Tween-80 (polyoxyethylene sorbitan monolaurate) per 50 ml of solution for 10 min, wash with SDW 5 times	Khan *et al.* (2010)
Shoot tips	Add 3 g $HgCl_2$ l^{-1} for 1 min, wash with SDW 3 times, treat with 5.25% NaOCl for 20 min	Ashrafuzzaman *et al.* (2012)

Continued

Table 6.1. Continued

Type of explant	Sterilization agents and method	Reference
Shoot tips, nodal cuttings	Wash under running tap water for 60 min, swab with 70% alcohol, wash with SDW 2 times, treat with 2.5% NaOCl with two drops of Tween-20 for 25 min, wash with SDW 7 times	Harb *et al.* (2015)
Shoot tips, nodal cuttings	Wash with soapy water + 1.0–2.5% NaOCl solution with or without 0.1% chloramphenicol	Miro and Acedo (2015)
Shoot tips	Wash under running tap water, wash with detergent soap and wash in tap water. Treat with 0.3% Kocide (antifungal) for 45 min, immerse in 70% ethanol for 2 min, wash with DW + 0.5% $HgCl_2$ for 5 min, wash with DW 3 times	Ali *et al.* (2017)
Shoot tips	Wash with water + 75% ethanol for 10 s, treat with 20% NaOCl for 10 min, wash with SDW 4 times	Okunade *et al.* (2018)
Seeds, shoot tips, nodal cuttings	Remove seedcoat, wash with 70% ethanol for 1 min, treat with 30%, 40% and 50% Clorox for 20 min each, wash with SDW, treat with 10%, 20% and 30% Clorox for 20 min each, wash with SDW 5 times	Kamal *et al.* (2018)
Nodal cuttings	Wash under running tap water, wash with 70% alcohol for 1 min, treat with 2.5%, 3.5% and 7% $Ca(OCl_2)$ with two drops of Tween-20 for 5, 10 and 15 min, respectively, wash with SDW for 5 min 3 times	Yapo *et al.* (2020)

DW, distilled water; $HgCl_2$, mercuric chloride; SDW, sterile distilled water; PVP, polyvinylpyrrolidone; NaOCl, sodium hypochlorite; $Ca(OCl_2)$, calcium hypochlorite.

Successful micropropagation requires appropriate explant selection and effective explant sterilization. The most used explant sources in jackfruit micropropagation are shoot tips, nodal cuttings, apical buds, young leaves, cotyledons and seeds (Table 6.1). Before culture initiation, the explants must be sterilized so they are contamination free. Various methods of surface sterilization have been proposed by researchers in jackfruit micropropagation (Table 6.1). In general, explants were cleaned thoroughly under running tap water and detergent soap, followed by rinsing in tap water before sterilization treatments. The explants were then sterilized. Various chemical agents are used to sterilize the plant tissues for *in vitro* cultures. Sodium hypochlorite (NaOCl), mercuric chloride ($HgCl_2$) and calcium hypochlorite ($Ca(OCl_2)$) solutions have been found to be effective by many researchers for surface sterilization of jackfruit (Table 6.1). Pretreatment with antibiotics, antioxidants (El-Zaher, 2008), fungicides (Ali *et al.*, 2017) and ethanol (Lee and Keng, 2005; Harb *et al.*, 2015; Ali *et al.*, 2017; Kamal *et al.*, 2018; Okunade *et al.*, 2018; Yapo *et al.*, 2020) minimized the infection rate. As sterilization agents are also toxic to plant tissues, appropriate concentrations of sterilizing agents, duration of sterilization and methods of use must be considered to minimize the damage to the explants. This procedure varies according to the type of explant and their morphological characters (Table 6.1).

The success of micropropagation is also generally dependent on the culture medium, which contains all the essential nutrient elements required for optimum growth and development of *in vitro* cultures. It contains macroelements, microelements, vitamins, organic compounds, plant growth regulators, carbon sources and other gelling agents. Murashige and Skoog (MS) medium is used for micropropagation of many plant species, including jackfruit (Table 6.2). However, Amany *et al.* (2007) did a comparison of B5 Gamborg medium, woody plant medium and MS medium for *in vitro* propagation of jackfruit. Their findings support the use of MS medium for *in vitro* propagation of jackfruit in

Table 6.2. Basal medium and growth regulators used in micropropagation of jackfruit.

Culture medium	Growth regulators	Plant survival rate (%)	pH	Result	Reference
MS medium	1.0 mg BAP l^{-1} and 0.5 mg kinetin l^{-1}	75	5.8	Shoot proliferation	Roy *et al.* (1990)
	1.0 mg NAA and IBA l^{-1}			Root induction	
MS medium	4.5, 9.0, 18.0, and 36.0 μM BA and kinetin	50	5.7	Culture initiation	Amin and Jaiswal (1993)
	4.5 and 9.0 μM BA and kinetin			Shoot proliferation	
	0.5, 5.0, 10.0, 25.0 μM IBA and NAA			Root induction	
MS medium	2.22–11.10 μM BA and 0.53–2.37 μM NAA	75	5.8	Shoot proliferation	Roy *et al.* (1993)
	2.46–9.86 μM IBA and 2.68–10.74 μM NAA			Root induction	
MS medium	2.5 mg BA l^{-1} and 0.5 mg NAA l^{-1}	75	5.8	Shoot proliferation	Roy and Roy (1996)
	1.0 mg NAA and IBA l^{-1}			Root induction	
WPM	1.5 mg BAP l^{-1}	80	5.7	Shoot proliferation	Ara *et al.* (2000)
	2 mg BA l^{-1}			Root induction	
MS medium	4.5 mg BA l^{-1}	70–80	5.7–5.8	Shoot proliferation	Lee and Keng (2005)
	0.5 mg NAA l^{-1} and 0.5 mg IBA l^{-1}			Root induction	
MS medium	1.5 mg IBA l^{-1}	–	–	Shoot proliferation	Mannan *et al.* (2006)
MS medium	3.0 mg BA l^{-1} and 0.1 mg NAA l^{-1}	90	–	Shoot proliferation	Amany *et al.* (2007)
	1.5 mg IBA l^{-1}			Root induction	
MS medium	5 mg BA l^{-1}, 0.5 mg NAA l^{-1} and 0.5 mg kinetin l^{-1}	–	5.7	Shoot proliferation	El-Zaher (2008)
	3 mg IBA l^{-1} and 2 mg NAA l^{-1}			Root induction	
MS medium	3.5 mg BAP l^{-1}	80	5.8	Shoot proliferation	Azam *et al.* (2009)
	1.2 mg IBA l^{-1}			Root induction	
MS medium	1.5 mg BAP l^{-1}	–	–	Shoot proliferation	Khan *et al.* (2010)
	3.0 mg IBA l^{-1}			Root induction	
MS medium	2.0 mg BAP l^{-1}	–	–	Shoot proliferation	Ashrafuzzaman *et al.* (2012)
	2.0 mg IBA l^{-1}			Root induction	

Continued

Table 6.2. Continued

Culture medium	Growth regulators	Plant survival rate (%)	pH	Result	Reference
MS medium	2.0 mg BA l^{-1} and 0.5 mg kinetin l^{-1}	60	5.7–5.8	Shoot proliferation	Harb *et al.* (2015)
	1.0 mg NAA and IBA l^{-1}			Root induction	
MS medium	3 mg BAP l^{-1}, 0.5 mg GA_3 l^{-1}, 0.1 mg IBA l^{-1}	–	–	Shoot proliferation	Miro and Acedo (2015)
MS medium	2.0 mg BAP l^{-1}	–	5.8	Shoot proliferation	Ali *et al.* (2017)
MS medium	3.0 mg BAP l^{-1}	80	5.75	Shoot proliferation	Okunade *et al.* (2018)
	2.5 mg IBA l^{-1}			Root induction	
MS medium	2.5 mg BAP l^{-1}	88.89	5.7	Shoot proliferation	Kamal *et al.* (2018)
	2.5 and 5.0 mg IBA l^{-1}			Root induction	
MS medium	2.0 mg BAP l^{-1}	–	5.8	Shoot proliferation	Yapo *et al.* (2020)
	1.5 mg GA_3 l^{-1}			Shoot elongation	
	2.0 mg IBA l^{-1}			Root induction	

MS, Murashige and Skoog; BAP, 6-benzylaminopurine; NAA, naphthaleneacetic acid; IBA, indole-3-butyric acid; BA, benzyladenine; WPM, woody plant medium.

terms of the average number of shoots, leaves per explant and shoot length. A report by Okunade *et al.* (2018) showed that MS medium induced better morphogenesis of *in vitro*-cultured shoot tips of jackfruit than the other basal media, including DKW (Driver and Kuniyuki Walnut) medium, C2D (Chee and Pool) medium, AR (Anderson's rhododendron) medium and N6 (Chu) medium supplemented with 30 g sucrose l^{-1}, 3.0 mg 6-benzylaminopurine (BAP) l^{-1} and 6.5 g agar l^{-1}. They recommend DKW medium as an alternative to MS media for shoot multiplication in jackfruit. The pH of the medium is generally adjusted to between 5.4 and 5.8 (Table 6.2), which critically affects *in vitro* culture and plant growth regulators. Auxins, cytokinins and gibberellins are the most used plant growth regulators. Cytokinins in the culture medium promote shoot proliferation, while auxins promote root formation. Therefore, the formation of adventitious shoots or roots of *in vitro* culture is regulated by changing the ratio of cytokinin:auxin. A high cytokinin:auxin ratio enhances the formation of the shoot, but root formation is inhibited; in contrast, a low ratio promotes root formation (Ashrafuzzaman *et al.*, 2012).

Increased shoot proliferation of jackfruit nodal cuttings was obtained with MS medium supplemented with 1.0 mg BAP l^{-1} and 0.5 mg kinetin l^{-1} (Roy *et al.*, 1990). Benzyladenine (BA) and kinetin (4.5–9.0 μM), either separately or together, supported shoot proliferation; higher concentrations of the cytokinins inhibited bud breaking and favoured callus formation at the explant bases (Amin and Jaiswal, 1993). Multiple shoot proliferation was obtained by culturing shoot tips and nodal explants of jackfruit plants *in vitro* by adding 1.5 mg BAP l^{-1} in the medium (Ara *et al.*, 2000). Lee and Keng (2005) documented that shoot explants of jackfruit cultured on MS medium supplemented with 4.5 mg BA l^{-1} formed the highest number of multiple shoots, followed by root formation on half-strength MS medium supplemented with 0.5 mg NAA l^{-1} and 0.5 mg IBA l^{-1}. The number of shoots formed from each explant was high when the shoot explants were cultured on MS medium supplemented with BA only, and IBA was not needed for the induction of multiple shoots from the shoot explants of jackfruit (Lee and Keng, 2005). Mannan *et al.* (2006), Ali *et al.* (2017), Kamal *et al.* (2018), Okunade *et al.* (2018) and Yapo *et al.* (2020) also reported that BA (1.5–3.0 mg l^{-1}) was the most effective cytokinin for *in vitro* shoot proliferation and multiplication of jackfruit in terms of the frequency of shoot induction, multiple shoot formation and the size of shoots.

The addition of GA_3 to the medium significantly influences the internode elongation and accelerates the growth of *in vitro* cultures of jackfruit. GA_3 at 1.5 mg l^{-1} was reported as the optimal concentration to activate the elongation of internode cells responsible for shoot length growth (Yapo *et al.*, 2020).

Rooting was induced on *in vitro*-proliferated shoots of jackfruit by culturing in half-strength MS (½ MS) medium supplemented with 1.0 mg l^{-1} each of NAA and IBA (Roy *et al.*, 1990). Application of auxin was essential for root induction but the effectiveness of the auxins IBA and NAA was not significantly different. Rahman and Blake (1988) demonstrated that *in vitro*-grown shoots of jackfruit could be rooted without auxin, but IBA was found to be more effective than NAA or 2,4-dichlorophenoxyacetic acid for rooting. Ara *et al.* (2000) and Kamal *et al.* (2018) also observed a superior rooting response of *in vitro*-cultured shoot tips and nodal explants on IBA-containing medium.

Rooted shoots can be successfully acclimatized in ½ MS liquid medium without hormones and sugar (Roy *et al.*, 1990). Plants at 6–7 cm tall were transferred to pots containing a mixture of soil, vermiculite and sand (2:1:1) in a greenhouse and later to the field. Roy *et al.* (1993) and Roy and Roy (1996) also demonstrated that apical and nodal buds of jackfruit trees can produce multiple shoots *in vitro* and can be rooted to form complete plantlets. They observed a 75% survival rate of *in vitro*-generated jackfruit plantlets. Optimum rooting was achieved with 2 mg IBA l^{-1}, and these plantlets showed about 80% survival rate when the rooted plantlets were transferred to a soil substrate (Ara *et al.*, 2000). Amin and Jaiswal (1993) reported 60–80% successful rooting of shoots using ½ MS medium and 10 μM IBA or NAA, and regenerated plantlets were transferred to the soil with 50% survival.

The time of collecting of explants is often a crucial factor in the establishment of *in vitro* cultures. Amin and Jaiswal (1993) reported that shoot tip explants of jackfruit collected from the 10–20-day-old trunk sprouts and from actively growing upper branches of 30–50-year-old field-grown trees during November–January showed satisfactory culture establishment and shoot proliferation. Mannan *et al.* (2006) stated that January is the best season for collecting shoot buds of jackfruit, and inoculation *in vitro* played a vital role in the survivability of shoot buds and their proliferation. This could be attributed to the active growth phase of jackfruit trees in November–January in India. However, Roy *et al.* (1993) suggested the period of June–August, when phenolic exudation was very low as the best season for taking axillary bud explants from jackfruit trees for rapid and increased *in vitro* multiplication.

During the past few decades, effective *in vitro* micropropagation protocols for jackfruit have been developed and recommended by a number of researchers. However, as jackfruit is a woody perennial, successful micropropagation has encountered several limitations, including contamination, phenolic exudation, vitrification, induction of rooting and acclimatization (Ashrafuzzaman *et al.*, 2012). Therefore, further understanding of the different physiological, biochemical and molecular processes of *in vitro* morphogenesis will be necessary to address the problems of *in vitro* propagation of jackfruit. In the future, further research will be required to develop and integrate cost-effective micropropagation protocols that enable large-scale commercial micropropagation of jackfruit from superior mother plants, leading to expansion of the areas that grow micropropagated jackfruit.

6.5 Conclusion

Jackfruit is propagated mainly by seeds. However, vegetative propagation methods are more important as these methods can retain the true-to-type characteristics of the mother plant. Although challenging, micropropagation also has received much attention to support the ever-increasing demand for jackfruit planting material.

References

Ali, J., Bent, K. and Feyissa, T. (2017) Protocol optimization for in vitro shoot multiplication of Jackfruit (*Artocarpus heterophyllus* L.). *African Journal of Biotechnology* 16(2), 87–90. DOI: 10.5897/AJB2015.15191.

Alila, P., Sanyal, D. and Kar, P.L. (1997) Propagation of jackfruit by mound layering. *Crop Research* 14, 433–436.

Alila, P., Chiesotsu, S., Kar, P.L. and Sanyal, D. (2000) Influence of growth regulators and etiolation on air-layering in jackfruit. *Horticultural Journal* 13, 9–13.

Amany, M.H., Ali, E.A.M. and Ehab, S.B. (2007) *In vitro* propagation of jackfruit. *Journal of Applied Scientific Research* 3, 218–226.

Amin, M.N. and Jaiswal, V.S. (1993) *In vitro* response of apical bud explants from mature trees of jackfruit (*Artocarpus heterophyllus*). *Plant Cell, Tissue and Organ Culture* 33(1), 59–65. DOI: 10.1007/BF01997599.

Aralikatti, G., Mokashi, A.N., Hegde, R.V., Patil, R.V. and Angadi, S.G. (2011) Softwood grafting in jackfruit. *Acta Horticulturae* 890, 101–106. DOI: 10.17660/ActaHortic.2011.890.11.

Ara, K.A., Sharifuzzaman, S.M., Hossain, M.M. and Talukder, A.S.M.H.M. (2000) Micropropagation of year-round jack fruit (*Artocarpus heterophyllus* Lam.) through *in vitro* culture. *Annals of Bangladesh Agriculture* 10, 85–91.

Aseef, M.R., Kavino, M. and Vijayakumar, R.M. (2018) Effect of age-old rootstocks on growth pattern of grafted scions in jackfruit (*Artocarpus heterophyllus* Lam). *International Journal of Chemical Studies* 6, 1951–1954.

Ashrafuzzaman, M., Kar, S., Khanam, D. and Prodhan, S.H. (2012) *In vitro* regeneration and multiplication of jackfruit (*Artocarpus heterophyllus* L). *Research Journal of Biology* 2, 59–65.

Asia-Pacific Association of Agricultural Research Institutions (2012) *Jackfruit Improvement in the Asia-Pacific Region: A Status Report*. APAARI, Bangkok.

Azam, F.M.S., Rahmatullah, M. and Zaman, A.U. (2009) Tissue culture of a year-round fruiting variety of *Artocarpus heterophyllus* L. in Bangladesh. *Acta Horticulturae* 806, 269–276. DOI: 10.17660/ActaHortic.2009.806.33.

Baskaran, A., Saraswathy, S. and Parthtoan, S. (2008) Standardization of propagation methods for jack (*Artocarpus heterophyllus*, Lam). *Asian Journal of Horticulture* 3, 361–363.

Basuchaudhuri, P. (2021) Auxins in rooting of cuttings. *Indian Journal of Plant Sciences* 10, 69–85.

Biswas, M. and Hossain, A. (1984) Successful method of vegetative propagation of jackfruit. *Bangladesh Journal of Agricultural Research* 9, 145–146.

Campbell, R.J. and Ledesma, N. (2003) *The Exotic Jackfruit*. Fairchild Tropical Botanic Garden, Coral Gables, Florida.

Chandel, K.P.S., Chaudhury, R., Radhamani, J.J. and Malik, S.K. (1995) Desiccation and freezing sensitivity in recalcitrant seeds of tea, cocoa and jackfruit. *Annals of Botany* 76(5), 443–450. DOI: 10.1006/anbo.1995.1118.

Chatterjee, B.K. and Mukherjee, S.K. (1980a) Effect of different media on rooting of cuttings of jackfruit (*Artocarpus heterophyllus* Lam). *Indian Journal of Horticulture* 37, 360–363.

Chatterjee, B.K. and Mukherjee, S.K. (1980b) Stooling potentiality of clones of jackfruit (*Artocarpus heterophyllus* Lam). *Science and Culture* 46, 55–56.

Costa, J.M., Heuvelink, E. and van de Pol, P. (2017) Propagation by cuttings. *Module in Life Sciences* 1, 1–11.

Desai, S.A. and Desai, A.G. (1989) Effect of age, length, defoliation and storage of scion sticks on success of softwood grafting in jackfruit (*Artocarpus heterophyllus* Lam). *Proceedings of the Indian Academy of Sciences* 99(6), 585–591. DOI: 10.1007/BF03053428.

Dhua, R.S. and Sen, S.K. (1984) Role of etiolation, auxinic and non-auxinic chemicals on root initiation of air-layers of jackfruit (*Artocarpus heterophyllus* Lam). *Indian Journal of Horticulture* 41, 116–119.

Dhua, R.S., Sen, S.K. and Bose T.K. (1983) Propagation of jackfruit (*Artocarpus heterophyllus*) by stem cuttings. *Punjab Horticultural Journal* 23, 84–91.

Dutta, P. and Mitra, S.K. (1992) Regeneration of jackfruit (*Artocarpus heterophyllus* Lam.) air layers by pre-treatment of stock plant with cycocel and ethrel. *Advances in Horticulture and Forestry* 2, 77–83.

El-Zaher, M.H.A. (2008) Studies on micropropagation of jackfruit. 1 – behaviour of the jackfruit plants through the micropropagation stages. *World Journal of Agricultural Sciences* 4, 263–279.

Faleiro, F.G., Junqueira, N.T.V., Junghans, T.G., Jesus, O.N. de, Miranda, D. *et al.* (2019) Advances in passion fruit (*Passiflora* spp.) propagation. *Revista Brasileira de Fruticultura* 41(2), e–155. DOI: 10.1590/0100-29452019155.

Gawankar, M.S., Haldankar, P.M., Salvi, B.R. and Jamadagni, B.M. (2019) Correlation study between moisture loss and germination of recalcitrant seeds of jackfruit (*Artocarpus heterophyllus* Lam.) under different storage conditions. *Emergent Life Sciences Research* 05(2), 28–34. DOI: 10.31783/elsr.2019.522834.

Gawankar, M.S., Haldankar, P.M., Salvi, B.R., Haldavanekar, P.C., Burondkar, M.M. *et al.* (2020) Critical moisture level for seed viability in jackfruit (*Artocarpus heterophyllus* Lam.). *Emergent Life Sciences Research* 06(1), 01–05. DOI: 10.31783/elsr.2020.610105.

Goldschmidt, E.E. (2014) Plant grafting: new mechanisms, evolutionary implications. *Frontiers in Plant Science* 5, 727. DOI: 10.3389/fpls.2014.00727.

Gunasena, H.P.M., Ariyadasa, K.P., Wikramasinghe, A., Herath, H.M.W., Wikramasinghe, P. *et al.* (1996) *Manual of Jack Cultivation in Sri Lanka*. Forest Department, Forestry Information Service, Sir Lanka.

Gunjate, R.T., Kolekar, D.T. and Limaye, V.P. (1980) Epicotyl grafting in jackfruit (*Artocarpus heterophyllus* L). *Proceedings of the Indian Academy of Science, Section A* 49, 667.

Haq, N. (2006) *Jackfruit:* Artocarpus heterophyllus *Crops for the Future*. Center for Underutilized Crops, University of Southampton, Southampton, UK.

Harb, E.M., Alhady, M.R.A.A. and Elsalam, N.A.A. (2015) *In vitro* rapid propagation of jackfruit (*Artcarpus heterophyllus* Lam). *American–Eurasian Journal of Agricultural & Environmental Sciences* 15, 147–153.

Harshavardhan, A. and Rajasekhar, M. (2012) Effect of pre-sowing seed treatments on seedling growth of jackfruit (*Artocarpus heterophyllus* Lam.). *Journal of Research ANGRAU* 40, 87–89.

Hartmann, H.T., Kester, D.E., Davies, F.T. and Geneve, R.L. (1997) *Plant Propagation, Principles and Practices*, 6th edn. Prentice Hall, Upper Saddle River, New Jersey.

Hatibarua, P., Gogoi, S. and Mazumder, A. (1997) Adventitious rooting in jackfruit (*Artocarpus heterophyllus* Lam.) air-layers induced with etiolation and root hormones. *Annals of Biology Ludhiana* 13, 155–159.

Hayes, W.B. (1953) *Fruit Growing in India*. Kitabistan, Allahabad, India.

Henrique, A., Campinhos, E.N., Ono, E.O. and Pinho, S.Z. de (2006) Effect of plant growth regulators in the rooting of Pinus cuttings. *Brazilian Archives of Biology and Technology* 49(2), 189–196. DOI: 10.1590/S1516-89132006000300002.

Hossain, M.A. and Kamaluddin, M. (2004) Effects of lateral shading on growth and morphology of shoots and rooting ability of jackfruit (*Artocarpus heterophyllus* Lam.) cuttings. *Journal of Applied Horticulture* 06(1), 35–38. DOI: 10.37855/jah.2004.v06i01.07.

Howard, B.H. (1968) The influence of 4(indolyl-3) butyric acid and basal temperature on the rooting of apple rootstock hardwood cuttings. *Journal of Horticultural Science* 43, 23–31.

Howard, B.H. (1971) Propagation techniques. *Scientific Horticulture* 23, 116–126.

Howard, B.H. and Garner, R.J. (1965) *High temperature storage of hardwood cuttings as an aid to improved establishment in the nursery*. In: *Annual Report of East Malling Research Station 1964*. East Malling Research Station, East Malling, UK, pp. 85–87.

Islam, S., Farooque, A.M., Rahim, M.A. and Haque, T. (2004) Effect of age of rootstock and scion defoliation on epicotyl grafting in jackfruit. *Bangladesh Journal of Crop Science* 13, 89–93.

Jagadeesh, S.L., Reddy, B.S., Swamy, G.S.K., Gorbal, K., Hegde, L. *et al.* (2007) Chemical composition of jackfruit (*Artocarpus heterophyllus* Lam.) selections of western ghats of India. *Food Chemistry* 102(1), 361–365. DOI: 10.1016/j.foodchem.2006.05.027.

Junqueira, N.T.V., Lage, D.A.D.C., Braga, M.F., Peixoto, J.R., Borges, T.A. and Andrade, S.R.M.D. (2006) Disease reaction and productivity of a passion fruit clone propagated by cuttings and grafting in wild Passiflora herbaceous cuttings. *Revista Brasileira de Fruticultura* 28, 97–100.

Kamal, N.H.M., Aziz, M.A., Kadzimin, S. and Rashid, A.A. (2018) In vitro mass multiplication of *Artocarpus heterophyllus* Lam var Tekam Yellow. *Pertanika Journal of Tropical Agricultural Science* 41, 1289–1314.

Kamaluddin, M., Ali, M. and Bhuiyan, M.K. (1997) Effect of auxin on rooting of cuttings and growth of stecklings of jackfruit (*Artocarpus heterophyllus* Lam). *Chittagong University Studies, Science* 20, 71–75.

Kelaskar, A.J., Desai, A.G. and Salvi, M.J. (1991) Effect of rootstock, leaf retention, shade and tying material on patch bud grafts of jackfruit. *Indian Journal of Plant Physiology* 34, 58–62.

Khan, F.R., Ur-Rahma, H., Abbasi, N.A., Ibrahim, M. and Abbas, G. (2010) *In vitro* shoot and root proliferation of jackfruit as affected by different concentrations of growth regulators. *Sarhad Journal of Agriculture* 26, 533–538.

Khan, M.L. (2004) Effects of seed mass on seedling success in *Artocarpus heterophyllus* L., a tropical tree species of north-east India. *Acta Oecologica* 25(1–2), 103–110. DOI: 10.1016/j.actao.2003.11.007.

Khatun, M.J.M., Islam, M.S., Haque, T. and Khan, N. (2008) Propagation of jackfruit by modified cleft grafting as influenced by time of operation. *Progressive Agriculture* 19(2), 67–74. DOI: 10.3329/pa.v19i2.16930.

Konhar, T., Murmu, S. and Maharana, T. (1990) A study on the budding methods of propagation of jackfruit. *Orissa Journal of Agricultural Research* 3, 115–119.

Krishnasamy, V. (1990) Assessment of critical moisture for seed viability in jack. *South Indian Horticulture* 38, 218–219.

Kumar, M., Lal, S., Rajkumar, C. and Kamal, S. (2009) Effect of IBA on the success of stooling in jackfruit (*Artocarpus heterophyllus* Lam.) cv Pant garima. *Haryana Journal of Horticultural Science* 38, 23–25.

Lee, C.L. and Keng, C.L. (2005) Micropropagation of *Artocarpus heterophyllus* Lamk. *The Planter* 81, 689–695.

Madhav, R.V. (1965) The jackfruit in India. *Farm Bulletin* 34, 18.

Maiti, S.C. and Biswas, P. (1980) Effect of scion variety and type of scion shoot on success of epicotyl grafting of mango. *Punjab Horticultural Journal* 20, 152–155.

Maiti, C.S., Wangchu, L. and Sen, S.K. (2003) Effect of pre-sowing seed treatments with different chemicals on seed germination and seedling growth of jackfruit (*Artocarpus heterophyllus* Lam). *Environment and Ecology* 21, 290–292.

Mannan, M.A., Nasrin, H. and Islam, M.M. (2006) Effect of season and growth regulators on *in vitro* propagation of jackfruit (*Artocarpus heterophyllus* Lam). *Khulna University Studies* 7, 83–87.

Mathew, P.A. and Joseph, J. (1982) Epicotyl grafting in nutmeg (*Myristica fragrans* Houtt). Short Communication. *Journal of Plantation Crops* 10, 61–63.

Medagoda, I. and Tennakoon, T.M.K. (2001) Some studies on exploration, conservation and evaluation of jack fruit (*Artocarpus heterophyllus* Lam.) in Sri Lanka. *Annals of the Sri Lanka Department of Agriculture* 3, 121–129.

Miro, C.B. and Acedo, V.Z. (2015) Development of micropropagation protocol supporting sustainable production of jackfruit (*Artocarpus heterophyllus* Lam.). *Acta Horticulturae* 1088, 505–508. DOI: 10.17660/ActaHortic.2015.1088.92.

Naik, K.E. and Kumar, C.S.R. (2018) Effect of different age of rootstock on grafting of jackfruit (*Artocarpus heterophyllus* L.) var. Palur-1. *International Journal of Current Microbiology and Applied Sciences* 7(8), 3994–3998. DOI: 10.20546/ijcmas.2018.708.413.

Okunade, O.A., Sinniah, U.R., Namasivayam, P., Kumara Swamy, M. and Lynch, P.T. (2018) Influence of seed position within the fruit on seedling quality and *in vitro* shoot tip production of jackfruit. *The Journal of Horticultural Science and Biotechnology* 93(5), 510–518. DOI: 10.1080/14620316.2017.1405746.

Prakash, M. (1998) Effect of plant growth regulators and chemicals on germination of jackfruit. *Annals of Plant Physiology* 12, 75–77.

Prakash, O., Kumar, R., Mishra, A. and Gupta, R. (2009) *Artocarpus heterophyllus* (jackfruit): an overview. *Pharmacognosy Reviews* 3, 353–358.

Pushpakumara, D.K.N.G. (2011) Reproductive biology. In: Valavi, S.G., Peter, K.V. and Thottappilly, G. (eds) *The Jackfruit*. Studium Press LLC, Houston, Texas, pp. 41–59.

Quasem, A. and Sharkur, M.A. (1984) *Propagation of Jackfruit. Annual Report*. Regional Agricultural Research Station, Hathazari, Chittagong, Bangladesh.

Rahman, M.A. and Blake, J. (1988) Propagation of jackfruit by stem cuttings: effect of ringing, etiolation and indole butyric acid treatment. *Bangladesh Horticulture* 16, 25–29.

Rajmohan, K. (2011) Sexual, vegetative and *in vitro* propagation. In: Valavi, S.G., Peter, K.V. and Thottappilly, G. (eds) *The Jackfruit*. Studium Press LLC, Houston, Texas, pp. 129–139.

Ranasinghe, R.A.S.N., Maduwanthi, S.D.T. and Marapana, R.A.U.J. (2019) Nutritional and health benefits of jackfruit (*Artocarpus heterophyllus* Lam.): a review. *International Journal of Food Science* 2019, 4327183. DOI: 10.1155/2019/4327183.

Rao, N.K.S., Narayanswamy, S., Chacko, E.K. and Doreswamy, R. (1981) Tissue culture of the jack tree. *Current Science* 50, 310–312.

Ravi, K., Prativa, N., Ra, R. and Nalini, P. (2021) Comparative evaluation of grafting and budding methods in jackfruit. *Indian Journal of Horticulture* 78(4), 398–403. DOI: 10.5958/0974-0112.2021.00057.8.

Relf, D. and Ball, E.C. (2009) Propagation by cuttings, layering and division. Virginia State University, Petersburg, Virginia: Virginia cooperative extension. Available at: www.pubs.ext.vt.edu/426/426-002/426-002.html (accessed 11 July 2022).

Roy, S.K. and Roy, P.K. (1996) In vitro propagation and establishment of a new cultivar of jackfruit (*Artocarpus heterophyllus*) bearing fruits twice yearly. *Acta Horticulturae* 429, 497–502. DOI: 10.17660/ActaHortic.1996.429.63.

Roy, S.K., Rahman, S.L. and Majuar, R. (1990) In vitro propagation of jackfruit (*Artocarpus heterophyllus* Lam). *Journal of Horticultural Science* 65, 355–358. DOI: 10.1080/00221589.1990.11516065.

Roy, S.K., Islam, M.S., Sen, J., Hossain, A.B.M.E. and Hadiuzzaman, S. (1993) Propagation of flood tolerant jackfruit (*Artocarpus heterophyllus*) by in *vitro* culture. *Acta Horticulturae* 336, 273–278. DOI: 10.17660/ActaHortic.1993.336.36.

Rymbai, H. and Reddy, G. (2010) Effect of IBA, time of layering and rooting media on air-layers and plantlets survival under different growing nursery conditions in guava. *Indian Journal of Horticulture* 67, 99–104.

Selvi, R., Kumar, N., Selvarajan, M. and Anbu, A. (2008) Effect of environment on grafting success in jackfruit. *Indian Journal of Horticulture* 65, 341–343.

Sengupta, S. and Thakur, S. (2000) Studies on the effect of growth regulators on rooting of air-layers of jackfruit (*Artocarpus heterophyllus* Lam). *Orissa Journal of Horticulture* 28, 22–24.

Shanmugavelu, K.G. (1971) Effect of plant growth regulators on jack (*Artocarpus heterophyllus* Lam). *Madras Agricultural Journal* 58, 97–103.

Sharma, N., Anand, R. and Kumar, D. (2009) Standardization of pomegranate (*Punica granatum* L.) propagation through cuttings. *Biological Forum* 1, 75–80.

Sheela, S. and Salim, N. (2010) Reserve mobilization during germination of jackfruit (*Artocarpus heterophyllus* Lam.) seeds. *International Journal of Plant Science* 5, 620–626.

Singh, R. (1969) *Fruits*. National Book Trust, New Delhi.

Singh, U.R., Pandey, I.C., Upadhyay, N.P. and Prasad, R.S. (1982) Propagation of jackfruit by budding. *Punjab Horticultural Journal* 22, 103–105.

Singh, M., Sharma, B.K. and Yadav, H.S. (1995) Response of plant growth regulators and layering dates on air-layering of jackfruit (*Artocarpus heterophyllus* Lam). *Advances of Plant Sciences* 8, 130–133.

Singh, D.K., Bhattacharya, B. and Mondal, K. (2002) Role of pre-sowing seed treatments with different chemicals on germination behavior and seedling growth of jackfruit (*Artocarpus heterophyllus* Lam). *Environment and Ecology* 20, 740–743.

Soepadmo, E. (1991) *Artocarpus heterophyllus* Lam. In: *Plant Resources in South East Asia*, Vol. 2. pp. 86–91.

Soepadmo, E. (1992) *Artocarpus heterophyllus* Lam. In: Verheij, E.W.M. and Coronel, R.E. (eds) *Plant Resources of Southeast Asia, No.2: Edible Fruits and Nuts*. PROSEA, Wageningen, Netherlands, pp. 86–91.

Sonwalkar, M.S. (1951) A study of jackfruit seed. *Indian Journal of Horticulture* 8, 27–30.

Swamy, G.S.K. and Melanta, K.R. (1994) Effect of age of rootstocks on the success of softwood grafting in jackfruit (*Artocarpus heterophyllus* Lam). *Karnataka Journal of Agricultural Science* 7, 471–473.

Tang, L.F. and Fu, J.R. (1993) A reasonable suggestion on moist storage of jackfruit (*Artocarpus heterophyllus* Lam.) seed. *Acta Scientiarum Naturalium Universitatis Sunyatseni* 32, 111–115.

Tanushree, S., Swamy, G.S.K., Honnabyraiah, M.K., Rao, V., Suresha, G.J. *et al.* (2019) Performance of red jackfruit (*Artocarpus heterophyllus* L.) genotypes for softwood graft success. *Journal of Pharmacognosy and Phytochemistry* 8, 1392–1395.

Tomar, Y.K. (2011) Effect of various concentrations of bio-regulators and time of air-layering on the multiplication of jackfruit (*Artocarpus heterophyllus* Lam). *International Journal of Current Research* 3, 316–318.

Tsipouridis, G., Thomidis, T. and Bladenopoulou, S. (2006) Seasonal variation in sprouting of GF677 peach × almond (*Prunus persica* × *Prunus aygdalus*) hybrid root cuttings. *New Zealand Journal of Crop and Horticultural Science* 34(1), 45–50. DOI: 10.1080/01140671.2006.9514386.

Witherup, C., Zuberi, M.I., Hossain, S. and Zerega, N.J.C. (2019) Genetic diversity of Bangladeshi jackfruit (*Artocarpus heterophyllus*) over time and across seedling sources. *Economic Botany* 73(2), 233–248. DOI: 10.1007/s12231-019-09452-5.

Yapo, S.E.S., Beugré, M.M., Konan, Y.K.F., Ayolie, K., Dien, G.F. *et al.* (2020) *In vitro* regeneration of jackfruit (*Artocarpus heterophyllus* Lam., Moraceae) in Côte d'Ivoire. *Asian Journal of Science and Technology* 11, 11313–11318.

7 Orchard Management

Ganesan Karunakaran[1]*, Anbazhagan Thirugnanavel[2], Selladurai Rajendiran[3], Manivannan Arivalagan[1], Thandavarajan Sakthivel[1] and Prakash C. Tripathi[1]
[1]ICAR-Indian Institute of Horticultural Research, Bengaluru, India; [2]ICAR-Central Citrus Research Institute, Nagpur, India; [3]Central Horticultural Experiment Station (ICAR-IIHR), Chettalli, India

7.1 Introduction

Jackfruit (*Artocarpus heterophyllus* Lam.) grows rapidly, reaching a height of 3 m (10 ft) and a canopy diameter of 2 m (6.6 ft) in 2 years. In the first few years, the apical growth reaches about 1.5 m year^{-1}, and thereafter the growth rate slows down to 36–60 cm year^{-1}. A 20-year-old tree can reach up to 17.5 m in height. Jackfruit seedlings grow slowly in heavy shade. The trees can live up to 100 years of age. The tree starts fruit production after 4–5 years of planting and the commercial yield is achieved after 8–10 years. Jackfruit has a huge proliferating tap root system that explores a large volume of soil to fulfil its nutrient requirement from fertile and organic matter-rich soils. However, the nutrient requirements of jackfruit depends on the soil fertility, age of the tree, growth and yield potential. During the initial years, the interspaces between the rows and plants can be utilized for growing intercrops, which not only provide additional income but also improve the physico-chemical properties of the soil and check weed growth. In this chapter, the methods of planting, pre- and post-planting care, and orchard management are discussed in detail.

7.2 Preparation

7.2.1 Land preparation

Jackfruit is cultivated in home gardens, as a shade tree in coffee plantations and orchards and sometimes as a multi-tier cropping system (Haq, 2006). It is intercropped with coconut in the Philippines (Elevitch and Manner, 2006) and with durian in Malaysia (Acedo, 1992). More recently, farmers have been planting jackfruit on a commercial scale. Although it grows well in an agroforestry landscape, it performs to its full potential in the open field with full sunshine. Land preparation is vital for jackfruit cultivation on a commercial scale as the crop stand can last for about 80–100 years. The field should be thoroughly ploughed and harrowed until a fine tilth is attained.

7.2.2 Spacing

A jackfruit plantation can be established in a square or rectangular planting system at a spacing of 7.5–12.0 m. In poor soils, a hexagonal system may be followed. The spacing varies based on the need: 5.5 × 5.5 m spacing is used

*Corresponding author: ganesan.karunakaran@icar.gov.in

DOI: 10.1079/9781800622319.0007

Table 7.1. Recommended spacing for jackfruit trees.

Country	Spacing	Reference
Philippines	7.5 × 7.5 m to 12 × 12 m	BPI (1991)
Australia	6 × 12 m	Alexander *et al.* (1983)
Bangladesh	10 × 10 m to 12 × 12 m	Haq (2006)
South Asia	12 × 12 m	APAARI (2012)
Tamil Nadu, India	8 × 8 m	Anon (2021)
Kerala, India	12 × 12 m to 15 × 15 m	Anon (2019)
Malaysia	8 × 8 m to 12 × 12 m	Anon (2007)
Vietnam	7 × 7 m to 12 × 12 m	Anon (2013)

for a wind break, while for timber production, a closer spacing of 2 × 3 m or 3 × 3 m is used (Elevitch and Manner, 2006), as this may minimize branching and encourage straight, long trunk growth. In general, a wider spacing of 12 × 12 m (70 plants ha^{-1}) is followed for direct seeded and transplanted seedlings in fertile soils, while 10 × 10 m (100 plants ha^{-1}) spacing is followed for poor soils in commercial orchards. For grafted plants, a spacing of 8–10 m is recommended. In total, 150 plants can be accommodated with 8 × 8 m spacing, which is suitable for high-density planting. For fruit production, spacing of about 7.5–12 m is used, with 4.6–7.6 m plant-to-plant spacing and 6.1–7.6 m row-to-row spacing for dwarf varieties and varieties with a compact canopy. To make the best quality knot-free wood for timber, it is necessary to prune the lower branches and fruit-bearing spikes. The spacing recommended in different countries is described in Table 7.1. The planting pit holes can vary in size from 0.6 to 1 m^3 or even 1.5 m^3. The pit holes should be dug at least 15 days before planting to aerate the soil for efficient establishment of sapling roots.

7.3 Planting

In India and other Asian countries, the optimal time to plant, whether by direct seeding or transplanting, is at the start of the rainy season. Direct seeding may be done in early summer if adequate irrigation facilities are available, so that seedlings are established before the rainy season begins (Morton, 1987; Haq, 2006). The rainy season provides ample water and creates ideal conditions for tree establishment in the field. Furthermore, rather than transplanting in the heat of the day, it is advisable to do so in the late afternoon or early evening (Morton, 1987). The optimal period to plant grafts or seedlings is from June to August, but with irrigation facilities, they may be planted until November. Plants may die if they are exposed to prolonged dry weather after they have been planted. To avoid plant loss, the ball of earth should not be disturbed while planting. In places where rainfall is evenly distributed throughout the year, planting can be done at any time. Otherwise, planting should take place at the start of the rainy season to avoid water stress.

Pits of 1.0 m^3 size should be dug out 15 days before planting to expose the pit to the sunlight to kill soil-borne pathogens. Each pit should be refilled with 30 kg well-rotted farmyard manure, 500 g superphosphate and 250 g muriate of potash. The pits may be drenched with chlorpyrifos (20 EC (3 ml/L water)) to avoid early insect attacks. After planting, the soil is pressed down firmly to avoid waterlogging during the rainy season. This is important as jackfruit cannot withstand waterlogging.

7.3.1 Direct seeding

For direct seeding, four to five freshly extracted seeds are collected from ripe fruits of desirable mother trees. The pits should be mixed with soil and 20–30 kg well-rotted farmyard manure. The seeds should be sown 2–3 cm deep into the pits at 30 cm apart in a triangular formation (APAARI, 2012). After sowing, the pits are mulched and

Fig. 7.1. (A) Jackfruit rootstock ready for grafting. (B) Grafted jackfruit plant. (C) Grafted jackfruit ready for field planting. Photograph courtesy of Dr G. Karunakaran.

watered regularly depending on the weather and soil moisture status. Excess moisture may cause rotting of germinating seedlings. Seeds take 10 days to germinate and within 35–40 days, 100% of seeds should germinate (Khan, 2004). The most vigorous seedling in each pit should be retained and the others removed (Haq, 2006). Young plants are vulnerable to animal grazing; hence, fencing is required. Jackfruit is highly cross-pollinated and exhibits a high level of variability. Furthermore, it takes more than 7 years to fruit. To maintain the genetic purity of the variety, planting of seedlings should be discouraged. However, for timber production, seedlings are preferred as they grow tall and produce a substantial trunk.

7.3.2 Transplanting

The grafted plants (Fig. 7.1) and nursery grown seedlings are transplanted in the pits (1.0 m^3 size) dug in the field. The pits are dug a minimum of 4 weeks before planting and kept exposed for 2 weeks to control soil-borne pathogens and insects (Haq, 2006) and then filled in. One-year-old grafted plants are planted in the middle of the pits with the root collar at ground level to ensure that they are planted at the same depth, and proper staking should be provided (Fig. 7.2). Jackfruit seedlings are sensitive to root injury and should therefore be planted before the primary root penetrates the polythene bag. During planting, the polythene bag should be removed as carefully as possible so as not to injure or disturb the roots. The soil around the newly planted seedling is compacted by pressing the soil, making sure that the seedlings are in a vertical position and are aligned properly.

Transplanting is usually done during the onset of rainy season (late May or early June) in Asia, but if rain is delayed beyond 24 h, irrigation should be carried out. If adequate irrigation facilities are available, transplanting can be done any time of the year, even during the dry season. Similarly, replanting to replace plants that have died may be done at any time if there are available irrigation facilities (Fig. 7.3); otherwise, it is best to replant during the onset of the next rainy season. For replanting purposes, it is necessary to have extra planting materials (about 10% of the population density) in one section of the orchard. Such plants should be rebagged in much bigger polythene bags so that they can grow continuously until replanting

Fig. 7.2. Planting and staking of jackfruit in the field. Photograph courtesy of Dr G. Karunakaran.

time. The transplanted plants should be given post-planting care.

7.3.3 Aftercare

Stray cattle and goats frequently damage the plants and adequate fencing is necessary for about 2 years. Irrigation of young plants for one or two summers is necessary for survival and optimum growth of the plants. Protection against frost for at least the first few years is important in cooler regions. Basin cleaning and ploughing between the rows and plants should be followed as a routine measure. Pruning and training are necessary to achieve normal plant growth (Figs 7.4 and 7.5).

Fig. 7.3. A jackfruit plant at an early stage, 1 year after planting. Photograph courtesy of Dr G. Karunakaran.

Mulching and weed control

Weed control is essential during the initial 2–3 years of orchard establishment. Weeding is best done manually where there are few trees in household gardens or small orchards. In larger plantations or orchards, inter-row and circle weeding is practised and can be done mechanically or using herbicides. The space between the trees may be ploughed and harrowed twice a year at the beginning and end of the rainy season (Haq, 2006). In established orchards, growing a leguminous cover in the tree basin, surface mulching, mowing and complete weed control with herbicides are the options available to control the weeds. Mulching at the base of the tree can be done inexpensively using black polythene sheets (Fig. 7.6) or with organic materials such as dry grass (Fig. 7.7), wood shavings or water hyacinth (*Eichhornia crassipes*). Mulching can provide the jackfruit tree with better soil moisture retention and nutrient status. Up to 15–20 cm year^{-1} of ground plant material, preferable composted for 6 months or more, can be added below the canopy of the tree (Campbell *et al.*, 1998).

Fig. 7.4. An unpruned jackfruit tree (2 years old). Photograph courtesy of Dr G. Karunakaran.

7.4 Water and Nutrient Management

Jackfruit grows well and gives a good yield on hill slopes with a warm humid climate and on plains with a hot and humid climate. The crop grows successfully from sea level up to an elevation of 1500 m with a temperature range of 22–35°C. At higher altitudes, fruit quality deteriorates and fruit yield is also affected. Although jackfruit grows in a variety of soils, deep rich alluvial or open-textured loamy soil or red laterite soil with slightly acidic conditions (pH 6.0–6.5) with good drainage is ideal for cultivation. However, the plants cannot tolerate frost or drought. Similarly, waterlogging also weakens plant growth due to restricted aeration, and prolonged stagnant water conditions lead to death of the plant. Therefore, providing drainage is very important for jackfruit cultivation in high-rainfall areas (Nath *et al.*, 2009).

Generally, this crop is cultivated under rain-fed conditions with the application of organic manures only. The application of inorganic fertilizers is not commonly practised by farmers. However, jackfruit needs a well-balanced nutrition to promote regular and good bearing. Prolonged dry spells during the flowering to fruiting periods affect plant production. Therefore, effective irrigation management along with nutrient management practices are very much needed to improve the productivity of jackfruit. In the following sections, the nutrient and water requirements of jackfruit, their management practices and the crop responses are described.

Fig. 7.5. A jackfruit tree after pruning and mulching of leaves on the tree base. Photograph courtesy of Dr G. Karunakaran.

7.4.1 Nutrient requirements

Jackfruit is harvested every year from February to June in India, while at higher elevations, the harvest extends to September. Certain genotypes grown on plains bear fruits during the off-season (October–December). The bearing age is different for different varieties. Varieties such as 'Singapore Jack' start bearing 2–3 years after planting, while vegetatively propagated trees start bearing from 4–5 years and dwarf varieties start to give fruits after 3 years (Fig. 7.8). However, some varieties take about 8–10 years. Regional differences in fruiting habits have also been recorded. In South India, fruiting starts from 6–7 years, while at higher elevations and in North India, it is delayed due to cool weather. Jackfruit attains its full bearing potential after 15 years (Fig. 7.9). A well-established 15-year-old tree may yield about 250–500 fruits annually and the fruit weight ranges from 3 to 25 kg. Under good growing conditions, the fruit weight can be as high as 55 kg (Ghosh, 1996). Farmers generally harvest 20–200 fruits per tree. The yield of jackfruit varies widely across the country. This variation in yield may be due to variation in plant age, varieties, seasons and locations.

Jackfruit has a huge proliferous tap root system that explores the large volume of soil to fulfil its nutrient requirement from fertile and organic matter-rich soils. However, the nutrient

Fig. 7.6. Mulching the tree basin with black polythene. Photograph courtesy of Dr A.Sarkar.

Fig. 7.7. Mulching the tree basin with dry grass. Photograph courtesy of Dr A.Sarkar.

Fig. 7.8. Three-year-old jackfruit bearing fruits. Photograph courtesy of Dr G. Karunakaran.

requirement of a crop mainly depends on its age, growth rate, and yield potential, as well as soil fertility status; the information about the nutrient requirement for jackfruit is scanty. Over the years, leaf nutrient diagnosis techniques and standard leaf nutrient concentrations are used to arrive at optimum fertilization rates. However, the nutrient content stability in leaves depends on the sampling technique, including canopy height, leaf age and time of sampling. The stability of leaf nutrient concentrations in jackfruit has been observed in 3–6-month-old leaves from the central canopy between April and May. Further deficiency, optimum and excess nutrient concentration ranges for jackfruit have also been determined by Sun *et al.* (2015). Nutrition and irrigation management are discussed in more detail in Chapter 8 (this volume).

7.4.2 Irrigation

Jackfruit is generally cultivated under rain-fed conditions. Water management in jackfruit is not particularly critical and there is no specific irrigation schedule available for jackfruit. However, during the early plant growth period of 2–3 years, jackfruit needs regular watering. As jackfruit is very sensitive to drought, supplementary irrigation should be given during long dry spells and during the summer. Nath *et al.* (2009) noted that an imbalance in soil moisture after fruit set led to initial fruit drop, and at the fruit development stage led to fruit cracking. Hence, irrigation during the flowering and fruiting period of the crop is necessary to achieve a higher yield. Moreover, mulching with grasses/paddy straw/organic waste has shown a marked effect on fruit set and fruit growth in eastern Indian conditions. During the pre-bearing stage of jackfruit, raising intercrops such as pulses and use of the dried leaves of pulses as a spread below the jackfruit trees serve as a mulch for moisture conservation. Care must be taken to ensure optimal moisture. Similarly, too much soil moisture will affect the quality of the fruit – the carpels will develop an insipid taste when there is excess soil moisture. Thus, a proper drainage system should be installed in high-rainfall areas to avoid water stagnation around or near to the tree.

Advanced micro-irrigation systems can be adopted to save water and manpower. In addition, fertigation techniques can also be practised

Fig. 7.9. A jackfruit tree showing its full bearing potential. Photograph courtesy of Dr G. Karunakaran.

to reduce fertilizer consumption by increasing the frequency of application, which results in the reduction of wastage of fertilizer nutrients and improvement in the nutrient utilization efficiency of the crop. As described by Crane *et al.* (2003), application of fertilizers that is split into four to six applications is more beneficial, and nutrients will be available for the crop throughout the crop growth period. By adopting water and nutrient management practice through fertigation, farmers can save fertilizers and water to a great extent compared with conventional practices. Moreover, the advantages of harvesting higher yields under drip and fertigation systems are largely proven, and thus micro-irrigation or fertigation practices can be employed in jackfruit to harness a more sustainable yield and to attain a higher farm income.

7.5 Pruning and Training

Jackfruit does not respond well to indiscriminate pruning and it is not commonly practised. Young trees do not need pruning in the first year. When older trees are not pruned, a strong central leader usually develops, which is desirable for its timber value. However, where fruit production is the main purpose, pruning of the first lateral branches should be carried out during the year to slow upward growth and enhance the spread of the canopy. One or two prunings of shoot tips during summer causes lateral bud break and makes the tree more compact. Grafted trees have a dwarfing tendency but produce a large number of branches from the start. These branches should be pruned continuously to get a reasonable trunk. Thereafter, branches may be allowed to remain but removal of vigorously growing upright shoots is recommended. Inner branches of the canopy should be removed to allow more light and air to penetrate (Haq, 2006).

Singh (2010) suggested training of jackfruit trees in the early stage to build a strong framework and avoid weak crotches. The plants should be trained on a single stem. Apical growth needs to be controlled within the first year of planting for canopy architecture. Plants are headed back to a uniform height of 70–80 cm from the ground level 3–4 months after planting to induce the emergence of new growth. Three to four well-spaced limbs are retained around the main stem to form the scaffold limbs of the plant. Additionally unwanted shoots are removed from time to time to give the plant the desired shape.

In older trees, light to medium pruning can be carried out to remove overcrowded branches (Muthulakshmi, 2003). Full-grown trees should receive pruning once a year after harvest. This is done by removing the branches that grow too close to the ground or parallel with the trunk, while keeping the branches that are at least 40 cm off the ground and 40–50 cm higher or lower than branches of other levels. Fruit thinning is also recommended to prevent damage to branches, as heavy fruit loads break branches and can result in death or stunting of the tree. Limiting the number of fruits per limb also improves the quality of the fruits and increases their size (Haq, 2006). Overall, 20–35% of the developing fruits are removed to ensure the long-term health of the tree (Campbell *et al.*, 1998).

7.6 Floor Management

Jackfruit requires a long time to occupy the land fully, so the space available during the initial years can be utilized by intercropping. Pulses, spices and vegetables can be grown as intercrops. The intercrops should not hinder the intercultural operations of jackfruit. Intercropping with leguminous cover crops helps to control weeds, conserve soil moisture and improve soil fertility.

7.6.1 Intercropping

Root distribution studies of the wild jackfruit (*Artocarpus hirsutus*) revealed that most of the physiologically active roots were concentrated within a radius of 75 cm and to a depth of 30 cm, although the tap root can reach even deeper. The tree roots seldom extend beyond 2.25 m laterally from the stem; hence, the effect of overlapping root zones and the associated competitive effects may not be a serious problem for growing intercrops during the first few years of tree growth (Jamaludheen *et al.*, 1997). It makes good sense for the space between the trees to be utilized by growing intercrops such as short-duration pulses, spices and vegetables, particularly if irrigation is available. The selection and growing of intercrops should ensure that there is no interference with orchard operations. If the cultivation of field crops is not possible, a leguminous cover crop can be planted to help control weeds and conserve soil fertility and moisture. As jackfruit takes about 8–10 years to attain full bearing, intercropping will benefit growers and allow them to receive a regular income (Haq, 2006). Different legumes such as pigeon pea, horse gram, black gram, cowpea, French bean, millets, oilseeds such as niger, groundnut, and fodder (e.g. *Stylosanthes*, deenanath grass) should be selected for intercropping under rain-fed conditions. After 10 years, shade-tolerant crops such as elephant foot yam, ginger (Fig. 7.10) or turmeric can be grown as intercrops.

7.6.2 Monocropping

Jackfruit grows well in equatorial to subtropical maritime climates at an altitude of 10–1600 m with an average of 100–240 cm of rainfall. The trees reach 6–25 m height with a canopy diameter of 3.5–6.7 m at 5 years of age. Jackfruit grows best in fertile and well-drained soils. It tolerates a wide range of soils, including shallow

Fig. 7.10. Ginger intercropped in a jackfruit orchard. Photograph courtesy of Dr H. Lembisana Devi.

limestone, sand and rocky substrates. The trees grow well in 30–50% sunlight and can tolerate moderate wind quite well. The tree regrows well, even after heavy pruning.

7.6.3 Jackfruit-based mixed cropping systems

Jackfruit-based mixed cropping systems utilize solar energy, soil resources and spacing, thus increasing the cropping intensity. The main crop, filler crop and intercrops are the three essential components occupying different tiers in the space of the production system. Mostly, fruit trees with a larger canopy serve as the main crop, which utilizes the uppermost layer of the multi-tier cropping system. Generally, the main crop uses 10–25% of the land for up to 10 years, and it takes 20–25 years to occupy the entire land.

Short stature, prolific, precocious-bearing fruit trees are used as filler crops that utilize the middle layer of the multi-tier cropping system. Filler trees are used to generate additional income for the same land for the initial 10 years (during the juvenile period of the main crop). They are generally hardy and have a shorter economic life. They are planted at closer spacing within the main crop. The filler plants can be removed after the main crops attain an adequate canopy size for yielding economically. A fruit-based mixed cropping system with jackfruit is established to improve productivity in hilly slopes, uplands and valleys. Even under stress conditions, fruit trees have the efficiency of providing better economic returns than annual crops.

The root distribution pattern in different tiers improves nutrient and water-use efficiency in the multi-tier system. Decomposition of fallen leaves from multi-tier trees increases the organic carbon content and soil microbial activity leading to environmental stability in the rhizosphere. The yield of the component crops can increase the total productivity of the land where poor soil fertility coupled with low water-holding capacity of the soil contributes towards low productivity of any crop. A high return per unit area under upland conditions is the ultimate result of a fruit-based mixed cropping system.

Suitable crop combinations should be decided based on soil and climatic conditions, the growth pattern of the trees, irrigation facilities and socio-economic requirements for laying out the fruit-based mixed cropping system. The selection of suitable varieties of different crops is the most critical factor contributing to total productivity of the multi-tier system.

Jackfruit is ideal for a multi-tier cropping system. For a multi-tier system, the pits for jackfruit (main crop) are dug at 10 × 10 m (100 plants ha^{-1}), whereas the pits of filler plants are marked at 5 × 5 m (300 plants ha^{-1}) between the rows and within the plants of the main crop. For planting main crops with more extensive canopies, a pit size of 3 m^3 is used, whereas for filler crops with medium-sized canopies, the appropriate pit size is 0.75 m^3. The pits are refilled with 15–20 kg farmyard manure, 1 kg neem cake and 300 g single superphosphate. Jackfruit grows well in fertile soils and demands moisture, particularly during the initial establishment (up to 18 months). Hence, the pits should be filled with 50% river silt or tank silt, or compost. Application of 50–100 g of vesicular–arbuscular mycorrhiza with jackfruit-based rhizosphere soil will be beneficial in facilitating the initial establishment of jackfruit grafts or seedlings. The planting is done after 1 or 2 rainy days.

7.6.4 Plantation-based cropping system

Jackfruit is used as a shade tree for coffee, pepper, betel nut and cardamom. Wide spacing (15 × 15 m) for jackfruit is recommended for intercropping. It can be intercropped with coconut, durian, mango and citrus. Banana, papaya and pulses can be grown as short-term crops. Jackfruit is well suited for the home garden for its fruits, fodder and timber. It is ideal in a multi-species windbreak and has withstood hurricane-force winds. The fruits can be borne on the main trunk; hence, damage due to wind is minimal. It can support climbers such as black pepper and yam.

Jackfruit + pepper-based cropping system

Jackfruit trees are grown as a live stand for pepper vines (Fig. 7.11). The jackfruit tree should bear on the upper branches, and it should have shy-bearing or low-bearing ability on the main trunk. Moreover, the tree should produce fruits of smaller to medium size. The bearing column of the pepper should be restricted to 5 m height from the ground for easy culture operation and for sunlight penetration to the pepper vine for more spike production.

Jackfruit + coconut-based cropping system

Growing of perennial crops (jackfruit) in association with matured coconut palm is referred to as a coconut-based cropping system.

Jackfruit + cardamom-based cropping system

Jackfruit is grown to provide shade in cardamom plantations. It is fast growing when compared with other conventional shade trees such as *Toona ciliata* (Chandhana vembu); *Acrocarpus fraxinifolius* (Korangatti), *Diospyros ebenum* (Sorakkali), *Eugenia jambolana* (Jamun), *Spathodea campanulata* (Kattu rani), *Swietenia mahagani* (Mahogany), *Vernonia arborea* (Karunai), and is preferred over other shade trees.

Pineapple and jackfruit-based agrisilviculture systems

Jackfruit and pineapple-based agroforestry is a common practice in household areas. In this system, the partial shade of the trees improves the pineapple's physical environment, ultimately enhancing the latter's yield and quality. The average annual net returns from the traditional pineapple–jackfruit practices were higher than a sole pineapple crop (Abedin and Quddus, 1991). The jackfruit saplings are planted (120 ha^{-1}) during July–August and the pineapple suckers (30,000 ha^{-1}) in September. The jackfruit trees provide suitable biophysical conditions for the understorey crops and meet other basic requirements of the growers, such as food, fodder, fuelwood and timber (Hasan *et al.*, 2008).

7.7 Jackfruit Avenues and Wind Breaks

Jackfruit is extensively used as an avenue plantation in peninsular regions of India, Bangladesh, Sri Lanka and China (Fig. 7.12). The suitability of jackfruit tree as an avenue tree is mainly due to its thick foliage and branching habit, which helps jackfruit to withstand heavy winds (Elevitch, 2004). The recommended spacing for avenue planting is 3 m in a single row on either side of the avenue, with subsequent thinning. Apart from the shade and aesthetic beauty, such plantations also function as effective wind breaks.

Fig. 7.11 A jackfruit tree used as a live stand for pepper vine. Photograph courtesy of Dr L.Wangchu.

Fig. 7.12. Jackfruit as an avenue plant in Yunnan Province, China. Photograph courtesy of Dr Lanxi Su.

7.8 Conclusion

Orchard profitability and sustainability are largely dependent on the proportion of crop that can be marketed as first quality. Planting high-yielding varieties that are propagated vegetatively and planted in a systematic manner is common for most major fruits. In the recent past, in countries such as Malaysia, Indonesia, Vietnam, Thailand, Sri Lanka and India, demand from industrial processing has greatly promoted the emergence of jackfruit plantations. Although research input into orchard management of jackfruit is meagre, information available from different growing countries shows that jackfruit responds well to management of soil, nutrition, irrigation, training and pruning, weeding and mulching.

References

Abedin, M.Z. and Quddus, M.A. (1991) Agroforestry system in Bangladesh with particular reference to economic and tenurial issues. In: *Agroforestry in Asia and the Pacific" No. 1995/5*. FAO Regional Office for Asia and the Pacific, Bangkok, pp. 13–33.

Acedo, A.L. (1992) *Jackfruit Biology, Production, Use, and Philippine Research*. Forestry/Fuelwood Research and Development Project. Winrock International Institute for Agricultural Development, Little Rock, Arizona.

Alexander, D.M.P., Scholefield, P.B. and Frodshan, A. (1983) Jackfruit. In: *Some Tree Fruits from Tropical Australia*. Report no.27. Commonwealth Scientific and Industrial Research Organization, Canberra.

Anon (2007) Report. Horticultural Research Centre, Malaysian Agricultural Research and Development Institute, Kuala Lumpur.

Anon (2013) *Newsletter*. International Tropical Fruit Network, Serdang, Malaysia.

Anon (2019) *Jackfruit*. Kerala Agricultural University, Kerala, India. Available at: www.celkau.in/crops/Fruits/jackfruit.aspx (accessed 11 July 2022).

Anon (2021) Jackfruit. Tamil Nadu Agricultural University, Tamil Nadu, India. Available at: www.agritech.tnau.ac.in/horticulture/horti_fruits_jack.html (accessed 11 July 2022).

Asia-Pacific Association of Agricultural Research Institution (2012) *Jackfruit Improvement in Asia – Pacific Region – A Status Report*. APAARI, Bangkok.

BPI (Bureau of Plant Industry) (1991) *Jackfruit (Nangka). BPI Circular*. Department of Agriculture, BPI, Manila, Philippines.

Campbell, R.J., El-Sawa, S.F. and Eck, R. (1998) *The Jackfruit (*Artocarpus heterophyllus *Lam)*. Fairchild Horticulture Series, Vol. 2. Fairchild Tropical Botanic Garden, Miami, Florida.

Crane, J.H., Balerdi, C.F. and Campbell, R.J.V. (2003) *The Jackfruit (Artocarpus heterophyllus Lam.) in Florida*. University of Florida Extension Services, Florida.

Elevitch, C.R. (2004) *The Overstory Book: Cultivating Connection with Trees*, 2nd edn. Permanent Agriculture Resources, Holualoa, Hawaii.

Elevitch, C.R. and Manner, H.I. (2006) *Artocarpus heterophyllus* (jackfruit). *Species Profiles for Pacific Island Agroforestry* 10, 1–25.

Ghosh, S.P. (1996) *Technical Report for Use of Underutilized Tropical Fruits in Asia Network*. Underutilized Tropical Fruits in Asia Network, Southampton University, Southampton, UK.

Haq, N. (2006) *Fruits for the Future 10: Jackfruit (Artocarpus heterophyllus)*. Southampton Centre for Underutilized Crops, University of Southampton, Southampton, UK.

Hasan, M.K., Ahmed, M.M. and Miah, M.G. (2008) Agro-economic performance of jackfruit-pineapple agroforestry system in Madhupur tract. *Journal of Agriculture & Rural Development* 6(1), 147–156. DOI: 10.3329/jard.v6i1.1672.

Jamaludheen, V., Kumar, B.M., Wahid, P.A. and Kamalam, N.V. (1997) Root distribution pattern of the wild jack tree (*Artocarpus hirsutus* Lam.) as studied by ^{32}P soil injection method. *Agroforestry Systems* 35(3), 329–336. DOI: 10.1007/BF00044462.

Khan, M.L. (2004) Effects of seed mass on seedling success in *Artocarpus heterophyllus* L., a tropical tree species of north-east India. *Acta Oecologica* 25(1–2), 103–110. DOI: 10.1016/j.actao.2003.11.007.

Morton, J.F. (1987) Jackfruit (*Artocarpus heterophyllus*). In: *Fruits of Warm Climates*. J.F. Morton, Miami, Florida, pp. 58–64.

Muthulakshmi, P. (2003) Genetic diversity and canopy management in jackfruit (*Artocarpus heterophyllus* Lam). PhD thesis, Kerala Agricultural University, Kerala, India.

Nath, V.V., Pandey, V., Pandey, D. and Kumar, D. (2009) *Fruits for the Future. Vol. 2. Lesser Known Tropical and Subtropical Fruits*. Satish Serial Publishing House, Delhi, India.

Singh, G. (2010) Jackfruit. In: *Canopy Management in Fruit Crops*. Department of Agriculture Cooperation, Ministry of Agriculture, Government of India, New Delhi.

Sun, Y., Yang, J., Wang, H., Zu, C., Tan, L. *et al*. (2015) Standardization of leaf sampling technique in jackfruit nutrient status diagnosis. *Agricultural Sciences* 06(2), 232–237. DOI: 10.4236/as.2015.62023.

8 Nutrition, Fertilization and Irrigation

Prabhat Kumar Ray*
Dr Rajendra Prasad Central Agricultural University, Samastipur, Bihar, India

8.1 Introduction

Mineral nutrients and water are essential natural resources for life-supporting systems of all plants. They play a major role in stability and enhancement of crop production, leading to self-sufficiency and sustainability. Impressive progress has been made in irrigation and fertilizer application technologies and their field-level management in different crops over the past six to seven decades, enriching our understanding of their uptake mechanisms and functions in plant metabolism. At the same time, there have also been marked advances in increasing crop yields by the supply of mineral nutrients through fertilizer application and scheduling irrigation at the right time, in the right amount, in the right manner and in the right place.

To achieve optimal yield and quality with maximum gains in water- and nutrient-use efficiency, precise efforts are required to meet the site-specific demand for nutrients and water. As this varies greatly as a result of variability in soil properties and land topography, rainfall, crop morphology, genetic makeup of the crop, and other crop- or soil-related factors (e.g. prevalence of insect pests, diseases and weeds, excessive salt in soil, shallow soil depth with a rocky layer underneath), crop management practices such as fertilization and irrigation need critical attention towards their optimization. This chapter brings together current information on various aspects of nutrient and irrigation management strategies for the optimal production of jackfruit trees under both optimal and suboptimal conditions.

8.2 The Jackfruit Tree

8.2.1 Canopy size

Jackfruit (*Artocarpus heterophyllus*) is a medium- to large-sized evergreen tree that typically attains a height of 10–20 m, occasionally reaching up to 30 m, with a dense crown. According to Coronel (1983), it is the largest tree-borne fruit in the world. The canopy shape is usually conical or pyramidal in young trees and becomes spreading and domed in older trees. The canopy diameter at 5 years ranges from 3.5 to 6.7 m and can reach 10–12 m or more in older trees. A medium-sized tree can typically reach 8–25 m in height and 30–80 cm trunk diameter within a period of 20–25 years (Elevitch and Manner, 2006).The bark is thick, somewhat scaly, and greyish-brown or dark grey in colour. The tree casts a very dense shade. Heavy side branching may be found close to the ground. The leaves are dark green, alternate, entire, simple, glossy, leathery, stiff, large (10–18 cm long and 5–10 cm wide), and oblong, elliptic or oval in form. The trees generally produce within 3–4 years

*pkraypusa@gmail.com

DOI: 10.1079/9781800622319.0008

and can live to 100 years old, but productivity declines with age (Love and Paull, 2011).

8.2.2 Root spread

In general, jackfruit is propagated sexually by seed and the tree has a deep taproot system. Propagation by vegetative means such as stem cuttings, air layering and grafting is possible but uncommon in practice. The trees have evolved the ability to grow very deep to find a lower water table and can withstand drought conditions fairly well. Many believe that the root distribution somehow mirrors the branching pattern of the aerial portion of the tree, or that roots extend out only to the edge of the tree crown, known as the 'dripline'. In fact, for most trees, the roots extend laterally far beyond the tree crown but do not extend to depths equivalent to the tree height. The two main functions of roots are to take up water and minerals, and to hold the tree up. The roots normally account for 20–30% of a tree's mass (Hartman *et al.*, 2000). However, the canopy and roots have roughly around the same mass. This helps put into perspective the relative value of the roots and leaves to each other – too few roots and the canopy suffers from lack of water, but too few leaves and the roots get insufficient food. Therefore, there has to be a balance between roots and shoots.

The taproot continues to grow as the tree matures, and a number of major 'framework roots' (five to seven) grow out sideways from the tap root. These laterals may descend obliquely down several tens of centimetres before growing horizontally. Although the lateral roots can be well over 30 cm diameter at the base of the tree (helping to cause the marked swelling at the base of the trunk, called the root collar), they quickly taper over the first metre to often less than 10 cm diameter, and down to just 2–5 cm over the first 3–4 m from the trunk, becoming less rigid and more rope like.

Trees spend their energy developing a diverse and dense root system. The long taproot generally ends up being about 2–3 m deep, while the other roots, which grow out of the taproot, grow much further down into the soil profile. The tree's extensive lateral root system produces a high volume of these roots. Water supply greatly influences vertical and horizontal spread or the total volume of roots in contact with the soil (Fig. 8.1). There are several million fine roots across the root zone. These are the most active parts of the whole root system. Those in this vast network of roots that are less than 2 mm in diameter grow and die off in waves as the tree matures and advances further in age. They cover a large soil surface area for maximum nutrient absorption.

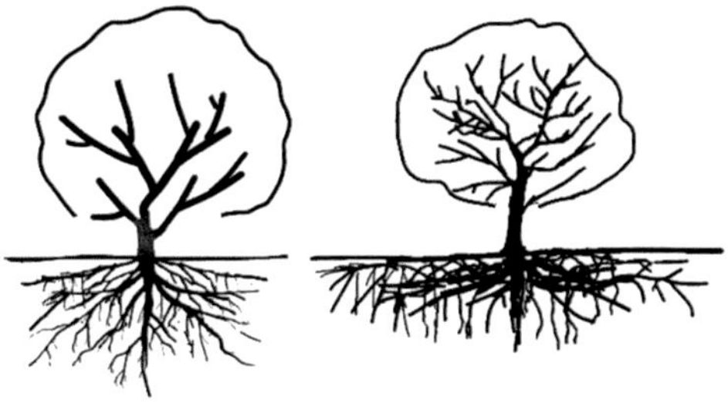

Fig. 8.1. Schematic diagram of jackfruit roots. Left: a tree with optimum water supply indicating balanced vertical and horizontal root growth. Right: a tree getting excess water, showing more horizontal growth reaching well beyond the dripline and a reduced canopy size (not to scale).

Horizontal root spread in mature trees is variable and ranges from 60 to 120 m^2, and usually exceeds that of the branches or the tree's dripline. If the water supply is plentiful, the horizontal spread is more, exceedingly well beyond the dripline (Fig. 8.1). The depth of root penetration or vertical spread varies from 2 to 8 m, commonly 2–3 m, and the zone containing majority of the functional roots is 0–60 cm deep. Practically, 0–30 cm or 0–45 cm depth is considered for drenching soil to field capacity when applying irrigation to the orchard. Atkinson *et al.* (1976) found that, in apple, over 60% of roots over 2 mm diameter existed at 0–25 cm depth, but only 35% of roots less than 2 mm diameter were present at this depth. The roots of a mature tree may grow in many directions, including upwards. The direction adopted may be under genetic control, as root systems of different species or varieties have characteristic morphologies owing, in part, to the degree by which the individual root deviates from the vertical position.

Root distribution studies of wild young jackfruit trees revealed that most of the physiologically active roots were concentrated within a radius of 75 cm and a depth of 30 cm (Jamaludheen *et al.*, 1997). Although the tap root might reach even deeper, the tree roots seldom extend beyond 2.25 m laterally from the stem, and thus the effect of overlapping root zones and associated competitive effects may not be a serious problem for intercropping during the first few years of tree growth (Jamaludheen *et al.*, 1997). As the jackfruit requires a long time to fully occupy the land provided during planting for future mature trees, it is desirable that the interspaces should be utilized by growing intercrops. In mature trees, most of the absorptive roots lie in the outer two-thirds of the circular area from the tree trunk to the dripline.

The rate of root growth, live root wood strength, rates of decay and site occupancy, particularly lateral root growth rate, or root distribution in both lateral as well as vertical directions, change with tree age, variety, rootstock and soil types, and can also be modified by soil management and orchard factors such as plant spacing, pruning and tillage. The relative density of roots in the soil increases with increasing density of planting (Atkinson, 1978, 1980, 1983; Hamido and Morgan, 2020). However, the root:shoot ratio remains unaffected by spacing in most cases, indicating a proportionate shrinkage in root spread with a reduction in canopy size (Coutts and Philipson, 1980; Williamson and Coston, 1990). Trees at higher planting densities make greater use of the subsoil at a much earlier point in their lives. The ability of different root types (white, thinner roots and woody, dark brown, thicker roots) to function in absorption is related to their anatomy. The effectiveness or contribution to nutrient and water uptake of the different root types will depend on the relative amounts present, inherent rates of absorption, contact with the soil and differential environmental effects on different roots (Atkinson, 1978; Atkinson and Wilson, 1980).

8.3 Nutrition

Soil serves as a reservoir of mineral nutrients and water for growth and reproduction of trees and also anchors their roots to keep them upright. It is the uppermost part of the Earth's crust, formed mainly by the weathering of rocks, formation of humus and material transfer. Soils vary widely in terms of origin, appearance, characteristics and production capacity. Jackfruit trees mainly use the topsoil (0–50 cm depth) as the most effective rooting volume to obtain water and nutrients, but they also use the lower subsoil layers up to 3 m or deeper to meet their metabolic requirements. Soil contains inorganic minerals, organic matter and living organisms, air and water in varying proportions. All play an important role in crop production.

8.3.1 Soil

The physical properties of soil largely determine the texture, structure, physical condition and tilth of the soil, whereas the chemical properties control soil pH and the exchange of nutrient cations and anions. These soil properties alter the resistance offered to root elongation and enlargement, proliferation, degree of aeration, living conditions for soil life, nutrient mobility and water uptake, which in turn affect plant nutrition. Fertile soils with a deep rooting zone of about 2–3 m are most suited for jackfruit trees. The soil volume should contain no stony or densely compacted layers or unfavourable chemical conditions that impede deep-root growing systems. Fine-textured soils such as clays, clay loam, silts and fine sands have good water-holding properties, whereas coarse-textured soils have low water-holding capacity but good drainage. Soil texture influences several components of soil fertility to a large extent, such as the level of nutrient reserves and the proportion of the available nutrient fraction. It also influences several properties such as aeration, pore-space distribution, water-holding capacity and drainage characteristics.

Sandy soils are low in plant nutrients, as there is nothing to bind the nutrients and they are easily leached away by heavy rain or excessive water supply. Loamy soils, such as sandy or silty loam, have intermediate properties and are generally the most suitable for growing jackfruit. Clay soils are often rich in nutrient reserves (but not necessarily in plant-available forms), have a high water-holding capacity because of the many medium and small pores, but soil aeration is restricted. Soils with a high clay component are highly productive, as they have innate fertility and hold many nutrients. Jackfruit is grown on a variety of soils: deep alluvial, sandy loam or clay loam of medium fertility, calcareous or lateritic soil, shallow limestone or stony soil, with a pH of 5.0–8.5. The tree prefers well-drained deep alluvial soils with moderate fertility with slightly acidic to neutral soil (pH 5–7.5) (Elevitch and Manner, 2006; Haq, 2006; Janick and Paull, 2008; Sidhu, 2012). In shallow soil, the growth of the tree can be slow and the tree may not be tall, robust or vigorous. Sometimes, jackfruit is grown on soils with serious chemical and physical constraints that have made them unsuitable for field crops. Such a soil must be given special considerations for its upheaval. It should be used for planting only after reclamation has been done satisfactorily.

Jackfruit trees exhibit moderate tolerance to saline soils (Elevitch and Manner, 2006; Haq, 2006; Janick and Paull, 2008). Red lateritic soils are also suitable for jackfruit trees if they have adequate depth, an open texture, free drainage and a neutral pH (Hussain, 1998; Pandey *et al.*, 1998). However, such soils formed from acidic igneous rocks in heavy rainfall areas in Kerala (South India) have certain inherent limitations, such as poor native fertility, resulting in zinc deficiency, low base status, high active acidity and exchangeable aluminium, and excessive leaching of nutrients. Potassium (K) deficiency occurs most in acidic red and lateritic soils or in organic soils that have few K-bearing minerals. This can be managed well with the use of suitable fertilizers and soil amendments. Jackfruit trees are also grown successfully at a high pH (8.5) in the calcareous soils of southern Florida (Crane *et al.*, 2007) and northern Bihar (India).

8.3.2 Plant nutrients

Like any other plant, jackfruit trees essentially need 16 nutrients for their proper growth, development and fruiting. These nutrients are indispensable as plant constituents, for biochemical reactions and for the production of organic materials referred to as photosynthates (e.g. carbohydrates, proteins, fats, vitamins) by photosynthesis. The essential nutrients required by plants are exclusively inorganic, a feature distinguishing these organisms from humans, animals and many species of microorganisms, which additionally need organic foodstuffs to provide energy. By contrast, plants absorb light energy from solar radiation and convert it to chemical energy in the form of organic compounds, while at the same time taking up mineral nutrients to provide the chemical elements also essential for growth (Bowling, 1980). Optimal crop nutrition is an important prerequisite for obtaining high yields and good-quality fruits.

Of these 16 essential elements, carbon (C) and oxygen are obtained from carbon dioxide (CO_2) and hydrogen (H) is obtained from water (H_2O). These three elements are required in large quantities for the production of plant constituents such as cellulose or starch. In general, oxygen, C and H make up 95% of plant biomass, and the remaining 5% is made up by all other elements. The other 13 elements are called mineral nutrients because they are taken up in mineral (inorganic) forms. They are traditionally divided into two groups, macronutrients (nitrogen (N), K, calcium (Ca), magnesium (Mg), phosphorus (P) and sulfur (S)), and micronutrients (chlorine (Cl), iron (Fe), boron (B), manganese (Mn), zinc (Zn), copper (Cu) and molybdenum (Mo)), according to the amounts required. Somewhat arbitrarily, a dividing line is drawn between those nutrients required in greater quantities (macronutrients) and those elements required in smaller quantities (micronutrients). The difference in plant concentration between macronutrients and micronutrients is enormous. The relative contents of N and molybdenum (Mo) in plants occur at a ratio of 10,000:1. Plants need about 40 times more magnesium (Mg) than Fe. These examples indicate the significant difference between macronutrients and micronutrients. This division does not mean that one nutrient element is more important than another, just that they are required in different quantities and concentrations. Regardless of the amount required, physiologically, all of them are equally important. These elements are taken up by plants in specific chemical forms regardless of their source. Several elements other than the essential nutrients (e.g. nickel, sodium

(Na), cobalt, silicon and aluminium) have been found to render beneficial influences in plants. Although not essential (as the plant can live without them), beneficial nutrients can improve the growth of plants in some respects.

Nitrogen

N is the most abundant mineral nutrient in plants. It constitutes 2–4% of plant dry matter. Plants absorb N either as a nitrate ion (NO_3^-) or an ammonium ion (NH_4^+). N is a part of the chlorophyll (the green pigment in leaves) and is an essential constituent of all proteins. It is responsible for the dark green colour of stem and leaves, vigorous growth, branching/tillering, leaf production, size enlargement and yield. Absorbed N is transported through the xylem (in the stem) to the leaf canopy as NO_3^-, or may be reduced in the root region and transported in an organic form, such as amino acids or amides. N is mobile in the phloem (the plant tissue through which the sap containing dissolved food materials passes downwards to the stem and roots); as such, it can be retranslocated from older to younger leaves under N deficiency and translocated from leaves to the developing seed or fruit. The principal organic forms of N in phloem sap are amides, amino acids and ureides. NO_3^- and NH_4^+ are not present in this sap.

N deficiency in plants results in a marked reduction in growth rate. N-deficient plants have a short stature and reduced vigour. As N is a constituent of chlorophyll, its deficiency appears as a yellowing or chlorosis of the leaves. This yellowness usually appears first on the lower leaves, while the upper leaves remain green as they receive some N from older leaves. In a case of severe deficiency, the leaves turn brown and die. As a result, crop yield is reduced significantly. The effects of N toxicity are less evident than those of its deficiency. Excess N can create excessive vigour and affect fruit-bearing ability and fruit quality. With an excessive supply of N, trees become overvigorous and sometimes such trees are unfruitful.

Phosphorus

P is much less abundant in plants (compared with N and K) having a concentration of about one-fifth to one-tenth that of N in plant dry matter. P is absorbed as the orthophosphate ion (either as $H_2PO_4^-$ or HPO_4^{2-}) depending on soil pH. As the soil pH increases, the relative proportion of $H_2PO_4^-$ decreases and that of HPO_4^{2-} increases. P is essential for growth, cell division, root lengthening, seed and fruit development, and early ripening. It is a part of several compounds including oils and amino acids. The P compounds adenosine diphosphate (ADP) and adenosine triphosphate (ATP) act as energy carriers within the plants.

P is readily mobile within the plant (unlike in the soil) in both xylem and phloem tissues. When the plant faces a P shortage (stress), P from the old leaves is readily translocated to young tissue. With such a mobile element, the pattern of redistribution seems to be determined by the properties of the source (old leaves and stems) and the sink (shoot tip, root tip, expanding leaves and later the developing seed). Plant growth is markedly restricted under P deficiency, which retards growth of the secondary and tertiary branches and root development, and delays ripening of fruits.

The deficiency symptoms usually start on older leaves. A bluish-green to reddish colour develops, which can lead to bronze tints and a red colour. A shortage of inorganic phosphate (PO_4^{3-}) in the chloroplast reduces photosynthesis. Because RNA synthesis is reduced, protein synthesis is also reduced. A decreased shoot:root ratio is a feature of P deficiency, as is the overall lower growth of tops. Extremely high levels of P can result in toxicity symptoms. These generally manifest as a watery edge on the leaf tissue, which subsequently becomes necrotic.

Potassium

K is the second most abundant mineral nutrient in plants after N. It is four to six times more abundant than the macronutrients P, Ca, Mg and S. K is absorbed as the monovalent cation K^+ and is mobile in the phloem tissue of the plant. K is involved in the working of more than 60 enzymes, in photosynthesis and the movement of its products (photosynthates) to storage organs (thick branches, roots and fruits), water economy and providing resistance against a number of pests, diseases and abiotic stresses (frost and drought). It plays a role in regulating stomatal opening and therefore in the internal water relations of the plant.

The general symptom of K deficiency is chlorosis along the leaf boundary, followed by scorching and browning of the tips of older leaves. The affected area moves inwards as the severity of deficiency increases. K-deficiency symptoms show on the older tissues because of the mobility of K. Affected plants are generally stunted and have shortened internodes. Such plants have slow and stunted growth, weak stalks and susceptibility to lodging, a greater incidence of pests and diseases, low yield, shrivelled grains and, in general, poor crop quality. Slow plant growth can be accompanied by a higher rate of respiration, which means a wasteful consumption of water per unit of dry matter produced. K-deficient plants may lose control over the rate of transpiration and suffer from internal drought.

Calcium

Ca ranks with Mg, P and S in the group of least abundant macronutrients in plants. It is absorbed by plant roots as the divalent cation Ca^{2+}. Ca is a part of the architecture of cell walls and membranes. It is involved in cell division, growth, root lengthening, and activation or inhibition of enzymes. Ca is immobile in the phloem.

Ca deficiency is seen first on growing tips and in the youngest leaves. This is the case with all nutrients that are not very mobile in the plant. Ca-deficiency problems are often related to the inability of Ca to be transported in the phloem. The problems occur in organs that do not transpire readily (i.e. large, fleshy developing fruits). Ca-deficient leaves become small, distorted, cup-shaped, crinkled and dark green. They cease growing, become disorganized and twisted and, under severe deficiency, die prematurely. Although all growing points are sensitive to Ca deficiency, those of the roots are affected more severely.

Magnesium

Mg ranks with Ca, P and S in the group of least abundant macronutrients in plants. Plants take up Mg in the form of Mg^{2+}. Mg occupies the centre spot in the chlorophyll molecule and thus is vital for photosynthesis. It is associated with the activation of enzymes, energy transfer, maintenance of electrical balance, production of proteins, and metabolism of carbohydrates. Mg is mobile within the plant. As Mg is readily translocated from older to younger plant parts, its deficiency symptoms first appear in the older parts of the plant.

A typical symptom of Mg deficiency is the interveinal chlorosis of older leaves in which the veins remain green but the area between them turns yellow. As the deficiency becomes more severe, the leaf tissue becomes uniformly pale, and then brown and necrotic. The leaves are small and break easily (brittle). Twigs become weak and leaves drop early.

Sulfur

S is required by crops in amounts comparable to P. The normal total S concentration in vegetative tissue is 0.12–0.35% and the total N:total S ratio is about 15. Plant roots absorb S primarily as the sulfate ion (SO_4^{2-}). However, it is possible for plants to absorb sulfur dioxide (SO_2) gas from the atmosphere at low concentrations. S forms a part of the amino acids cysteine, cystine and methionine. Hence, it is essential for protein production. S is involved in the formation of chlorophyll and in the activation of enzymes. Under low S conditions, mobility is low, as the S in structural compounds cannot be translocated. As the S status of the plant rises, so does its mobility. This pattern of mobility means that in plants with adequate S, SO_4^{2-} is preferentially translocated to young, actively growing leaves.

As the supply of S becomes more limiting, young leaves lack S and hence show deficiency symptoms. In many ways, S deficiency resembles that of N. It starts with the appearance of pale yellow or light green leaves. Unlike N deficiency, S-deficiency symptoms in most cases appear first on the younger leaves, and are present even after N application. Plants deficient in S show a retarded growth with delayed flowering and fruiting. S toxicity can occur under highly reduced conditions, possibly as a result of hydrogen sulfide (H_2S) injury. Most plants are susceptible to high levels of atmospheric SO_2. Normal SO_2 concentrations range from 0.1 to 0.2 mg SO_2 m^{-3}, and toxicity symptoms are observed when these exceed 0.6 mg SO_2 m^{-3}.

S-toxicity symptoms appear as necrotic spots on leaves, which then spread over the whole leaf.

Boron

B is probably taken up by plants as undissociated boric acid (H_3BO_3). It appears that much of the B uptake mainly follows water flow through roots. B in a plant is like the mortar in a brick wall, the bricks being the cells of growing parts such as tips (meristems). Key roles of B relate to: (i) membrane integrity and cell wall development, which affect permeability, cell division and extension; and (ii) pollen tube growth, which affects seed/fruit set and hence yield. B is relatively immobile in plants, and frequently the B content increases from the lower to the upper parts of plants.

B deficiency usually appears on the growing points of roots, shoots and the youngest leaves. Young leaves are deformed and arranged in the form of a rosette. There may be cracking and cork formation in the stalks, stem and fruits, thickening of stem and leaves, shortened internodes, withering or dying of growing points, and reduced bud, flower and fruit setting. Other symptoms are deformed misshapen fruits, and thickening and development of brownish necrotic patches on roots.

Symptoms of B deficiency include leaf distortion, death of growing points, poor ovule fertilization, fruit set, cracking of fruits and development of deformed fruits. B-toxicity can arise under excessive B application in arid or semi-arid areas, where irrigation water is rich in B content (more than 1–2 p.p.m.). B-toxicity symptoms are yellowing of the leaf tip, followed by gradual necrosis of the tip and leaf margins, which spreads towards the midrib (central vein). The leaves become scorched and may drop early.

Chlorine

Cl is absorbed as the chloride anion (Cl^-). It is thought to be involved in the production of oxygen during photosynthesis, in raising cell osmotic pressure and in maintaining tissue hydration. Deficiency of Cl leads to chlorosis in younger leaves and overall wilting as a consequence of the possible effect on transpiration. Typical Cl-toxicity symptoms are burning of the leaf tips or margins, bronzing, premature yellowing and shedding.

Copper

Cu is taken up as Cu^{2+}. Its uptake appears to be a metabolically mediated process. However, Cu uptake is largely independent of competitive effects and relates primarily to the levels of available Cu in the soil. Cu is involved in chlorophyll formation and forms a part of several enzymes such as cytochrome oxidase. As much as 70% of the Cu in plants may be present in the chlorophyll, largely bound to chloroplasts. It participates in lignin formation and protein and carbohydrate metabolism, and is possibly required for symbiotic N fixation. Cu is a part of plastocyanin, which forms a link in the electron transport chain involved in photosynthesis. Cu is not readily mobile in the plant and its movement is strongly dependent on the Cu status of the plant.

Cu-deficiency symptoms are first visible in the form of narrow, twisted leaves and pale white shoot tips. At maturity, panicles/ears are poorly filled and even empty where the deficiency is severe. In fruit trees, dieback of terminal growth can occur. Cu-toxicity symptoms are highly variable with species and less established than its deficiency symptoms. Excess Cu induces Fe deficiency and therefore chlorosis is a common symptom.

Iron

Fe is absorbed by plant roots as Fe^{2+} and to a lesser extent as Fe chelates. For efficient utilization of chelated Fe, separation between Fe and the organic ligand has to take place at the root surface, after the reduction of Fe^{3+} to Fe^{2+}. Absorbed Fe is immobile in the phloem. Fe is generally the most abundant of the micronutrients with a dry-matter concentration of about 100 p.p.m. It plays a role in the synthesis of chlorophyll, carbohydrate production, cell respiration and chemical reduction of NO_3^- and SO_4^{2-}, and in N assimilation.

Fe deficiency appears on younger leaves first. Otherwise, its deficiency symptoms are somewhat similar to those of Mn, as both Fe and Mn lead to failure in chlorophyll production. Yellowing of the interveinal areas of leaves (commonly referred to as iron chlorosis) occurs. In severe deficiency, the leaves become almost pale white because of the loss of chlorophyll.

Manganese

Mn is taken up by plants as the divalent ion Mn^{2+}. It is known to activate several enzymes and functions as an autocatalyst. It is essential for splitting the water molecule during photosynthesis. It has certain properties similar to Mg. It is also important in N metabolism and in CO_2 assimilation. Like Fe, it is generally immobile in the phloem. Mn-deficiency symptoms resemble those of Fe and Mg deficiency where interveinal chlorosis occurs in the leaves. However, Mn-deficiency symptoms are first visible on the younger leaves, whereas in Mg deficiency, the older leaves are affected first. Mn deficiency in oats is characterized by 'grey speck' where the leaf blade develops grey lesions but the tip remains green, the base dies and the panicle may be empty. The younger leaves develop chlorotic patches between the veins (somewhat resembling Mg deficiency). Mn-toxicity symptoms lead to the development of brown spots, mainly on older leaves and an uneven green colour. Mn toxicity leads to internal bark necrosis in apple trees.

Molybdenum

Mo is absorbed as the molybdate anion, $MO_2O_4^-$, and its uptake is controlled metabolically. Mo is involved in several enzyme systems, particularly nitrate reductase, which is needed for the reduction of NO_3^- and nitrogenase. Thus, it is involved directly in protein synthesis. Mo appears to be moderately mobile in the plant. This is suggested by the relatively high levels of Mo in seeds and because deficiency symptoms appear in the middle and older leaves. Mo deficiency in plants can resemble N deficiency because of its role in N fixation. Mo deficiency can cause marginal scorching and rolling or cupping of leaves and yellowing and stunting in plants.

Zinc

Zn is taken up as the divalent cation Zn^{2+}. Zn is required directly or indirectly by several enzymes systems, auxins and for protein synthesis, seed production and rate of maturity. Zn is believed to promote RNA synthesis, which in turn is needed for protein production. The mobility of Zn is low. The rate of Zn mobility to younger tissue is particularly depressed in Zn-deficient plants. Common symptoms of Zn deficiency are stunted plant growth, poor fruit setting and fruit growth, and development of light green, yellowish, bleached or brown rusty spots on leaves. In fruit trees, the shoots may fail to extend and the small leaves may bunch together at the tip in a rosette-type cluster. Little-leaf condition is also a common symptom. The internodes are short. Flowering, fruiting and maturity can be delayed. Shoots may die off and the leaves can fall prematurely. Zn toxicity can result in a reduction in root growth and leaf expansion followed by chlorosis.

8.3.3 Determining nutritional needs

A number of methods are used to determine the nutritional needs of plants. The objective of all of these methods is to assess the available nutrient status of soils and plants so that corrective measures can be taken to ensure optimal plant nutrition and minimum depletion of soil fertility throughout the tree's productive lifespan. When considering tree nutrition, it is essential to start with both standard soil and tissue diagnostics. These tools are more valuable when used together as an integrated approach with visual diagnosis (i.e. nutrient deficiency symptoms appearing on the trees).

Looking for subnormal growth

The most direct way to determine if fertilizers are needed is to observe the growth rate of the tree and the colour of the leaves. The jackfruit leaf is large, thick and green. If the tree shows slow, stunted growth and there is unusual yellowing of the leaves, especially between the veins, it may be a sign of nutritional deficiency. Nutrient deficiency symptoms can be confusing as deficiency of more than one element can produce almost similar symptoms. Also, several biotic factors may be responsible for the development of symptoms similar to those of some nutrient deficiencies. However, a few common

Table 8.1. Soil test values indicating different fertility status of the orchard soil. From Anon (2011).

Soil nutrients	Soil fertility ratings		
	Low	Medium	High
Organic carbon as a measure of available nitrogen (%)	<0.5	0.5–0.75	>0.75
Available N as peralkaline permanganate method (kg ha^{-1})	<280	280–560	>560
Available P by Olsen's method in alkaline soil (kg ha^{-1})	<10	10–24.6	>24.6
Available K by neutral N, ammonia acetate method (kg ha^{-1})	<108	108–280	>280

Table 8.2. Critical limits for diethylenetriaminepentaacetic acid-extractable micronutrients. From Anon (2011).

Micronutrient level	Availability (μg g^{-1} soil)			
	Zn	Cu	Fe	Mn
Very low	0–0.5	0–0.1	0–2	0–0.5
Low	0.5–1	0.1–0.3	2–4	0.5–1.2
Medium	1–3	0.3–0.8	4–6	1.2–3.5
High	3–5	0.8–3	6–10	3.5 – 6
Very high	>5	>3	>10	>6

visual symptoms of nutrient deficiencies such as small chlorotic leaves, dead areas of leaf tips and margins or between veins, dieback of stem tips and twigs, bark lesions and excessive gum formation indicate that the orchard trees might be suffering from nutritional deficiency.

Soil testing

Soil testing helps in understanding the inherent fertility status of the soils. However, its usefulness depends on a number of factors, such as representative and correct soil sampling, analysis of the sample using a validated procedure, and correct interpretation of the analytical data to make practical recommendations. Soils are naturally variable; therefore, proper soil sampling is critical in assessing the soil fertility level of the orchard. Samples should also be taken from within the drip line of the tree canopy and from within the existing tree row. Samples drawn from topsoil (0–15 cm depth) and subsoil (15–45 cm depth) are tested separately. Soil testing is generally done in a laboratory following standard analytical methods for estimating levels of pH, organic matter, electrical conductivity as an indicator of salinity, P, K, Ca, Mg, B, Zn and Cu. Once the soil test report is available, a comparison of the test report with predefined nutrient availability standards in the soil (Tables 8.1 and 8.2) will indicate if the elements evaluated are present in adequate, deficient or excessive amounts. The test results are interpreted professionally by experts to prescribe fertilizers for managing the soil fertility.

Leaf analysis

Leaf analysis is used to directly measure the amount of nutrients in the trees. It generally provides more current plant-based information than soil testing but is more costly and requires greater efforts in sampling, sample handling and analysis. Ideally, both tools should be used as they complement each other. As nutrient concentration varies among tree fruit tissues and changes over the season, the standards must be sensitive enough to clearly alert the growers when nutrient levels become deficient or excessive. The value of leaf analysis depends on sampling technique, including canopy height, leaf age and time of sampling.

Sun *et al.* (2015a) standardized the leaf sampling technique for jackfruit to calculate fertilizer needs. The best period of drawing leaf samples was April and May. Three- to 6-month-old, fully expanded leaves taken from the middle portion of bearing and non-bearing current-season shoots (Fig. 8.2) lying near the

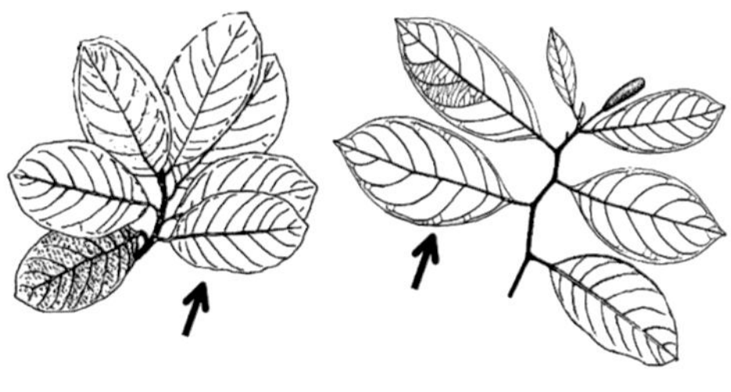

Fig. 8.2. Leaves (indicated by arrows) are selected from the shoot for tissue testing. Redrawn from Vozzo (2002) and Haq (2006).

central canopy region were found to be the most suitable to develop foliar nutrient standards for jackfruit. The recommended values of nutrient concentrations in leaf tissue are summarized in Table 8.3. Uniformity of the samples with respect to vigour, age and maturity of the twig/shoot is important. Caution should also be taken to avoid contaminationof the leaf samples by fungicides (e.g. containing Cu or Zn), or foliar nutrient applications (i.e. B, Ca). Samples should not be taken within 15 days of an application containing nutritional elements.

Nutrient removal

There exists a common belief that if the quantity of nutrients removed from the orchard by harvested fruits, fallen leaves and pruned twigs is worked out, their application to the tree through fertilizers will help maintaining normal growth and productivity. This concept has encouraged many researchers (Salazar-García and Lazcano-Ferrat, 2001; Xiuchong *et al.*, 2001; Palmer and Dryden, 2006; Mellado-Vázquez *et al.*, 2019) to assess the impact of replacement to the soil of nutrients removed by the crops as a method of crop fertilization. However, research data on annual nutrient removal from a perennial framework of jackfruit trees in terms of fruits, leaves, dead roots, etc. are not available.

As jackfruit is a tropical, evergreen, woody tree, its need for N, P and K is comparable to the needs of other evergreen fruit trees such as mango, avocado, olive, sapote and lychee. Therios (2009) reported that, in general, the amounts of nutrient absorbed from orchard soils are 0.50 kg N 100 kg^{-1} fruit and 1.0 and 0.76 kg N 100 kg^{-1} of leaves and wood, respectively. For K, the quantities are 0.95, 0.56 and 0.39 kg, respectively. For P, the quantities required every year by an olive orchard for the production of 100 kg fruit, 50 kg leaves and 50 kg wood are 0.120, 0.120 and 0.150 kg, respectively (Therios, 2009). The large fruit load, the significant number of leaves and wood removed every year by pruning and the leaf drop due to diseases and water stress entail, as a consequence, the loss of a large amount of nutrients from trees.

Fernández-Escobar *et al.* (2015) observed that the amounts of nutrients removed annually from 12-year-old olive trees by fruit harvest and trimmed pruning materials were (kg ha^{-1}

Table 8.3. Suggested leaf nutrient concentrations for jackfruit. Modified from Sun *et al.* (2015a).

Nutrient	Deficiency	Low	Optimum	High	Excess
N (%)	<2.24	2.24–2.41	2.42–2.76	2.77–2.94	>2.94
P (%)	<0.14	0.14–0.15	0.16–0.18	0.19–0.20	>0.20
K (%)	<0.96	0.96–1.37	1.38–2.24	2.25–2.66	>2.66
Ca (%)	<0.47	0.47–0.74	0.75–1.31	1.32–1.59	>1.59
Mg (%)	<0.01	0.01–0.14	0.15–0.41	0.42–0.55	>0.55
S (%)	<0.15	0.15–0.16	0.17–0.23	0.24–0.25	>0.25
Fe (mg kg^{-1})	<74.19	74.19–88.44	88.45–116.99	117.00–131.25	>131.25
Mn (mg kg^{-1})	<94.59	94.59–239.70	239.71–529.95	529.96–675.07	>675.07
Cu (mg kg^{-1})	<2.24	2.24–6.29	6.30–14.40	14.41–18.46	>18.46
Zn (mg kg^{-1})	<27.15	27.15–32.72	32.73–43.87	43.88–49.45	>49.45
B (mg kg^{-1})	<23.42	23.42–34.03	34.04–55.26	55.27–65.88	>65.88

year^{-1}): Ca 57.9, N 54.4, K 45.5, P 6.87, Mg 3.79, Cu 0.12, B 0.11, Mn 0.08 and Zn 0.05. These data show that olive trees remove small amounts of nutrients and therefore the need for fertilization is relatively low. Salazar-García and Lazcano-Ferrat (2001) estimated annual nutrient removal from yield data of mature avocado trees. A crop of 20 t ha^{-1} of Hass avocado removed 52, 21 and 94 kg N, P_2O_5 and K_2O, respectively. Estimated mean fruit nutrient removal rates for a 70 t ha^{-1} picked yield of apples were (kg ha^{-1} year^{-1}): K 82, N 31, P 7, and Ca and Mg 4 (Palmer and Dryden, 2006).

Jackfruit fresh foliage is rich in mineral nutrients such as N (2.15%), P (0.11%) and K (1.26%) (Kumar *et al.*, 1998). Hence, incorporation of foliage mulch to the soil can substantially improve the soil nutrient status. Furthermore, tree biomass accumulation is considerably higher than many tree species, and as a result of this higher biomass accumulation, the nutrient accumulation (dry weight basis) in the various aboveground tissue types for jackfruit (stem wood, branch wood, twigs and foliage) is fairly high: 958.3 kg N ha^{-1}, 30.9 kg P ha^{-1} and 386.2 kg K ha^{-1} (Kumar *et al.*, 1998). Considering the level of major nutrients present in leaves and fruits of jackfruit (Swami *et al.*, 2012; Amadi *et al.*, 2018; Ranasinghe *et al.*, 2019), removal of around 40, 20 and 75 kg N, P_2O_5 and K_2O, respectively, by trees producing an average fruit yield of 15 t ha^{-1} year^{-1} may be a fair approximation. K removal is much higher than that of N and P.

Field trials

This is a proven and effective method for assessing nutrient needs. It is also essential for the calibration of soil test and plant analysis results with crop response. Fertilizer experiments are time consuming but provide direct information on the amounts of nutrients required for a crop in a specific environment at a given yield level. These are carried out by researchers to generate data needed for making practical recommendations for farmers. Fertilizer experiments are required to reach conclusions on soil nutrient supply, optimal rates and combinations of nutrients, best timing, application techniques, etc. This is considered a more appropriate approach because it recommends nutrient applications based on the direct economic response of the crop. Several field trials have been carried out to determine nutrient applications for jackfruit trees. The outcomes of these trials are discussed in greater detail in sections 8.4.1 to 8.4.4.

Diagnosis and Recommendation Integrated System

The Diagnosis and Recommendation Integrated System (DRIS) is an interpretation technique for diagnosing the nutrient concentration or equilibria in selected plant tissue and has been used successfully in variety of crops (Mourão Filho, 2004; Bangroo *et al.*, 2010; Raghupathi *et al.*, 2014; Naik and Bhatt, 2017). It is impossible to diagnose imbalance of any particular nutrient in isolation in plant tissues, as large numbers of nutrients either increase or decrease together. As in the case of mango, nutrient concentration build-up and depletion are very gradual in leaves due to the perennial and deep-rooted nature of the plant. The nutrient concentration variation in leaves appears to be an overall orchard phenomenon rather than an individual tree phenomenon (Raghupathi *et al.*, 2004). The response in terms of the increase in nutrient concentration in the leaf due to fertilizer application in mango is reported to be very mild (Sukthumrong *et al.*, 2000). Because of the many interactions between nutrients and other inputs, more sophisticated indicators than just individual nutrient concentrations have been suggested, such as simple or complex nutrient ratios derived through the approach called DRIS.

DRIS provides a means of ordering nutrient ratios or products into a meaningful expression called the DRIS index. Essentially, the nutrient index is the mean of deviations of the ratios containing a given nutrient from their respective optimum or norm value. All indices are balanced around zero. Therefore, nutrient indices sum to zero. The more negative an index, the more lacking is the nutrient it represents relative to other nutrients used in the diagnosis. Alternatively, a large positive index indicates that the corresponding nutrients are present in relative excess (Raghupathi *et al.*, 2004).

DRIS helps in simultaneous identification of nutrient imbalances, deficiencies and excesses of a crop nutrient and ranks them in order of their importance for taking remedial steps. The DRIS norms based on foliar composition can be developed in any crop and at any stage of its development. It provides a mathematical means of ordering large numbers of nutrient ratios into nutrient indices that can easily be interpreted for better inference. Savita *et al.* (2016) considered DRIS as a holistic approach for formulation of nutrient management strategies.

The DRIS approach uses the principle of nutrient interrelationships in determining the order of nutrient requirement (Mourão Filho, 2004; Hundal *et al.*, 2005; Raghupathi *et al.*, 2014). It is based on the theory about that, as plants approach their growth potential, their nutrient status becomes less variable (Sharma *et al.*, 2005; Agbangba *et al.*, 2011; Savita *et al.*, 2016). DRIS index values that describe the relative status of each nutrient are generated during a DRIS assessment. This approach uses nutrient concentration ratios rather than nutrient concentrations, to give a rating of relative balances among nutrients (DRIS index). The DRIS approach has been used in many fruit crops, such as mango (Naik and Bhatt, 2017), grape (Sharma *et al.*, 2005) and mandarin (Srivastava and Singh, 2008).

Sun *et al.* (2015b) working on DRIS indices for low-yielding (fruit yield <60 t ha^{-1}) and high-yielding (fruit yield ≥90 t ha^{-1}) jackfruit trees in Hainan (China) developed standard DRIS indices (grade) of leaf nutrients. Based on the mean of nutrient concentration and standard deviation from the high-yielding population, the indices were classified on the availability parameters as deficient, low, optimum, high and excessive to serve as a diagnostic guide (Table 8.4). The nutrient imbalance in low-yielding population was explained using DRIS indices. The order of nutrient requirement for low-yielding jackfruit trees was determined as Ca > K > Fe > P > N > S > Mg > B > Mn, where Ca, K and Fe were the most deficient nutrients (Table 8.5). For active (mobile) nutrients such as Ca and Fe, the nutrient content of trees should be considered more carefully, as their immobilization due to certain factors can make the total concentration values misleading.

Yield Targeting Concept

When the nutrient supply is suboptimal, the leaf growth rate, and thus the LAI or total vegetative frame of the tree, can be limited by low rates of net photosynthesis or insufficient cell expansion, or both of these factors. This is particularly evident with a suboptimal supply of N and P. Furthermore, the yield response is strongly modulated by interactions between mineral nutrients and seasonal growth factors in perennial tropical trees. Under field conditions, the interactions between water availability and nutrient supply are of particular importance. The depressions in yield that accompany a large supply of nutrients in combination with low soil moisture levels are reflected as considerable deterioration in both quantitative and qualitative components of the fruits. Premature ripening of fruits imposed by water or nutrient deficiency is another yield-limiting factor.

The Yield Targeting Concept is a technique used to estimate fertilizer requirement for a crop

Table 8.4. Standards of DRIS index grade for jackfruit leaf nutrients. Adapted from Sun *et al.* (2015b).

Nutrient	Deficiency	Low	Optimum	High	Excess
N	<–4.61	–4.61 to –1.10	–1.09 to 5.95	5.96 to 9.47	>9.47
P	<–11.50	–11.50 to –3.23	–3.22 to 13.35	13.36 to 21.63	>21.63
K	<–8.29	–8.29 to –3.05	–3.04 to 7.44	7.45 to 12.68	>12.68
Ca	<–16.50	–16.50 to –3.32	–3.31 to 23.08	23.09 to 36.27	>36.27
Mg	<–45.00	–45.00 to –30.68	–30.67 to –2.02	–2.01 to 12.30	>12.30
S	<–12.38	–12.38 to –3.70	–3.69 to 13.67	13.68 to 22.35	>22.35
Fe	<–18.71	–18.71 to –2.83	–2.82 to 28.97	28.98 to 44.86	>44.86
Mn	<–53.82	–53.82 to –37.06	–37.05 to –3.52	–3.51 to 13.24	>13.24
B	<–23.12	–23.12 to –12.07	–12.06 to 10.05	10.06 to 21.10	>21.10

Table 8.5. DRIS indices and order of nutrient requirements for low-yielding orchards. Adapted from Sun *et al.* (2015b).

	N	P	K	Ca	Mg	S	Fe	Mn	B
Index	19.55	3.08	–222.41	–359.14	143.04	94.35	–22.88	197.60	146.82
Order of requirement	Ca > K > Fe > P > N > S > Mg > B > Mn								

to obtain the targeted yield from a particular soil and climatic conditions. The technique is based on soil tests and requires knowledge of the yield that can be achieved from a crop with manipulation of soil fertility with fertilizers. Soil test laboratories generally do not consider the yield goal when making recommendations for fertilizers on the basis of soil test results. However, the research institutes working on the crop develop basic equations that are used for the determination of nutrient requirements to obtain average or higher yield levels of the crop. Unfortunately, yield targeting equations that can be applied to jackfruit production with varied input supplies in a particular locality are not available and are yet to be worked out by research stations working on this crop.

8.4 Fertilization

8.4.1 Fertilizer recommendations

Traditionally, many growers do not apply manures or fertilizers, believing that jackfruit trees are well adapted to medium- and low-fertility soils and able to produce well without any additional application of fertilizers. This perspective is due to the fact that the crop has yet not acquired prominence or economic importance like crops such as mango, guava or lychee. It is still grown as a minor crop cultivated mostly without much attention. Growers often have a few trees of jackfruit in their mango or other fruit orchards. In many areas, it is a prominent, highly preferred agroforestry crop (Kunhamu, 2011). However, large orchards with only jackfruit trees are uncommon. Thus, important cultural practices have not been standardized as they have for other perennial tree fruit crops. The repercussions are clearly reflected in a wide range of fruit harvests (1.4–27 t ha^{-1}) from jackfruit orchards in the eight major jackfruit-producing countries in Asia (Haq, 2006). As well as genetic variability and some climatic factors, poor orchard management, particularly with respect to nutrition and irrigation, is responsible for the wide yield gap in this crop. With good management practices, fruit yield as high as or exceeding 90 t ha^{-1} has been reported from jackfruit orchards in Hainan, China (Sun *et al.*, 2015b). Given the individual fruit's notably large size and its high pulp contents, a good yield from jackfruit trees requires an adequate and well-balanced supply of nutrients.

Fertilizers should be used according to the dose calibrated from the nutritional needs of jackfruit trees, specific to the orchard and their past cropping history (fruit yield, quality and consistency of production). Recommendations for a tree vary considerably from its juvenile pre-bearing phase to the fully fledged production phase. As the pace of tree growth depends largely on variety, soil and climate, region-specific nutrient requirements of jackfruit need to be worked out precisely. It is known that young plants, during the pre-bearing phase, have exceedingly high nutrient uptake, although not particularly large quantities of nutrients are required due to the smaller tree size. During this time, the plants develop the fruit-bearing structure (canopy) and a thorough, extensive root system not only for anchorage and nutritional support but also for water.

Jackfruit is known for its fast growth, attaining a height of 3 m and a canopy diameter of 2 m in 2 years. During the early years, the mean annual increment in height is about 1.5 m $year^{-1}$, declining to 36–60 cm $year^{-1}$ towards maturity (Acedo, 1992). The tree attains a height of 15–17 m and 20–34 cm trunk diameter at the age of 20 years (Kunhamu, 2011). There is clearly a high demand for nutrients to sustain the rapid, possibly exponential growth of aboveground biomass, and relatively little recycling occurs because new tissue is being developed. Large soil reserves of nutrients are mobilized to the aboveground biomass and developing roots during this period.

When the trees have reached maturity, there is essentially no substantial net biomass increase, or it is very small relative to that occurring during the vegetative pre-bearing phase. The nutrient status during this stage is characterized by extensive recycling of nutrients already in the biomass. During this stage, there is relatively less dependence on soil nutrients and much greater emphasis on nutrient recycling and nutrient content already in the biomass. Measurement of soil nutrients may not be of much practical help during this stage because of the extensive recycling occurring in the plant tissues. The new tissues derive nutrients from those already in the plant, perhaps stored in older tissue, and thus estimating the amounts of nutrients to supply during this phase through fertilizers is difficult. However, the quantity of fertilizers required at the maturity stage is determined mainly on the basis of leaf nutrient status. Nutrients are added to increase the foliar level of nutrients from the existing level (presumably deficient) to a predetermined adequate level.

8.4.2 Amount of fertilizer

Fertilizer requirement (or rate per tree) denotes the quantity of additional nutrients required by a jackfruit tree to increase its growth and yield to the optimal level in a given set of soil and climatic conditions. At the time of planting, farmyard manure or well-rotted cattle manure at a rate of 10–20 kg per pit is usually mixed with the native soil. If the fertility of the native soil is good or average, this application serves the plant's needs at the establishment stage in the first year. Subsequent applications are required from the second year onward. However, when soil test data reveal a poor fertility status of the soil, small quantities of P_2O_5 (100–125 g per pit) and K_2O (150–200 g per pit) are also added along with farmyard manure to optimize initial growth of the young saplings. Hore (2011) suggested thorough mixing of the pit soil with 30 kg well-rotted farmyard manure, 500 g single superphosphate and 250 g muriate of potash and then refilling with the mixture. From the second year onward, N, P and K are applied two to three times annually to the developing plants. Many workers (Janick and Paull, 2008; Almeida *et al.*, 2009; Goswami and Chacrabati, 2016; Amadi *et al.*, 2018) have observed very high contents of K and fairly good amounts of Ca and Mg in the fruits. As they are lost every year from tree reserves through the harvests, their replenishment needs to be duly considered when deciding fertilizer recommendations.

Fertilizer needs have not been well studied for jackfruit. The tree seems to perform well, even on moderately fertile soils. A large variation in annual application rate from 0 to 100 kg N ha^{-1} and from 0 to 70 kg P and K ha^{-1} has been reported (Haq, 2006). The University of Agricultural Sciences, Dharwad (India), recommended 50 kg farmyard manure, 600 g N, 300 g P_2O_5 and 240 g K_2O $year^{-1}$ per tree. Mukherjee *et al.* (1983) reported that supply of farmyard manure at 37 kg per plant, 750 g ammonium sulfate ($(NH_4)_2SO_4$), 625 g single superphosphate and 300 g muriate of potash $year^{-1}$ per plant in the second and third year of planting with an increased dose of 938 g $(NH_4)_2SO_4$, 781 g single superphosphate and 375 g muriate of potash increased growth, flowering and fruiting in jackfruit trees.

The use of manure is better than fertilizers. Application of 80 kg farmyard manure ha^{-1} was found to be sufficient to sustain growth and yield of mature trees (Samaddar, 1985). Annual applications of 80 kg farmyard manure per tree would provide approximately the equivalent of 0.4 kg N, 0.2 kg P and 0.4 kg K per tree, as its N, P and K contents are about 0.5%, 0.25% and 0.5% respectively. If soil fertility is good, regular application of 80 kg farmyard manure would be able to sustain production. However, when fertility is low or average, it may be necessary to supplement the manure with fertilizers, as the nutrients in the manure are made available rather slowly, and the quantity applied this way may not be sufficient to meet the plant's requirement.

Coronel (1983) recommended a commercial fertilizer regime as 100–150 g $(NH_4)_2SO_4$ (20–0–0 = 20% N, 0% P, 0% K) per tree in the first year, increasing in pre-bearing years, and then 0.5–1.0 kg of 14–14–14 fertilizer per tree increasing with age and size, with a fully grown

Table 8.6. Fertilizer recommendations for jackfruit from planting to fully bearing trees in North Bihar (India).

Nutrients (per tree year^{-1})	Planting year	Non-bearing years			Bearing years				
	Year 1	Year 2	Year 3	Year 4	Year 5	Year 6	Year 7	Year 8	Year ≥9
Farmyard manure (kg)	20	25	30	35	40	45	45	50	50
Nitrogen (g)	–	125	250	375	500	625	750	875	1000
Phosphorus (g)	–	50	100	150	200	250	300	350	400
Potash (g)	–	100	200	300	400	500	600	700	800

[a]Phosphorus as P_2O_5 and potassium as K_2O (P.K. Ray, unpublished data).

15–20-year-old tree receiving 2–3 kg complete fertilizer. He suggested that the use of nutrient-rich organic mulches such as prunings from fast-growing, nitrogen-fixing trees can reduce or eliminate the use of commercial fertilizer for the jackfruit trees under agroforestry ecosystem. Saha *et al.* (2011) recommended application of 50.0 kg farmyard manure, 5.0 kg wood ash, 2.0 kg bone meal, combined N, P and K at the rate of 2.0 kg, and 50.0 g borax per mature tree (≥10 years) for a genotype bearing fruits throughout the year. Tarai *et al.* (2018), for a full-bearing jackfruit tree, suggested 50 kg farmyard manure, 750 g N, 400 g P, 500 g potash and 5 kg ash for an average production of about 40–50 t fruits ha^{-1}.

The trees grow well and yield satisfactorily (20–25 t ha^{-1}) in the calcareous, sandy loam soils of North Bihar (India) with average fertility status and pH 8.4, electrical conductivity 0.36 dS m^{-1}, organic carbon 0.27% and free calcium carbonate ($CaCO_3$) 30.2%. The jackfruit trees of various age groups growing in this soil are fertilized as shown in Table 8.6. In Malaysia, the fertilizer recommendation is that N, P, K, and Mg are applied in the ratio 8:4:2:1 at 30 g per tree at 6 months, doubling every 6 months up to 2 years. Older trees receive 1 kg per tree of a 4:2:4:1 ratio every 6 months. Higher rates of 2–3 kg per tree are suggested in the Philippines. Application occurs before and at the end of the wet season, and is done around the outer canopy dripline. In Hawaii, trees are usually given a quarterly application of 8–8–8 fertilizer (Love and Paull, 2011). In the Konkan region of Maharashtra (India), mature jackfruit trees fertilized with 50 kg farmyard manure, 0.75 kg N, 0.40 kg P and 0.50 K year^{-1} per tree produced 50 t ha^{-1} (Priya Devi *et al.*, 2014). The NPK dose (g year^{-1} per tree) for mature trees is 750:400:500 in Tamil Nadu, 800:480:1050 in Madhya Pradesh, 600:300:240 in Karnataka and 210:160:1000 in Assam. These nutrients are added to the orchard soil along with farmyard manure ranging from 30 to 100 kg year^{-1} per tree (Tandon, 1987). With 20 t ha^{-1} of farmyard manure, about 100 kg N ha^{-1} is added. In the first year, 20–30% of this N is utilized, but up to 40–50% can be utilized by the second year, including the residual effect. Thus, the next year, a similar dose of farmyard manure can be skipped altogether or a separate dose may be considered (Hartman *et al.*, 2000).

In Bangladesh, where jackfruit has been assigned the status of the national fruit, some farmers use 10 kg rotted cow dung, 250 g urea, 250 g triple superphosphate and 250 g muriate of potash during planting of jackfruit saplings and thereafter do not apply any fertilizers (Hasan *et al.*, 2008). However, evidence shows significant improvements in flowering and fruiting of jackfruit with application of manures and fertilizers. Mondal (2000) recommended annual application of 25–50 kg cow dung, 500–700 g urea, 300–600 g triple superphosphate and 300–400 g muriate of potash per tree falling in the age group of 6–10 years, and 60–110 kg cow dung, 1.0–1.2 kg urea, 0.8–1.0 kg triple superphosphate and 1 kg of muriate of potash per tree for those older than 10 years. Hossain and Haq (2006) advocated application of 10 kg farmyard manure, 200 g urea, 250 g triple superphosphate, 100 g muriate of potash and 100 g gypsum year^{-1} per tree for 2–4-year-old trees. This dose was gradually increased with advancement of tree age up to to 30 kg farmyard

Table 8.7. Yearly amounts of manure and fertilizers used for jackfruit in Bangladesh. Modified from Hossain and Haq (2006).

	Recommended dose for trees in different age groups (quantity to be applied year^{-1} per tree)				
Manure/fertilizer	Years 2–4	Years 5–7	Years 8–10	Years 11–20	Years >20
Farmyard manure (kg)	10	15	20	25	30
Urea (g)	200	250	300	350	500
Triple superphosphate (g)	250	250	500	750	1000
Muriate of potash (g)	100	200	250	300	500
Gypsum (g)	100	100	200	300	500

Table 8.8. Fertilizer recommendations for jackfruit in Florida. Modified from Crane *et al.* (2007).

Year	Times year^{-1}	Amount per tree (lbs)[a]	Total amount per tree (lbs)[b]	No. of minor element sprays year^{-1}[c]	Chelated iron soil drenches (year^{-1} oz per tree)
1	6	0.25–0.5	1.5–3.0	6	0.5–0.75
2	6	0.5–1.0	3.0–6.0	6	0.75–1.0
3	6	1.0–1.5	6.0–9.0	6	1.0–1.5
4	2–3	1.5–2.5	9.0–10.0	6	1.5–2
5	2–3	2.5–3.5	10.0–14.0	4–6	2–4
6	2–3	3.5–4.0	14.0–16.0	4 – 6	2–4
7	2–3	4.0–4.5	16.0–18.0	4–6	2–4
8	2–3	4.5–5	18.0–20.0	4–6	2–4

[a]Amount of NPK per tree per application; use 6–6–6 or 8–3–9 or slow-release fertilizer material.
[b]Total amount year^{-1} per tree.
[c]Foliar sprays should contain zinc, manganese, boron, molybdenum and possibly magnesium.
[d]Chelated iron soil drenches may prevent iron deficiency; foliar iron sprays are not generally effective.

manure, 500 g urea, 1 kg triple superphosphate, 500 g muriate of potash and 500 g gypsum year^{-1} per tree for 20-year-old or older mature trees (Table 8.7).

In the Philippines, the recommended annual fertilizer schedule for plants of different age groups are: 200–300 g complete fertilizer (14–4–14) per tree at planting time, followed by 300–500 g complete fertilizer (14–14–14) plus 200–300 g urea (45–0–0) per tree for young trees, and 1.5–3 kg complete fertilizer (14–4–14) plus 200–300 g muriate of potash (0–0–60) for fully bearing trees (Yap, 1972; BPI, 1991). In Florida, for jackfruit trees growing in acid to neutral soils or in alkaline calcareous soils, 113 g fertilizer (6–6–6 with minor elements) and 30% N from organic sources per tree were used after planting when new growth begins. The fertilizer amount is increased gradually as the trees grow (Table 8.8). For mature trees, fertilizer should be applied two to three times from bloom to right after harvesting and pruning. Fe applications (in chelated forms) are most effective from May to September, and foliar sprays from April to September (Crane *et al.*, 2007).

Integration of organic sources of nutrients other than farmyard manure has been tried recently. At a density of 100 plants ha^{-1}, 8-year-old plants produced 76.3 kg fruit yield with application of N:P:K at a ratio of 500:300:300 g year^{-1} per tree. When similar plants were given only vermicompost at 4.00 kg year^{-1} per tree, they produced an average fruit yield of 56.3 kg per tree (i.e. about 25% lower than the inorganic fertilizers) (Laishram and Ghosh, 2018). Briones and Lina (2020) evaluated the effects of organic and inorganic fertilizers on the physicochemical and sensory acceptability of jackfruit and observed that fruit yield per tree remained comparable for both organic and inorganic treatments but use of vermicast at 75 kg per tree in the organic treatment produced the best postharvest quality of fruits.

Although partial substitution of chemical fertilizers by organic manure for improving soil quality and crop production has always been appreciated, there is limited information regarding the complex soil–microbial–microfauna interactions derived from different organic or inorganic fertilizer application rates in the rhizosphere of jackfruits. Recently, Su *et al.* (2021) reported the impact of various proportions of organic manure and chemical fertilizers on the soil properties and yield of jackfruit. They found that soil amended with more organic manure (70% organic manure plus 30% chemical fertilizer or 100% organic manure) presented a more positive correlation with beneficial nematodes and microorganisms, but a negative correlation with the plant parasite *Pratylenchus* spp. and the dominant fungal taxon Chytridiomycota.

There is scope for arbuscular mycorrhizal fungi symbiosis in jackfruit. Boberg and Dyberg (2002) found that jackfruit trees growing on an acid sulfate soil had arbuscular mycorrhizal fungi colonies and exhibited wide variations in the levels of arbuscular mycorrhizal fungi colonization between trees, despite the low pH and high soil nutrient levels. In Indonesia, a microbial biofertilizer consortium patented as BIOVAM-LIPI has been found to be very effective in controlling tree growth, particularly plant height and diameter (Lekatompessy *et al.*, 2016). Its utilization is environmentally friendly and it is expected to reduce dependence on chemical fertilizers. A judicious combination of mineral fertilizers with organic and biological sources of nutrients is better, as such integrated applications are not only complementary but also synergistic, as organic inputs have long-term beneficial effects on soil fertility beyond their nutrient content (Roy *et al.*, 2006).

8.4.3 Micronutrients

Application of micronutrients is recommended only when soil or leaf test data suggest corrective application. B deficiency is very common in jackfruit. Irregular or deformed fruit shape and size, high rates of dropping of female spikes and lower fruit set are important symptoms of B deficiency. The shape and size of the fruit is directly related to pollination and the growth-regulating processes of the trees. It has been found that B deficiency restricts the germination of pollen grains. If pollination and fertilization fail to occur within 3–6 days after anthesis, the fruits become twisted and shrivelled in shape (Bose, 1990).

Boron may stimulate the normal flow of hormones and enhance pollen grain and pollen-tube formation for proper fertilization. It also increases the stickiness of the stigma for receiving the pollen grains (Samaddar and Yadav, 1982; Tessy *et al.*, 1996). Mature fruits show the symptoms of developing spongy and corky tissues along with whitish pockets in the fruit mesocarp (Tarai *et al.*, 2018). Several studies (Konsaeng *et al.*, 2005; Farid *et al.*, 2007; Halder *et al.*, 2008; Singh *et al.*, 2011; Kumar and Kumar, 2016) have reported that application of B at 15 g per tree in 20-year-old trees significantly corrected fruit deformity in jackfruit. The lowest percentage of deformed fruits (24%) was recorded with this treatment, and significant improvements in the sex ratio (female:male) led to an increased number of normal fruits having greater fruit weight. Tarai *et al.* (2018) suggested that deficiency of B in the trees could be managed by spraying them with 1% solution of borax at monthly intervals from January to May or by adding borax at 250 g per tree along with fertilizer application. Mousavi and Motesharezadeh (2019) found that B contributes positively to and facilitates the assimilation and distribution of necessary nutrients for normal fruit development without deformations or external defects.

In some jackfruit orchards in North India, copper deficiency symptoms were also noticed and could be corrected by the foliar application of $CuSO_4$ at 0.1 g l^{-1} water (Agarwala *et al.*, 1991). In the case of Fe deficiency in young trees, foliar sprays of $FeSO_4$ applied every 2 weeks reduced the deficiency symptoms significantly but not completely, and increased the plant growth rate. This effect began to occur after about 6 weeks of application of the $FeSO_4$ (Grange and Kriesing, 2000). The rates for chelated Fe for mature trees ranges from 20 to 100 g per tree one to two times

Table 8.9. General recommended doses of micronutrient fertilizer. From Dwivedi *et al.* (2015).

Micronutrient	Material and doses for application	
	Soil application ha^{-1}	Foliar spray on plant
Zinc	Zinc sulfate: 25–50 kg[a]	0.5% zinc sulfate + 0.25% lime
Iron	Ferrous sulfate: 25–50 kg[a]	1–2% ferrous sulfate + 0.5–1% lime
Copper	Copper sulfate: 10 kg	0.1% copper sulfate + 0.05% lime
Manganese	Manganese sulfate: 10 kg	1% manganese sulfate + 0.25% lime
Boron	Borax: 10 kg	0.1–0.2% borax

[a]25 kg ha^{-1} for light soils and 50 kg ha^{-1} for heavy soils.

a year. Zn is widely deficient in calcareous soils of North Bihar (India) and in orchards established in coarse-textured sandy loam or loamy soils. Here, an application of $ZnSO_4$ of 62.5 kg ha^{-1} once every 2–3 years is suggested. Zn deficiency can also be corrected by spraying 0.5% $ZnSO_4$ two to three times at 15-day intervals. A list of suitable chemicals that are generally used in micronutrient deficiencies is given in Table 8.9.

8.4.4 Fertilizer application

Timing of application

The timing of nutrient applications is very important for improving the efficiency of nutrient use by the trees. It is determined and affected by a number of factors including climate, rainfall, soil type, and the physiological and developmental state of the trees. In determining the quantity of fertilizers to supply, the age of the tree should be considered as well as what is desired to be achieved. If the tree is young and the owner wants it to grow vegetatively as rapidly as possible, then extra food is called for. In contrast, for mature bearing trees where further architectural growth is restricted to a greater extent, balanced fertilization is required to have desired vegetative as well as reproductive growth. It is often suggested that orchardists use less N if vegetative growth is excessive and if flowering and fruit setting are not adequate.

During the early years of establishment, frequent application is desirable. Young plants of jackfruit grow very fast if soil water and nutrients are not limiting. In Florida, Crane *et al.* (2007) recommended six split applications of the recommended dose at bimonthly intervals up to 3 years of age, while from year 4 onwards, the number is reduced to two to three applications a year from bloom to right after harvesting and pruning. In India and Bangladesh, only two applications – one before the rainfall in June–July and the second after the rainfall in September–October – have been prescribed from the planting year until maturity of the plants and thereafter (Tandon, 1987; Mondal, 2000; Haq, 2006; Hossain and Haq, 2006; Hasan *et al.*, 2008; Hore, 2011; Priya Devi *et al.*, 2014).

The best time to fertilize is a short period of 1 or 2 weeks immediately after fruit harvest. This provides sufficient time to accentuate the lost internal energy and, in turn, helps trees to prepare well for the next fruiting. In India and Bangladesh, this period coincides with the rainy season, and danger of washing away the applied nutrients cannot be ignored. However, considering the tree's requirements, mineral fertilizers may be applied in the rainy season but strictly not on rainy days or soon after the rain. They should be applied at a time when it is possible to work the soil immediately after application. The gap between two rainfalls can provide a favourable situation. Usually, whole farmyard manure, a half dose of N, a full dose of P and a half dose of K are applied at this stage. The remaining half of the N and K are applied in October after the end of the rainy season. The trees perform better if the two N and K applications are applied in three equal splits, one soon after harvest, the second in October and the third immediately after fruit setting in March. For established bearing trees, two or three feeds will suffice, but for developing, pre-bearing young trees, more than two are practised in Brazil, Australia, USA and China. Due to the high labour cost, splitting of the dose into several applications during the early establishment phase is not followed in India, Nepal and Bangladesh.

Methods of application

There are a number of methods of applying nutrients: surface broadcasting, deep-root feeding (also called the punch bar system, auger holes or soil drill holes), liquid soil injection, injection or placement of nutrients in holes (implants) in tree trunks, foliar spraying and addition to irrigation water. Each application method serves a specific role depending on the orchard site, plant condition and availability of equipment. The method that is most efficient under a given set of conditions depends largely on the objectives (mode) of fertilization, the soil conditions and economics. Although many of the more sophisticated methods available today may work equally well, it is foolish to waste money on complicated techniques that give no better results than simple ones (Hartman *et al.*, 2000).

The area to be fertilized beneath the canopy is governed by the location of the absorptive roots. It was once thought that most tree roots occurred within the dripline but now it is known that tree roots commonly exist beyond the dripline. In surface applications, farmyard manure plus fertilizer mixture, or fertilizers alone, are spread over the soil surface and incorporated into the soil by light spade digging. It is generally recommended that 50–60% of the outward, circular area that extends from the trunk to the dripline be covered for spreading the fertilizers. Generally, a 60–90 cm circular band, 2–3 m away from the trunk is used to place the fertilizers. The use of surface application is appropriate where water penetration is not limited by compaction or excessive slopes. If the applied fertilizers are not worked into the soil properly, they (particularly P and K fertilizers) seldom reach the deep-feeding roots of jackfruit plants where they can be utilized.

The radial trench method is commonly practised to overcome this problem. Here, plant nutrients are placed nearer to the root system for better utilization. One or two trenches are dug in areas radiating out from the trunk to the dripline. When two trenches are dug, their dimensions are is 30 cm wide and 30 cm deep. If the trees are planted 10 m apart and the radius of their crown is 5 m, the two trenches are positioned one after another at a distance of 3 and 4 m away from the trunk. When a single trench is dug, it should be 60 cm wide and 30 cm deep and placed 3.5 or 4.0 m away from the trunk. The trenches are opened far enough away from the trunk to enhance absorption of nutrients and minimize damage caused by digging to the major thick roots near the trunk base.

The punch bar system and soil drill holes and application through soil or tree injection are not popular as they require costly machines so there is a significant difference in the economic aspects of application by these methods. Foliar application is generally practised for overcoming micronutrient deficiencies. For nutrients such as Fe, B, Zn, Cu and Mn, foliar application is far superior to soil application. As micronutrients are required in extremely small amounts and absorbed quickly through the leaves, their deficiencies can usually be quickly and safely corrected by foliar sprays. Sometimes N is also applied by foliar application. The cost of application can be reduced by combining nutrient and pesticide applications.

Fertilizers may also be applied by fertigation, i.e. along with irrigation water. This allows improved efficiency of irrigation and nutrient use as there is a substantial reduction in the quantity of water applied per irrigation or per season and the cost of fertilizer application is also saved. This method has been widely used in many fruit crops and can be adopted successfully for jackfruit orchards. With this method, the placement of mobile nutrients such as N can be regulated in the soil profile by controlling the amount of water applied per tree. Furthermore, the applied nutrients are readily available for rapid plant uptake. Due to a better soil nutrient and moisture regime, good aeration, and improved and steady plant physiological processes, there is a yield gain of 20–30% in different fruit crops with fertigation (Solaimalai *et al.*, 2005; Rupali *et al.*, 2014; Srivastava and Malhotra, 2017). Unfortunately, there is a dearth of scientific literature on fertigation in jackfruit, but its adoption should remarkably improve fertilizer and water-use efficiency of the orchard.

Impact of overfertilization

Fertilizers are sometimes used to excess, causing unnecessary expense, occasional damage to tender or sensitive plant parts, and potential water pollution. Before deciding whether to fertilize, the orchardist should understand the growth and bearing status of the trees and determine their nutritional needs. Fertilizers should be applied only if the crop responds positively. Excess application should be avoided as it may reduce yield and quality by overstimulating vegetative growth at the cost of flowering and fruit setting in jackfruit trees. It is also responsible for delaying fruit maturity. In addition, excess N results in luxuriant plant growth, which makes them more attractive to insects, particularly leaf-feeding insects, and more susceptible to disease.

The toxic effects of N on plant growth may be caused by a particular form of N, excessively high levels of N, or both. NH_4^+ supplied exogenously as a substrate presents a different situation from NH_4^+ generated internally from NO_3^- reduction and photorespiration. The release of gaseous ammonia (NH_3) from fertilizer bands can also result in NH_3 toxicity. Excess N causes many plants to grow vigorously and be dark green, but the plants also have some characteristic changes in developmental patterns such as elongation of the vegetative phase, overvigorous trees, reduction in fruiting, delayed maturity and increased succulence.

Excess N may also change the biochemistry of plants. There is little evidence to indicate that urea itself may be toxic; rather, urea toxicity may be inflicted indirectly through NH_4^+ and nitrite (NO_2^-) (Goyal and Huffaker, 1984). When too much N fertilizer (in excess of crop needs) is used, it acidifies the orchard soil. Acidity is caused by differential cation–anion uptake by the trees (Wallace, 2008). This acidity must be compensated for by liming in order to maintain an optimal pH range for better nutrient availability and microbial activity. However, the acidifying effect of N fertilizers may result in additional mobilization of nutrients such as Fe, Mn and Zn.

To be readily available for the plant, P must be in a suitable chemical form and physically accessible. Despite being present in soils in large quantities, in top soils (0–30 cm layer) as well as in the subsoil, most P is not readily available to plants. Although application of P fertilizers enhances availability, it also adds to the large pool of P already present in the soil that can be detrimental for growth of the plants. The subsoil not only provides P to the plant but also serves as a sink for excess fertilizer P above plant requirements (Siebers *et al.*, 2021). The rate of P removal is higher from jackfruit orchards compared with those of other fruits, and thus the supply seldom becomes in excess. However, excessive P held locked in the soil is a potential threat to water quality (Kulkarni and Goswami, 2019). With an excess K supply, Ca deficiency can be accentuated, so an adequate Ca supply is particularly important where a large application of K is made. Similarly, excess K application can also decrease the supply of Mg. Kulkarni and Goswami (2019) reported that excess use of fertilizers may also lead to heavy-metal accumulation, eutrophication and accumulation of PO_4^{3-} and NO_3^-. Policies and rules for encouraging effective and environment friendly use of fertilizers and its supply can improve the soil quality and prevent the adverse effects on humans (Roy *et al.*, 2006).

8.4.5 Nutrient recycling

Perennial woody fruit trees such as jackfruit generally require less fertilizer per year for their upkeep and maintenance than many annual field crops. There are two main reasons for this. First, a high level of fertilizer application is not required once trees acquire a large canopy consisting of numerous thick and thin branches and start bearing, as excess fertilizer can stimulate vegetative growth that favours low fruit setting and yield. Second, the trees are able to store a large quantity of nutrients in their trunk and recycle them across phenological growth stages to match the higher requirements for enhanced metabolic activities. Furthermore, the expansive and deep perennial root systems make them capable of absorbing nutrients from the deeper soil throughout the year. Nutrients stored or locked in the plant tissues are recycled and used when internal demand is high. Physiological adaptations such as mycorrhizae and bark storage proteins allow perennial fruit trees to take up and recycle a large proportion of P and

N (Boberg and Dyberg, 2002). Taken together, these physiological adaptations allow fruit growers to maintain productive plantings while using less fertilizer than most annual field crops. Despite this, application of fertilizers annually is vital to reduce yearly yield fluctuation and prevent them from becoming old and unproductive at an early stage.

Nutrient return via litter decay has long been recognized as the primary pathway for nutrient cycling in tree-based ecosystems. The role of leaf litter as a source of nutrients, however, is dependent on its decomposition rate, which in turn, controls the release of the tissue-held mineral ions. The bioelemental concentration in jackfruit litter is 0.91% N, 0.13% P, 0.6% K and 15.2% lignin (Kumar *et al.*, 1998; Kunhamu, 2011). Its role in improving soil fertility has been reported through increased earthworm activity in the soil beneath jackfruit trees (Isaac and Nair, 2006). Jackfruit litter decays much faster than that of some other trees, with almost 80% of the litter mass decomposing in 5 months. Allowing fallen leaves to rot in the orchard soil will greatly help reducing the dependence on added fertilizers.

8.5 Irrigation

Irrigation is one of the most important factors that has a marked influence on the health and bearing potential of jackfruit trees. Lack of water can lead to declining tree health, lower yield and poor-quality fruits. In fact, every aspect of tree growth is affected by water stress. The long-term viability of orchards therefore depends on the availability of water, particularly during the critical stages of the tree's production cycle. Water constitutes over 50% of the fresh weight of woody perennial trees (Kramer and Boyer, 1995) and a restriction in the supply of water may upset the physiology of the plant, resulting in various nutrient and metabolic disturbances leading to a decline in tree health, erratic flowering, lower yield and poor-quality fruits (Jackson *et al.*, 2000; Fernández and Cuevas, 2010). To ensure a higher profitability, it is important to safeguard jackfruit trees against drought at all phenological stages.

8.5.1 Determining irrigation requirements

Irrigation is a costly input. The aim of obtaining the maximum possible yield with a given amount of water requires thinking about three main driving variables: (i) how much water the root zone of the crop can hold; (ii) how much water infiltrates into the soil; and (iii) how much water the crop is using. The jackfruit tree possesses a huge crown where there is constant trade-off between vegetative and reproductive growth. Manipulation in cultural practices can help improve the balance between vegetative flushing and fruiting, thereby overcoming wide yearly fluctuations in yield. The total amount of water needed by the trees will vary with the climatic conditions under which they are grown, as well as with their age and size, and with the depth and type of soil and the rooting habits of the trees.

To determine the irrigation requirement of a crop, it is essential to know the rate of evapotranspiration (ET) of the concerned crop. Only a small fraction of the water (1%) is used in the plant's metabolism (photosynthesis and other biochemical reactions) and most of the absorbed water (as much as 99%) escapes as vapour in the process of transpiration from the plant canopy (Hillel, 1987). Thus, if we can estimate precisely the rate of water loss through transpiration from the plant plus evaporation from the soil, we would be able to predict the water requirement of the crop more accurately. ET is a key factor in irrigation scheduling of any crop. There are several approaches to determine the crop's irrigation requirement. The basic idea behind all approaches is to measure the level of water available in the soil (root zone) and/or in plant issues and replace the deficiency if it occurs. Obviously, when the measured value is below the desired threshold for optimum production, irrigation is applied.

Soil-based measurements

Soil characteristics play a major role in irrigation scheduling. Coarse-textured soils such as sands hold less crop-available water than fine-textured soils such as clays. Soil texture refers to the proportion of sand, silt and clay in a soil. If the texture of a field is in question, it can best be determined through particle size distribution tests by a soil test laboratory. The lowest moisture level at which plant roots are able to extract moisture from the soil (before they wilt beyond recovery) is called the 'permanent wilting point'. The highest moisture level that a soil can hold against gravity (when the excess has drained after saturation) is called field capacity. Plant-available soil water (ASW) is the water available between the permanent wilting point and the field capacity. The water content of a saturated soil (immediately after flooding and surface runoff) is about twice the field capacity and about four times the permanent wilting point. The relationships among saturation percentage, field capacity and permanent wilting point generally hold consistent for all soils, from clay loams to sandy soils (Jones, 2012).

Considerable differences occur in the amount of ASW present at various depths in non-uniform soils. This suggests that deep rooting can sometimes compensate for a limited supply of available water in the surface soil. An increase in the amount of organic matter, either from root growth or by addition of manure, usually increases the water-holding capacity of coarse-textured soils and improves the aeration of fine-textured soils. A perennial jackfruit tree, by virtue of its deep root system, absorbs water from deeper soil layers, so even if the water potential of the upper soil layer (0–30 cm) drops below –1.5 MPa, the trees may not show wilting symptoms immediately. However, if these conditions continue for some weeks, they will adversely affect the subsequent growth and fruiting of the trees. In traditional irrigation by surface flooding of the tree basin, ASW is allowed to deplete up to 50% in the top 30 cm soil profile before irrigation, whereas in drip irrigation systems, only a maximum of 10% depletion in ASW is permitted. A problem with soil-based techniques is that plants depending on their inherent qualities take considerable time to show a reaction to water stress in the root zone. Plant-based measurements discussed below reflect soil moisture stress early.

Plant-based measurements

Several plant-based measuring techniques have been examined for their potential use in irrigation scheduling. Detection of early signs of thirst through visual observations is the simplest of all and can prevent any significant reduction in yield by early action. However, in some instances it is too late when the visual symptoms become apparent. The tree most often suffers a hidden thirst. The young leaves are very sensitive and so are a good indicator of water stress. They begin to curl or become flaccid, or change their angle of orientation with a shortage of water in the soil. When this happens, irrigation is required; otherwise, the stress may become severe enough to reduce the yield. Nevertheless, many people consider this to be subjective and argue that various discrepancies such as differences in the orientation of the stem, leafage, inherent colour differences and lamina thickness leave sufficient scope for human bias.

Various other methods have been tried over the years to monitor water levels in plant tissues. These include: determination of leaf tissue hydration (relative water content) with punched discs or intact leaves, monitoring stem diameter, sap flow, measurements of leaf-water potential (LWP), determination of free proline or abscisic acid accumulation in the leaves, leaf extension rate, observations of stomatal aperture, stomatal conductance, CO_2 exchange and transpiration rate. Nearly all of these techniques require specialized instrumentation and trained personnel and are difficult to carry out on a routine basis. Hence, the majority of them are generally impractical in ordinary irrigation management and are more suitable for academic research on water relations.

Although an infrared thermometer is a handy tool for measuring canopy temperature for irrigation scheduling in many plants, determining water stress has proved problematic because of the effects of extreme atmospheric conditions, e.g. situations of very low vapour saturation deficit (Jones, 1999). Measurement of LWP has received by far the most attention of the plant-based measurement techniques in irrigation scheduling. Use of a pressure chamber (Scholander type) is the simplest and perhaps the most rapid practical field method for such estimations. For mango, a predawn LWP of –0.3 MPa has been observed as optimal for good growth under irrigated condition, while the values of non-irrigated trees tend to decrease over time (Larson *et al.*, 1989). As diurnal variation in LWP is very high, ranging from –0.2 to –2.0 MPa, the predawn LWP is preferred. The predawn values of LWP are almost stable and are free from the influences of weather. However, differences in the age of the sampled leaf and its position in the tree's canopy (height, interior or exterior of the canopy) induces considerable variation, even in the predawn LWP values.

Rodrigues *et al.* (2010) measured the predawn LWP (ΨL), stomatal conductance (g_s) and transpiration (E) in well-watered (normal) and water-stressed jackfruit trees. The value of ΨL in well-watered trees was –0.5 MPa while in stressed plants it was –0.6 MPa. ΨL values at midday of well-watered and stressed trees oscillated between –0.35 and –0.83 MPa on a typical summer day. Stomatal conductance and transpiration rate showed typical isohydric behaviour. The rates of g_s and E measured at 7.00 a.m. for well-watered trees were 0.8 mol m^{-2} s^{-1} and 1.2 mmol m^{-2} s^{-1}, respectively. In the afternoon at 2.00 pm, there was a sharp reduction in the rates of both g_s and E with respective values 0.1 mol m^{-2} s^{-1} and 0.2 mmol m^{-2} s^{-1}.

Because many plants have a marked capacity for passive and active adjustment of cell solute concentration in response to varying water status, LWP is not always a good measure to indicate the exact nature or degree of stress experienced by the plants (Jones, 2004). This highly dynamic nature of LWP (i.e. its dependence on many environmental as well as plant internal factors) poses great difficulties in reaching a definite conclusion. Earlier attempts to identify a threshold value of LWP at which irrigation should be applied in fruit crops have been challenged because there cannot be a single threshold for all the seasons in a year. Also, the threshold may not indicate the same degree of water stress in all agroclimatic situations. At present, there is little information available on the LWP of jackfruit under various degrees of soil-water stress, and no threshold value to mark the need for irrigation has been indicated. To judge the hardiness of the leaves to drought, Wu *et al.* (2008) observed jackfruit leaves under a scanning electron microscope and reported a thick horny layer of epidermis cells, two- or four-layer palisade tissue in the mesophyll and cell gaps in spongy tissues. There were many tannin parenchyma cells between the mesophyll and the vein, and the xylem of the vein was highly developed. These attributes imparted strong drought-tolerant and insect-resistant qualities to the jackfruit leaves.

Atmosphere-based approaches

Atmosphere-based approaches apply thermodynamic limits to the amount of water that can evaporate from a cropped surface under a particular set of environmental conditions. Early work (Doorenbos and Pruitt, 1977, 1979) was based on evaporation measured from a pan together with crop coefficients, which were then used to estimate ET. De Jager and van Zyl (1989) proposed a breakdown of the single crop coefficient into vegetation and soil subcomponents. Each of these was further broken down to account for canopy cover and plant- and soil-water status. de Jager and van Zyl, 1989 used evaporation from a reference crop to represent atmospheric evaporative demand. This theoretical formulation of crop evaporation provided the basis for weather-based water-balance models used in many irrigation-scheduling tools and/or integrated soil-, plant- and atmosphere-based crop irrigation models.

8.5.2 Scheduling irrigation

Irrigation scheduling is the term used to describe the procedure by which an irrigator determines the timing and quantity of water application. The objective is to provide the crop with the right amount of water, when the crop needs it, at the lowest cost and with the least impact on the environment. Accordingly, the two classical questions of irrigation scheduling are: (i) when to irrigate; and (ii) how much water to apply. In conventional low-frequency irrigation by surface flooding or sprinkle methods, the answer to the first question is generally when the reservoir of ASW in the root zone is nearly depleted. In practice, this implies when the crop is near the point of experiencing distress. In perennial woody tree crops, it is at around 50% of ASW. With high-frequency irrigation, as in drip irrigation, the grower does not need to worry about soil moisture depletion as such a situation is avoided entirely.

The question of when to irrigate can now be answered: as frequently as possible, even daily. For the second question of how much water to apply, the answer is enough to meet current evaporative demand and prevent salinization of the root zone (Hillel, 1997). However, the evaporative demand is a variable imposed by weather conditions, which fluctuate over time, and the real problem is how to translate pan evaporation into an estimate of the crop's ET, and in turn into actual irrigation requirements. Actual field measurements of specific crop responses to varying amounts of irrigation under local conditions should provide more reliable guidance on optimal irrigation amounts, but the methods of monitoring relevant weather variables (e.g. temperature, wind, atmospheric humidity, solar radiation, pan evaporation, rain), developed during the late 1980s, have also helped in calculating irrigation requirements quite precisely. Thus, the potential ET was determined and then by applying some functional equations or formulae, the crop ET could be estimated.

The modified Penman–Monteith method is one of the most widely used mathematical approaches for ET estimation. Several publications (Doorenbos and Pruitt, 1977, 1979; Hillel, 1987; Allen *et al.*, 1998) have provided detailed information on the equations used to calculate crop ET (ET_c) using reference ET (ET_o) and the crop factor (K_c). Hillel (1997) suggested a simple equation to estimate irrigation requirement as:

$$I = 0.3E_{pan}(1 + C) - R$$

where I is the irrigation requirements, E_{pan} is pan evaporation, C is the fractional ground cover of the crop, varying from 0 (when the crop is just sown or planted) to 1 (when the crop stand is full), and R is the rainfall that occurred since the previous irrigation.

According to Pereira *et al.* (2015) and Rallo *et al.* (2017), successful adoption of the United Nations Food and Agriculture Organization (FAO) Penman–Monteith reference ET (ET_o) and K_c approaches (proposed by Allen *et al.*, 1998) is owed primarily to the simplicity yet relatively high level of robustness of the procedures and to transferability and repeatability of the K_c method. Trajkovic and Kolakovic (2010) observed that the simplified pan-based equations to work out ET estimates can be a successful alternative to the FAO-56 Penman–Monteith equation for estimating reference ET using the following simple equation:

$$ET_o = ET_p \times K_c$$

where ET_p is potential ET. Nowadays, ET_o is preferred over ET_p for calculating crop water requirements as ET_o intends to avoid some ambiguities that existed in the definition of ET_p. By adopting a reference crop (grass), consistent crop coefficient selection has become easier and more practical, and thus, as a result, actual ET_c estimates in new areas are more reliable (Djaman *et al.*, 2018). However, it appears from the available literature that a large number of empirical or semi-empirical equations have been developed for assessing ET_o from numerous meteorological data. Records of such weather variables are often incomplete or not always available for many locations, which is a shortcoming of these complex models. Cobaner (2013) examined the efficiency of a wavelet regression (WR) model in estimating ET_o based on only Class A pan evaporation and compared the WR model estimates against those of the FAO-56 Penman–Monteith equation. The results showed that the WR model is capable of accurately predicting the ET_o values as a product of pan evaporation data. Following the atmosphere-based methodology (Hillel, 1997; Allen *et al.*, 1998), the water requirement was calculated for jackfruit plants of different age groups grown under the warm subtropical climate of North Bihar, India (Table 8.10). For 1-, 2- and 3-year-old young jackfruit trees, the water requirement for the warm summer months has been calculated as 24, 94 and 212 l $week^{-1}$ per tree, respectively. During winter (November–February), the requirement is reduced drastically. The highest water requirement during May–June for a mature tree of 10 years or more in age is estimated to be 1885 l $week^{-1}$ per tree.

Table 8.10. Water requirements of young jackfruit trees (l $week^{-1}$ per plant) in North Bihar (India). Data from P.K. Ray (unpublished).

Month	Year 1	Year 2	Year 3
March–June	24	94	212
July–October	15	61	123
November–February	4	18	40

Table 8.11. Ranges in available water capacity and intake rate for various soil textures. From Tan (1990).

	Available soil water (mm water cm^{-1} soil depth)		Water intake rate (mm h^{-1})	
Soil texture	Range	Average	Range	Average
Sand	0.5–0.8	0.65	12–20	16.0
Loamy sand	0.7–1.0	0.85	7–12	9.5
Sandy loam	0.9–1.2	1.05	7–12	9.5
Loam	1.3–1.7	1.50	7–12	9.5
Silt loam	1.4–1.7	1.55	4–7	5.5
Silty clay loam	1.5–2.0	1.75	4–7	5.5
Clay loam	1.5–1.8	1.62	4–7	5.5
Clay	1.5–1.7	1.60	2–5	3.5

8.5.3 Soil moisture monitoring

Here, the critical step is to choose the allowable depletion of soil water in the root zone. As indicated earlier, for surface and overhead irrigation of perennial fruit trees, the optimal soil moisture range is usually from field capacity (100% ASW) down to 50% of ASW, while for drip irrigation the optimal range is from field capacity down to 90% of ASW in the irrigation zone. When the allowable soil-water depletion reaches the predetermined level, irrigation is applied to bring ASW back to 100%. At this point, the 'balance' on the water budget is reset to the maximum ASW (or field capacity) in the effective rooting depth (30–45 cm). The soil moisture available to the crop is dependent on the storage of water in the soil, which varies according to the texture of the soil. The level of ASW and water absorption or intake rate for a range of soil textures is presented in Table 8.11.

Irrigation scheduling is traditionally based on either soil-water measurements, where the soil-water status is measured directly to determine the need for irrigation, or on soil-water balance calculations. The soil-water balance calculation is done using a simple equation:

$$P + I + W - \mathrm{ET} - R - D = \pm[\Delta S]$$

where P is precipitation, I is irrigation, W is the contribution from water table upwards, R is surface runoff, D is drainage and ΔS is soil water stored in the soil layer where the roots are active to supply water to the plant (Rana and Katerji, 2000). Units for all the terms are mm per unit time. Rana and Katerji (2000) further mentioned that, as it is often difficult to accurately measure all the terms of the above equation, it is often expressed in its simplified form:

$$P + I - \mathrm{ET} = \pm[\Delta S]$$

This method calculates ET as the residual term within a water-balance equation (Taylor and Gush, 2007). The water-balance technique has been found to be adequate over a wide range of conditions. Measuring the response of the plant to water deficits rather than sensing the soil moisture status directly has also been used in practice (Jones, 1990).

8.5.4 Recommendations for irrigation

Recommendations for adoption of irrigation in an area are generally based on the objectives. In young plantations, the main objective is to maximize growth so that trees can mature as fast as possible. To achieve this, the soil is kept moist all year round. Irrigation is applied at 12–15-day intervals in the summer months to avoid water stress. Mulch around the base of the tree will help keep the root zone moist for longer (Ghosh and Bera, 2006; Crane *et al.*, 2007). Older trees are hardier and less sensitive to drought. In many areas, jackfruit is often not irrigated and is cultivated as a rain-fed crop with low output. The crop is reported to be successful in tropical climates with an average annual rainfall of 1000–2400 mm, mean maximum temperature of hottest month as 32–35°C, mean minimum temperature of the coldest month as 16–20°C and mean annual temperature of 24–28°C

(Elevitch and Manner, 2006; Haq, 2006; Patil, 2013). The annual gross irrigation needs of jackfruit vary between 800 and 900 mm, and part of the water requirement is met by rainfall. In the drier and warmer climatic zones, this value may reach over 1000 mm (de Costa and Rozana, 2000). Trees at lower elevations (300–600 ft) in South Kona, Hawaii, were given 15 min of water daily from a half-gallon water bubbler. Jackfruit trees in another block were given 10 min of water three times a week from a 1 gallon h^{-1} drip emitter (Love and Paull, 2011).

The bearing trees are sensitive to drought and respond well to irrigation between flowering and fruiting (Haq, 2006; Pushpakumara, 2006; Patil, 2013; Choudury *et al.*, 2017; Kishore, 2018). The demand is distinctly higher at this stage because the tree experiences enhanced physiological and metabolic activities compared with the postharvest period. However, to determine the actual irrigation needs in an area, it would be necessary to carry out a water budget that takes into account soil-water data, the effective rainfall and the actual ET_c. Also, care needs to be taken when the postharvest period is devoid of rain for a long time and the soil water has dropped to less than 40% ASW. Excessive water stress should be avoided in order to avoid reductions in the fruit load of the return bloom, which might limit subsequent yields. de Costa and Rozana (2000) reported the effects of shade and water stress on growth of jackfruit and found that soil moisture ranging between 30% and 40% of ASW resulted in a marked reduction in yield and fruit quality. Watering in the summer months is essential to produce a good yield from the orchard. Requirements during the early years of establishment are low (Table 8.10), but from year 6 onwards, watering at the rate of 800 l $week^{-1}$ per tree is required. An adult tree needs about 1900 l $week^{-1}$ per tree during the warmest summer months under the warm, subtropical agroecological conditions of North Bihar (India). Water use for a mature tree varied from 10.5 to 40 mm $week^{-1}$ in different months of the year when the crop factor for winter months was 0.6 and for rest of the year was 0.75. (Table 8.12).

Water availability is crucial in flowering and fruit production (Fathin *et al.*, 2021). Both female and male inflorescences are more sensitive to water deficits than the surrounding leaves. Under a normal soil-water regime, flowering synchrony occurs well between sexes within the same tree. The sensitivity of fruit set to water deficits is also quite high, and excessive fruit drop may occur during the early stages of fruit growth if water stress is induced by lack of irrigation. Fruit size is affected by water deficits during the growth period, which lasts for about 4 months after fruit set. Water stress during fruit development in the later stages of fruit growth negatively affects the quality of mature fruit. When trees are frequently irrigated, the K, Ca and P concentrations in the fruit increase and this also helps in preventing deformities in fruit shape (Fathin *et al.*, 2021).

When irrigation is practised in a traditional way, enough water is applied in a circular bordered basin around the tree trunk to return the soil moisture to field capacity. This economizes the use of water as the entire orchard floor is not

Table 8.12. Quantity of water to be applied (l $week^{-1}$ per plant) to jackfruit trees of different age groups at different phenological stages in a North Bihar climate. Data from P.K. Ray (unpublished).

	Tree age				≥10 years	
Months (phenological stage)	4–5 years	6–7 years	8–9 years	≥10 years	Crop factor	Water use (mm $week^{-1}$)
November–February (flowering)	89	186	319	397	0.60	10.5
March–April (fruit development)	260	558	929	1157	0.75	24.5
May–June (fruit development)	424	885	1513	1885	0.75	40.0
July– October (vegetative flushing)	282	590	1008	1256	0.75	26.6

flooded. Alternatively, drip irrigation can be used and is preferred over basin flooding as it greatly economizes water use. So far, little information is available on drip irrigation of jackfruit. The young jackfruit tree is intolerant to poor drainage (Soepadmo, 1992) and roots fail to grow and proliferate under flooded conditions. The soil at the base of the plant should be raised to prevent water stagnation in areas with excessive rainfall and temporary water stagnation.

Ghosh and Bera (2006) found that mulching of the tree basin with dry grasses and watering at 30 l per plant at 30-day intervals starting from early November resulted in the greatest fruit retention (83%) and fruit yield (100 kg per plant versus 30 kg per plant in the control). The highest edible flake recovery (55.3%) was obtained when irrigation was applied with mulching at 15-day intervals. Mulching jackfruit trees by placing a 5–15 cm thick layer of bark, wood chips, fallen dried leaves, paddy straw or a similar mulch material at 20–30 cm away from the tree trunk not only retains soil moisture and reduces weed problems in the orchard but also improves the top soil (Crane *et al.*, 2007).

8.5.5 Deficit irrigation

The pressure on availability of water for varied multiple usage is increasing fast. Decreased water availability due to global warming along with uneven rainfall patterns has increased the requirements for optimum and efficient use of irrigation. This led to the concept of ‘regulated deficit irrigation’ (RDI) in the early 1980s. This practice aims to reduce water use by imposing water deficits at certain developmental stages of growth without causing a proportionate loss in yield (Fereres and Soriano, 2007). Nowadays, deficit irrigation (DI) is a common practice in many areas of the world, especially in dry regions, where it is more profitable for a farmer to maximize crop water productivity than to maximize the harvest per unit land (Geerts and Raes, 2009; Pérez-Pastor *et al.*, 2009).

RDI has been researched extensively with fruit crops over the last three decades, revealing the potential of RDI as a way of reducing water use in tree crops with little or no impact on yield and fruit quality (Fereres and Soriano, 2007; Iniesta *et al.*, 2009; Ruiz-Sanchez *et al.*, 2010). RDI is designed mainly to restrict water when the sensitivity of plant to water stress is the minimum. Three strategies, namely RDI, partial root-zone drying (PRD) and sustained deficit irrigation (SDI), are practised under the ambit of deficit irrigation (Ruiz-Sanchez *et al.*, 2010).

RDI consists of applying water in quantities below those necessary to satisfy ET_c during certain periods of the crop cycle when production and crop quality are hardly affected, and in the application of all the water needed during the rest of the cycle, especially at critical periods of the cycle when the yield and/or quality would be most affected by a lack of water. The PRD technique consists of keeping approximately half of the root system sufficiently watered while the other part is allowed to dry. This is done by irrigating one side or alternate rows of the trees. For this method, the two halves are watered alternately.

In several species, it has been observed that when part of the root system is in the process of drying, a root–shoot signalling mechanism is triggered; these signals molecules are transported via the xylem to the leaves, where they translate into partial stomatal closure (Dodd, 2005; Egea, 2008). Such stomatal regulation reduces water loss through transpiration with little effect on photosynthesis, thus increasing transpiration efficiency (i.e. the dry matter produced per unit of water transpired) (Stoll *et al.*, 2000). In some fruit species, PRD has given similar results to those obtained with RDI, with no clear advantage of the former when the volumes of water applied in both methods were the same (Egea *et al.*, 2010).

SDI is based on distributing the water deficit uniformly over the whole crop cycle to avoid the occurrence of severe water stress at any particular moment, which might have unfortunate results. When PRD (irrigation at 50% of ET_c, alternating irrigation from one-half of the tree to the other every 2–3 weeks) was compared with SDI in the same peach cultivar, no differences in yield between the two deficit irrigation treatments were found, although the yield was lower than in fully irrigated trees (Tapia *et al.*, 2009). PRD resulted in a lower reduction in root growth than that obtained with SDI with respect to the full irrigation treatment (Abrisqueta *et al.*, 2008).

As jackfruit is often grown without irrigation in many places, the RDI strategies can be a recommendable practice for irrigation of jackfruit trees when the available water is insufficient and irrigation is essential to optimize maximum yields.

8.5.6 Critical time of irrigation

The water need of the fruit tree is governed by the annual phenological and soil-water–plant relationship. Jackfruit trees require frequent irrigation during fruit development period (i.e. from fruit setting to fruit maturity). Irrigation during the rapid fruit growth phase can reduce fruit drop, increase final fruit size and reduce deformities in fruit shape (Falcão and Clement, 2001; Fathin *et al.*, 2021). In addition, an adequate water supply during early vegetative growth helps in developing a good framework for heavy bearing in later years. Irrigation prior to the start of flowering or at the fruit-bud differentiation stage is not recommended as it interferes with the ontogeny and morphology of the fruits. Once fruit setting is over, adequate soil moisture needs be maintained in the root zone through irrigation.

8.5.7 Methods of irrigation

Irrigation efficiency depends on the size and uniformity of fields, climatic conditions and the manner in which the irrigation is delivered, as water may be lost through deep percolation, runoff and evaporation. The duration of the water application depends on the amount of water to be applied and the water intake rate of the soil. Common irrigation methods include traditional flood systems (e.g. borders and level basins) and pressurized systems (e.g. overhead and undercanopy microsprinklers) and low-pressure drip systems. The physical properties of the soil will largely determine the suitability of the various irrigation systems in an orchard. Using a flood system, irrigation is applied two to four times per month, depending on weather and water availability, while sprinkler and microsprinkler systems apply irrigation two to three times a week; drip systems are used daily or on alternate days to apply water. Fruit trees irrigated with drip systems were larger and produced better yields than those irrigated by surface flooding, microsprays or any other irrigation system (Bryla *et al.*, 2005; Ray *et al.*, 2006; Kumar *et al.*, 2008; Pérez-Blanco *et al.*, 2019). These differences are attributed to irrigation efficiency, irrigation frequency and water placement within each system.

Orchards irrigated with inefficient systems will require more water to maintain a favourable water status and sustain good yields, than an orchard irrigated with an efficient system where little water is lost through percolation, evaporation, runoff or through the action of wind. Micro- and drip irrigation deliver irrigation more efficiently than sprinklers. The efficiency of microsprinklers is about 90–100%, while drippers are almost 100% efficient and can reduce water stress between irrigations (Tan, 1995). Microsprinklers have the potential to limit chemical and water loading to deep soil layers and groundwater, while sustaining fruit yield and quality on a variety of soils.

The placement of the water is another critical component of efficient irrigation. A concern with drip irrigation is that the limited wetted soil volumes will induce tree water deficits if the system stops functioning even for a short period in summer months. Human and economic factors such as labour and management skills, cost of installation, operation (energy costs) and maintenance of the system in relation to expected returns from the orchard greatly influence the system of irrigation adopted by a grower. There is no best system for a crop, different soils and varied orchard sizes. What we should seek is not the best system but an optimal one in a particular situation (Hillel, 1989).

8.5.8 Over-irrigation and waterlogging

With any irrigation system, applying too much water to the soil will result in runoff or loss of water below the root zone. Excess water can damage the crop due to poor aeration, rotting of finer roots, loss of soil nutrients through leaching, denitrification and increased soil erosion. Flooded young jackfruit seedlings are damaged severely by flood within 2–3 days, showing symptoms including epinasty, wilting, chlorosis, shoot browning and necrosis. Some seedlings died quickly and could not withstand the flood, even for a day (Tiradhanapiboon, 1989).

Over-irrigation slows root growth, increases the potential for iron chlorosis in alkaline soils, and leaches N, S and B out of the root zone leading to nutrient deficiencies. Leached NO_3^- can enter water bodies or become denitrified under anaerobic conditions within the soil profile. Such conditions can exist in pockets or compact zones within an otherwise aerated soil. Waterlogging causes loss of N through denitrification of NO_3^-. However, in upland soils, NO_3^- levels are often quite high, such that periodic waterlogging by heavy rainfall as in a monsoon-type climate or excess irrigation can result in a large loss.

As free-draining soils become waterlogged less readily, this risk is greatest on high clay, fine-textured soils. Under very high soil moisture and high temperature conditions, volatilization of NH_3 can be appreciable. Sometimes, an excessive water supply can induce excessive vegetative vigour that is not desired for good fruiting. Excess shoot vigour might compete for carbohydrates and other tree reserves available for bud break in November (Uddin *et al.*, 2005; Kishore, 2018). Irrigation scheduling should be such that in no case are plants are exposed to an excess supply for long. As this situation may also occur due to heavy rainfall, there must be a provision for efficient drainage of excess water from the field.

8.5.9 Effect of salinity and alkalinity

In saline and alkaline soils, exchangeable Na is present in very large amounts compared with exchangeable Ca and K. Na is not an essential plant nutrient. The performance of jackfruit trees under high levels of salinity is poor. However, jackfruit can be grown in soils showing electrical conductivity values of 4–6 dS m^{-1} after using suitable soil amendments (Ray, 1998). Soils with higher electrical conductivity values are not good and the yield reduction may be up to 60% or even more. This results from osmotic stress caused by the total soluble salt concentration and from toxicities or nutrient imbalances created when specific salts become excessive, or from reduction of water penetration into the soil profile due to poor structure (Hoffman and Shalhevet, 2007). Irrigation with saltwater should always be avoided, especially when the orchard soil possesses excess salt. Reclamation of saline soils is possible through the addition of Ca salts (calcium sulfate or gypsum) or through the use of acid-forming substances, such as S, iron sulfate, pyrite (FeS_2) and organic matter. To reclaim control of salinity, it is necessary to reduce the soluble salt concentration to acceptable limits (<4 dS m^{-1}) by leaching through excessive watering.

8.5.10 Effects of water stress

Water stress can cause severe damage to plants when inflicted during the early years of orchard establishment. Root proliferation will not be very extensive and the plants will start to show wilting of the shoot apex or drooping of leaves during midday hours if there is water stress during the summer months.

As the tree grows older, there is gradual increase in its ability to withstand the stress. When the tree is not bearing fruits, moderate water stress does not affect the morphological and physiological processes of the tree, but if irrigation is not provided to mature trees during the fruiting period, a decline in yield is almost certain. However, there also exists some potential to use drought stress as a horticultural tool (Syvertsen, 1985).

A limited soil-water supply or creating a drought-like situation by withholding irrigation or the water supply at the time of fruit-bud differentiation (November–December) helps in getting a good fruit set. Enhanced water-use efficiency (yield per unit of water) can be achieved due to greater diversion of the sink to the limbs and improved biochemical conditions for flowering fruit setting. Once the fruit setting is over, the soil moisture regime should be brought back to the normal level. Prolonging the stress may cause the trees to experience rapid dehydration and can be disastrous. Even a moderate drought stress at this stage can substantially increase the percentage of fruit drop. In jackfruit, irrigation of mature fruiting trees in India is virtually stopped from harvesting until fruit setting takes place in February–March. This period coincides with plenty of rainfall during the monsoon season (July–September) and comparatively mild weather from October onwards until flowering.

8.6 Water- and Fertilizer-Use Efficiency

The efficiency of nutrient or water application is measured as the increase in yield obtained per unit of applied nutrient or water. Many aspects of crop management influence the actual yield level and the response to the applied nutrients and water in the field. In relation to water supply, nutrient-use efficiency will improve appreciably if regulated or measured irrigation is provided. This minimizes fertilizer losses (leaching or denitrification) from the effective root zone of the trees. The nutrient-use efficiency can also be improved by ensuring that lack of water does not retard crop growth or nutrient uptake appreciably at any stage. Excess water can be a cause of nutrient losses, and insufficient water at a critical stage can also limit growth and yield. It is important that all other production inputs and management factors are adequate. The best way to achieve higher nutrient- and water-use efficiency will depend on enhancement of the soil fertility status and maintaining a proper soil-water regime in a given production system.

The various DI (deficit irrigation) strategies are now widely recognized water-saving irrigation methods in fruit trees. Cheng *et al.* (2021) studied different DI regimes in seven crops and observed that yields under DI regimes were remarkably lower than those under traditional full irrigation, while the corresponding water-use efficiency was significantly higher. However, they inferred that it is critical to weigh up the cost of yield penalty, the economic benefit of saving water and the cost of implementing an irrigation method before choosing an effective water-saving irrigation method.

Addition of organic matter into the soil and applying mulch on the surface greatly help in improving the water-use efficiency (Crane *et al.*, 2007). Splitting the fertilizer dose into two to three applications improves their utilization efficiency. Whatever the level of fertilizer used, the factor most often limiting production is adequacy and time of water supply. Timing of irrigation, particularly at the most critical stages of crop growth, influences nutrient-use efficiency considerably. The peak time of nutrient requirement must be supplemented with an adequate water supply. In jackfruit, the reproductive phase has been found to be most critically affected by moisture deficiency. Matching mineral nutrition and irrigation with the tree's phenological demand will lead to a more efficient cultural practice for improved productivity. Excessive irrigation, resulting in the passage of water through the soil profile through deep percolation, will carry with it soluble nutrients, particularly NO_3^-, SO_4^{2-} and B. Thus, the extent of nutrient losses must be considered when determining nutrient application rates in areas with higher rainfall during the monsoon season. The adoption of drip or undercanopy microsprinklers and fertigation improves the water- and nutrient-use efficiencies of the orchards significantly.

8.7 Conclusion

In this chapter, efforts have been made to discuss the various aspects of fertilization and irrigation scheduling for higher productivity and fruit quality of jackfruit through better nutrient- and water-use efficiency. In spite of its high food value and market potential, jackfruit is one of the most poorly researched crops and there is a dearth of scientific information about its cultural practices. The literature available on fertilization and plant water use is incomplete and often unsubstantiated.

Detailed studies are required on plant water relations of this crop for fine-tuning of fertilizer and irrigation rates and their application timings. Such studies should take into consideration the nutrient availability in the plant body, residual nutrients in the soil from the applications, local soil and climatic conditions and the long-term impact of these inputs on soil and environmental health.

The below-ground root dynamics of the jackfruit have not been studied to any degree. Such information is of vital significance for a better understanding of the tree interactions with increases and decreases in water supply and nutrient availability.

In Indonesia, a microbial biofertilizer consortium found to be highly effective in controlling tree growth has been patented (Lekatompessy *et al.*, 2016). Development of such a microbial consortium exploiting the native and natural microbial synergisms in other production areas would not only help in enriching the soil in a value-added form but would also be a popular eco-friendly method of managing soil fertility constraints occurring within the rhizosphere.

As soil moisture and mineral nutrient uptake are interrelated, there is a need to obtain accurate and more reliable information on water- and nutrient-use efficiency. There is a lack of definitive information on crop ET values and the separate contribution of transpiration and soil evaporation to these estimations. Estimation of crop water use from weather data and potential or reference ET is not very accurate. More direct observations of transpiration rate are required. Integrating the soil and plant-based measurements for water-balance calculations will provide better opportunities for selecting the critical stages for irrigation.

Plant growth and fruiting of jackfruit under different DI strategies (RDI, PRD and SDI) need to be examined thoroughly as they have been proven to be highly effective in reducing water use in other fruit tree crops with little or no impact on yield and fruit quality.

Large-scale adoption of improved fertilization and irrigation technologies will open up new prospects in the cultural management of jackfruit orchards, allowing a reduction of input without negative influence on yield, fruit quality, the environment and human health.

References

Abrisqueta, J.M., Mounzer, O., Álvarez, S., Conejero, W., García-Orellana, Y. *et al.* (2008) Root dynamics of peach trees submitted to partial rootzone drying and continuous deficit irrigation. *Agricultural Water Management* 95(8), 959–967. DOI: 10.1016/j.agwat.2008.03.003.

Acedo, A.L. (1992) *Jackfruit Biology, Production, Use and Philippine Research.* Monograph no.1, Forestry/ Fuelwood Research and Development (F/FRED) Project. Winrock International Institute for Agricultural Development, North Little Rock, Arkansas.

Agarwala, S.C., Nautiyal, B.D., Shaama, C.P. and Chatterjee, C. (1991) Studies on copper deficiency in mango, guava and jackfruit. *Indian Journal of Horticulture* 48, 192–200.

Agbangba, E.C., Olodo, G.P., Dagbenonbakin, G.D., Kindomihou, V., Akpo, L.E. *et al.* (2011) Preliminary DRIS model parameterization to access pineapple variety Perola nutrient status in Benin (West Africa). *African Journal Of Agricultural Research* 6(27), 5841–5847. DOI: 10.5897/AJAR11.889.

Allen, R.G., Pereira, L.S., Raes, D. and Smith, M. (1998) *Crop Evapotranspiration – Guidelines for Computing Crop Water Requirements.* FAO Irrigation and Drainage Paper no.56. Food and Agriculture Organization, Rome.

Almeida, M.M.B., de Sousa, P.H.M., Fonseca, M.L., Magalhaes, C.E.C., Lopes, M. de F.G. *et al.* (2009) Evaluation of macro and micro-mineral content in tropical fruits cultivated in the northeast of brazil. *Cienciae-Tecnologia de Alimentos* 29, 581–586. DOI: 10.1590/S0101-20612009000300020.

Amadi, J.A., Ihemeje, A. and Afam-Anene, O.C. (2018) Nutrient and phytochemical composition of jackfruit (*Artocarpus heterophyllus*) pulp, seeds and leaves. *International Journal of Innovative Food, Nutrition and Sustainable Agriculture* 6, 27–32.

Anon (2011) *Methods Manual. Soil Testing in India.* Department of Agriculture and Cooperation, Ministry of Agriculture, Government of India, New Delhi.

Atkinson, D. (1978) The use of soil resources in high density planting systems. *Acta Horticulturae* 65, 79–90. DOI: 10.17660/ActaHortic.1978.65.12.

Atkinson, D. (1980) The distribution and effectiveness of the roots of tree crops. *Horticulture Review* 2, 424–490.

Atkinson, D. (1983) The growth, activity and distribution of the fruit tree root system. *Plant and Soil* 71(1–3), 23–35. DOI: 10.1007/BF02182638.

Atkinson, D. and Wilson, S.A. (1980) The growth and distribution of fruit tree roots: some consequences for nutrient uptake. In: Atkinson, D., Jackson, J.E., Sharples, R.O. and Waller, W.M. (eds) *Mineral Nutrition of Fruit Trees.* Butterworths, London, pp. 137–150.

Atkinson, D., Naylor, D. and Coldrick, G.A. (1976) The effect of tree spacing on the apple root system. *Horticulture Research* 16, 89–105.

Bangroo, S.A., Bhat, M.I., Ali, T., Aziz, M.A., Bhat, M.A. and Wani, M.A. (2010) Diagnosis and recommendation Integrated System (DRIS) – a review. *International Journal of Current Research* 10, 84–97.

Boberg, J. and Dyberg, A. (2002) *The Effect of Fertiliser Regimes on Mycorrhizal Colonisation of Mangosteen Growing on an Acid Sulphate Soil in Southern Vietnam*. Minor Field Studies No.194. International Office, Swedish University of Agricultural Sciences, Uppsala, Sweden.

Bose, T.K. (1990) Jackfruit. In: *Fruits of India: Tropical and Subtropical*. Noya Prakash, Kolkata, India, pp. 487–495.

Bowling, D.J.F. (1980) Uptake and transport of nutrients. In: Atkinson, D., Jackson, J.E., Sharples, R.O. and Waller, W.M. (eds) *Mineral Nutrition of Fruit Trees*. Butterworths, London, pp. 151–160.

Briones, R.L. and Lina, D.P. (2020) Effects of organic and inorganic fertilizer application on the postharvest quality of jackfruit (*Artocarpus heterophyllus* Lam.) var EVIARC Sweet. *Mindanao Journal of Science and Technology* 18, 259–276.

Bryla, D.R., Dickson, E., Shenk, R., Johnson, R.S., Crisosto, C.H. *et al*. (2005) Influence of irrigation and scheduling on patterns of soil and tree water status and its relation to yield and fruit quality in peach. *HortScience* 40(7), 2118–2124. DOI: 10.21273/HORTSCI.40.7.2118.

Bureau of Plant Industry (1991) *Jackfruit (Nangka)*. BPI, Department of Agriculture, Manila, Philippines.

Cheng, M., Wang, H., Fan, J., Zhang, S., Liao, Z. *et al*. (2021) A global meta-analysis of yield and water use efficiency of crops, vegetables and fruits under full, deficit and alternate partial root-zone irrigation. *Agricultural Water Management* 248, 106771. DOI: 10.1016/j.agwat.2021.106771.

Choudury, A.G., Bhutia, S.O., Hasan, M.A. and Das, B.C. (2017) Fruiting behaviour of different jackfruit genotypes. *Trends in Bioscience* 10, 2847–2848.

Cobaner, M. (2013) Reference evapotranspiration based on class A pan evaporation via wavelet regression technique. *Irrigation Science* 31(2), 119–134. DOI: 10.1007/s00271-011-0297-x.

Coronel, R.E. (1983) *Promising Fruits of the Philippines*. College of Agriculture, University of the Philippines Los Banôs, Laguna.

Coutts, M.P. and Philipson, J.J. (1980) Mineral nutrition and tree root growth. In: Atkinson, D., Jackson, J.E., Sharples, R.O. and Waller, W.M. (eds) *Mineral Nutrition of Fruit Trees*. Butterworths, London, pp. 122–149.

Crane, J.H., Balerdi, C.F. and Maguire, I. (2007) *Jackfruit Growing in the Florida Home Landscape*. Document no. HS882. IFAS Extension, University of Florida Bulletin, Gainesville, Florida.

de Costa, W.A.J.M. and Rozana, M.F. (2000) Effects of shades and water stress on growth and related physiological parameters of the seedlings of five forest tree species. *Journal of the National Science Foundation of Sri Lanka* 28(1), 43. DOI: 10.4038/jnsfsr.v28i1.2934.

de Jager, J.M. and van Zyl, W.H. (1989) Atmospheric evaporative demand and evaporation coefficient concepts. *Water South Africa* 15, 103–110.

Djaman, K., Lombard, K., Koudahe, K., Allen, S. and O'Neill, M. (2018) Variability of the ratio of alfalfa to grass reference evapotranspiration under semiarid climate. *Irrigation and Drainage Systems Engineering* 7, 1–6.

Dodd, I.C. (2005) Root-to-shoot signalling: assessing the roles of 'up' in the up and down world of long-distance signalling in planta. *Plant and Soil* 274(1–2), 251–270. DOI: 10.1007/s11104-004-0966-0.

Doorenbos, J. and Pruitt, W.O. (1977) *Crop Water Requirements*. FAO Irrigation and Drainage Paper no.24. United Nations Food and Agriculture Organization, Rome.

Doorenbos, J. and Pruitt, W.O. (1979) *Yield Response to Water*. FAO Irrigation and Drainage Paper no.33. United Nations Food and Agriculture Organization, Rome.

Dwivedi, B.S., Sharma, G.D., Amule, F.C., Tagore, G.S., Dwivedi, A.K. *et al*. (2015) *Laboratory Manual on Soil and Plant Analysis*. Department of Soil Science and Agricultural Chemistry, Jawaharlal Nehru Krishi Vishwa Vidyalaya, Jabalpur, India.

Egea, G. (2008) Characterization and modelling of the agronomic and physiologic responses of almond trees to water deficit. PhD thesis, Universidad Politécnica de Cartagena, Cartagena, Spain.

Egea, G., Nortes, P.A., González-Real, M.M., Baille, A. and Domingo, R. (2010) Agronomic response and water productivity of almond trees under contrasted deficit irrigation regimes. *Agricultural Water Management* 97(1), 171–181. DOI: 10.1016/j.agwat.2009.09.006.

Elevitch, C.R. and Manner, R.I. (2006) *Artocarpus heterophyllus* (jackfruit), *Moraceae* (mulberry family). Species profiles for pacific island agroforestry, permanent agriculture resources. Available at: https://fdocuments.net/document/artocarpus-heterophyllus-jackfruit.html?page=1 (accessed 13 July 2022).

Falcão, M.D.A. and Clement, C.R. (2001) Phenology and yield of breadfruit (*Artocarpus altilis*) and jackfruit (*Artocarpus heterophyllus*) in Central Amazonia. *Acta Amazonica* 31, 179–191. DOI: 10.1590/1809-43922001312191.

Farid, A.T.M., Halder, N.K. and Shahjahan, M. (2007) Effect of boron for correcting the deformed shape and size in jackfruit. *Indian Journal of Horticulture* 64, 144–149.

Fathin, A.N., Astuti, D., Winarni, W.W. and Ratnaningrum, Y.W.N. (2021) Flowering and fruiting phenology of jackfruit (*Artocarpus heterophyllus* Lam.) from Sumatra landraces in *ex situ* conservation area in Karangmojo, Yogyakarta. *Earth and Environmental Science* 914(1), 012052. DOI: 10.1088/1755-1315/914/1/012052.

Fereres, E. and Soriano, M.A. (2007) Deficit irrigation for reducing agricultural water use. *Journal of Experimental Botany* 58(2), 147–159. DOI: 10.1093/jxb/erl165.

Fernández, J.E. and Cuevas, M.V. (2010) Irrigation scheduling from stem diameter variations: a review. *Agricultural and Forest Meteorology* 150(2), 135–151. DOI: 10.1016/j.agrformet.2009.11.006.

Fernández-Escobar, R., Sánchez-Zamora, M.A., García-Novelo, J.M. and Molina-Soria, C. (2015) Nutrient removal from olive trees by fruit yield and pruning. *HortScience* 50(3), 474–478. DOI: 10.21273/HORTSCI.50.3.474.

Geerts, S. and Raes, D. (2009) Deficit irrigation as an on-farm strategy to maximize crop water productivity in dry areas. *Agricultural Water Management* 96(9), 1275–1284. DOI: 10.1016/j.agwat.2009.04.009.

Ghosh, S.N. and Bera, B. (2006) Effect of watering and mulching on yield and fruit quality of jackfruit grown in laterite soil. In: *Proceedings of the National Symposium on Production, Utilization and Export of Underutilized Fruits with Commercial Potentialities, 22–24 November 2006, Kalyani, West Bengal, India*. Bidhan Chandra Krishi Viswavidyalaya, Mohanpur, India, pp. 150–153.

Goswami, C. and Chacrabati, R. (2016) Jackfruit (*Artocarpus heterophylus*). In: Simmonds, M.S.J. and Preedy, V.R. (eds) *Nutritional Composition of Fruit Cultivars*. Academic Press, London, pp. 317–335.

Goyal, S.S. and Huffaker, R.C. (1984) Nitrogen toxicity in plants. In: Hauck, R.D. (ed.) *Nitrogen in Crop Production*. American Society of Agronomy, pp. 97–118, Madison, Wisconsin, pp. 97–118.

Grange, I. and Kriesing, C. (2000) Effects of foliar sprays on iron deficient jackfruit (*Artocarpus integrus* Merr.) growing in calcareous soils. *Warasan Din Lae Pui* 22, 43–51.

Halder, N.K., Farid, A.T.M. and Siddiky, M.A. (2008) Effect of boron for correcting the deformed shape and size of jackfruit. *Journal of Agriculture and Rural Development (Gazipur)* 6, 37–42.

Hamido, S.A. and Morgan, K.T. (2020) Effect of various irrigation rates on growth and root development of young citrus trees in high-density planting. *Plants* 9(11), E1462. DOI: 10.3390/plants9111462.

Haq, N. (2006) *Jackfruit,* (Artocarpus heterophyllus). Southampton Centre for Underutilised Crops, University of Southampton, Southampton, UK.

Hartman, J.R., Pirone, T.P. and Sail, M.A. (2000) *Pirone's Tree Maintenance*, 7th edn. Oxford University Press, New York.

Hasan, M.K., Ahmed, M.M. and Miah, M.G. (2008) Agro-economic performance of jackfruit-pineapple agroforestry system in Madhupur tract. *Journal of Agriculture & Rural Development* 6(1), 147–156. DOI: 10.3329/jard.v6i1.1672.

Hillel, D. (1987) *The Efficient Use of Water in Irrigation: Principles and Practices for Improving Irrigation in Arid and Semi-Arid Regions*. Technical Paper No.64. World Bank, Washington, DC.

Hillel, D. (1989) Adaptation of modern irrigation methods to research priorities of developing countries. In: *Technological and Institutional Innovation in Irrigation. Proceedings of a Workshop held at the World Bank, 5–7 April, 1988*, World Bank, Washington, DC, pp. 88–93.

Hillel, D. (1997) *Small-Scale Irrigation for Arid Zones: Principles and Options*. United Nations Food and Agriculture Organization, Rome.

Hoffman, G.J. and Shalhevet, J. (2007) Controlling salinity. In: Hoffman, G.J., Evans, R.G., Jensen, M.E., Martin, D.L. and Elliott, R.L. (eds) *Design and Operation of Farm Irrigation Systems*, 2nd edn. American Society of Agricultural and Biological Engineers, St Joseph, Michigan, pp. 160–207.

Hore, J.K. (2011) Agronomy, soils and climate requirements of jack tree. In: Valavi, S.G., Peter, K.V. and Thottappilly, G. (eds) *The Jackfruit*. Studium Press LLC, Houston, Texas, pp. 141–156.

Hossain, A. and Haq, N. (2006) *Jackfruit,* Artocarpus heterophyllus. *Field Manual for Extension Workers and Farmers*. Southampton Centre for Underutilised Crops, University of Southampton, Southampton, UK.

Hundal, H.S., Singh, D. and Brar, J.S. (2005) Diagnosis and recommendation integrated system for monitoring nutrient status of mango trees in submountainous area of Punjab, India. *Communications in Soil Science and Plant Analysis* 36(15–16), 2085–2099. DOI: 10.1080/00103620500194460.

Hussain, M.S. (1998) Red soils of Bangladesh: concept, distribution, utilization and future prospects. In: Sehgal, J., Blum, W.E. and Gajbhiye, K.S. (eds) *Red and Lateritic Soils, Vol. 2. Red and Lateritic Soils of the World*. A.A. Balkema, Rotterdam, Netherlands, pp. 35–44.

Iniesta, F., Testi, L., Orgaz, F. and Villalobos, F.J. (2009) The effects of regulated and continuous deficit irrigation on the water use, growth and yield of olive trees. *European Journal of Agronomy* 30, 258–265. DOI: 10.1016/j.eja.2008.12.004.

Isaac, S.R. and Nair, M.A. (2006) Decomposition of wild jack (*Artocarpus hirsutus* Lamk.) leaf litter under subcanopy and open conditions. *Journal of Tropical Agriculture* 42, 29–32.

Jackson, R.B., Anderson, L.J. and Pockman, W.T. (2000) Measuring water availability and uptake in ecosystem studies. In: Sala, O.E., Jackson, R.B., Mooney, H.A. and Howarth, R.W. (eds) *Methods in Ecosystem Science*. Springer, New York, New York, pp. 199–214.

Jamaludheen, V., Kumar, B.M., Wahid, P.A. and Kamalam, N.V. (1997) Root distribution pattern of the wild jack tree (*Artocarpus hirsutus* Lam.) as studied by ^{32}P soil injection method. *Agroforestry Systems* 35(3), 329–336. DOI: 10.1007/BF00044462.

Janick, J. and Paull, R.E. (2008) Jackfruit. In: *The Encyclopaedia of Fruit and Nuts*. CAB International, Wallingford, UK, pp. 481–485. DOI: 10.1079/9780851996387.0000.

Jones, H.G. (1990) Plant water relations and implications for irrigation scheduling. *Acta Horticulturae* 278, 67–76. DOI: 10.17660/ActaHortic.1990.278.4.

Jones, H.G. (1999) Use of infrared thermometry for estimation of stomatal conductance as a possible aid to irrigation scheduling. *Agricultural and Forest Meteorology* 95(3), 139–149. DOI: 10.1016/S0168-1923(99)00030-1.

Jones, H.G. (2004) Irrigation scheduling: advantages and pitfalls of plant-based methods. *Journal of Experimental Botany* 55(407), 2427–2436. DOI: 10.1093/jxb/erh213.

Jones, J.B. Jr (2012) *Plant Nutrition and Soil Fertility Manual*. CRC Press, Boca Raton, Florida. DOI: 10.1201/b11577.

Kishore, K. (2018) Phenological growth stages of jackfruit (*Artocarpus heterophyllus*) according to the extended BBCH scale. *Annals of Applied Biology* 172(3), 366–374. DOI: 10.1111/aab.12427.

Konsaeng, S., Dell, B. and Rerkasem, B. (2005) A survey of woody tropical species for boron retranslocation. *Plant Production Science* 8(3), 338–341. DOI: 10.1626/pps.8.338.

Kramer, P.J. and Boyer, J.S. (1995) *Water Relations of Plants and Soils*. Academic Press, Waltham, Massachusetts.

Kulkarni, S. and Goswami, A. (2019) Effect of excess fertilizers and nutrients: a review on impact on plants and human population. In: *Proceedings of International Conference on Sustainable Computing in Science, Technology and Management (SUSCOM), 26–28 February 2019, Amity University, Rajasthan, Jaipur, India*. p. 6. DOI: 10.2139/ssrn.3358171.

Kumar, B.M., Georgo, S.J., Jamaludheen, V. and Suresh, T.K. (1998) Comparison of biomass production, tree allometry and nutrient use efficiency of multipurpose trees grown under three age series in Kerala, India. *Forest Ecology and Management* 112, 145–163.

Kumar, R. and Kumar, V. (2016) Physiological disorders in perennial woody tropical and subtropical fruit crops – a review. *Indian Journal of Agricultural Sciences* 86, 703–717.

Kumar, M.D., Turral, H., Sharma, B., Amarasinghe, U. and Singh, O.P. (2008) Water saving and yield enhancing micro irrigation technologies in India: when and where can they become best bet technologies. In: Kumar, M.D. (ed.) *Managing Water in the Face of Growing Scarcity, Inequity and Declining Returns: Exploring Fresh Approaches. Proceedings of the 7th Annual Partners Meet, IWMI TATA Water Policy Research Program, ICRISAT, 2–4 April 2008, Patancheru, Hyderabad, India*, Vol. 1. International Water Management Institute (IWMI), South Asia Sub Regional Office, Hyderabad, India, pp. 1–36.

Kunhamu, T.K. (2011) Jack and agroforestry. In: George, T., Peter, K.V. and George, S.V. (eds) *The Jackfruit*. Studium Press LLC, Houston, Texas, pp. 177–189.

Laishram, M. and Ghosh, S.N. (2018) Nutrient management in jackfruit (*Artocarpus heterophyllus* Lam.) under rainfed condition. *Journal of Horticultural Sciences* 13(1), 97–102. DOI: 10.24154/JHS.2018.v13I01.011.

Larson, K.D., Schaffer, B. and Davies, F.S. (1989) Flooding, carbon assimilation and growth of mango trees. In: *Program and Abstracts: 86th Annual Meeting of the American Society for Horticultural Science, 29 July–3 August 1989, Tulsa, Oklahoma*, American Society for Horticultural Science, p. 126.

Lekatompessy, S.J.R., Nurjanah, L. and Sukiman, H. (2016) Application of biofertilizers BIOVAM-LIPI to promote plant growth of jackfruit. In: *Proceedings of 6th International Symposium for Sustainable Humanosphere, 15–16 November 2016, IPB International Convention Center, Bogor, Indonesia*, pp. 230–235.

Love, K. and Paull, R.E. (2011) *Jackfruit.* Fruit, Nut and Beverage Crops, Folder No.19. University of Hawaii, Honolulu, Hawaii.

Mellado-Vázquez, A., Salazar-García, S., Goenaga, R. and López-Jiménez, A. (2019) Survey of fruit nutrient removal by mango (*Mangifera indica* L.) cultivars for the export market in various producing regions of Mexico. *Terra Latinoamericana* 37(4), 437. DOI: 10.28940/terra.v37i4.528.

Mondal, F.M. (2000) *Production and Storage of Fruits (in Bangla)*. Mrs Afia Mondal, BAU campus, Mymensingh, Bangladesh.

Mourão Filho, F.D.A.A. (2004) DRIS: concepts and applications on nutritional diagnosis in fruit crops. *Scientia Agricola* 61(5), 550–560. DOI: 10.1590/S0103-90162004000500015.

Mousavi, S.M. and Motesharezadeh, B. (2019) Boron deficiency in fruit crops. In: Srivastava, A.K. and Hu, C. (eds) *Fruit Crops: Diagnosis and Management of Nutrient Constraints*. Elsevier, Amsterdam, Netherlands, pp. 191–209.

Mukherjee, S.K., Rao, D.P., Mukherjee, S., Ray, M. and Mohanty, B.B. (1983) Manurial treatment at planting on growth and development of jackfruit (*Artocarpus heterophyllus* Lam). *Indian Journal of Horticulture* 40, 150–154.

Naik, S.K. and Bhatt, B.P. (2017) Diagnostic leaf nutrient norms and identification of yield-limiting nutrients of mango in eastern plateau and hill region of India. *Communications in Soil Science and Plant Analysis* 48, 1574–1583. DOI: 10.1080/00103624.2017.1374396.

Palmer, J.W. and Dryden, G. (2006) Fruit mineral removal rates from New Zealand apple (*Malus domestica*) orchards in the Nelson region. *New Zealand Journal of Crop and Horticultural Science* 34(1), 27–32. DOI: 10.1080/01140671.2006.9514384.

Pandey, R.M., Palaniappan, R. and Raja, M.E. (1998) Suitability of red and lateritic soil landscapes for fruit crops. In: Sehgal, J., Blum, W.E. and Gajbhiye, K.S. (eds) *Red and Lateritic Soils, Vol.1, Managing Red and Lateritic Soils for Sustainable Agriculture*. A.A. Balkema, Rotterdam, Netherlands, pp. 191–201.

Patil, P. (2013) Jackfruit. In: Dhillon, W.S. (ed.) *Fruit Production in India*. Narendra Publishing House, New Delhi, pp. 287–301.

Pereira, L.S., Allen, R.G., Smith, M. and Raes, D. (2015) Crop evapotranspiration estimation with FAO56: past and future. *Agricultural Water Management* 147, 4–20. DOI: 10.1016/j.agwat.2014.07.031.

Pérez-Blanco, C.D., Hrast-Essenfelder, A. and Perry, C. (2019) Irrigation technology and water conservation: from panaceas to actual solutions. In: *Chapman Conference on the Quest for Sustainability of Heavily Stressed Aquifers at Regional to Global Scales, 21–24 October 2019, Valencia, Spain*, AGU, Washington, DC, pp. 1–44.

Pérez-Pastor, A., Domingo, R., Torrecillas, A. and Ruiz-Sánchez, M.C. (2009) Response of apricot trees to deficit irrigation strategies. *Irrigation Science* 27(3), 231–242. DOI: 10.1007/s00271-008-0136-x.

Priya Devi, S., Talaulikar, S., Gupta, M.J., Thangam, M. and Singh, N.P. (2014) *A Guide on Jack Fruit Cultivation and Value Addition*. Technical Bulletin No.41. Indian Council of Agricultural Research (ICAR), Research Complex, Goa.

Pushpakumara, D.K.N.G. (2006) Floral and fruit morphology and phenology of *Artocarpus heterophyllus* (Moraceae). *Sri Lankan Journal of Agricultural Sciences* 43, 83–106.

Raghupathi, H.B., Reddy, Y.T.N., Reju, M.K. and Bhargava, B.S. (2004) Diagnosis of nutrient imbalance in mango by DRIS and PCA approaches. *Journal of Plant Nutrition* 27(7), 1131–1148. DOI: 10.1081/PLN-120038540.

Raghupathi, H.B., Ganeshamurthy, A.N. and Ravishankar, H. (2014) Comparison of DRIS ratio norms of selected fruit crops. *Indian Journal of Horticulture* 71, 168–175.

Rallo, G., González-Altozano, P., Manzano-Juárez, J. and Provenzano, G. (2017) Using field measurements and FAO-56 model to assess the eco-physiological response of citrus orchards under regulated deficit irrigation. *Agricultural Water Management* 180, 136–147. DOI: 10.1016/j.agwat.2016.11.011.

Rana, G. and Katerji, N. (2000) Measurement and estimation of actual evapotranspiration in the field under Mediterranean climate: a review. *European Journal of Agronomy* 13(2–3), 125–153. DOI: 10.1016/S1161-0301(00)00070-8.

Ranasinghe, R.A.S.N., Maduwanthi, S.D.T. and Marapana, R.A.U.J. (2019) Nutritional and health benefits of jackfruit (*Artocarpus heterophyllus* Lam.): a review. *International Journal of Food Science* 2019, 4327183. DOI: 10.1155/2019/4327183.

Ray, P.K. (1998) Orchard management. In: Bose, T.K., Mitra, S.K., Farooqui, A.A. and Sadhu, M.K. (eds) *Tropical Horticulture, Vol. 1.* Naya Prokash, Kolkata, pp. 39–101.

Ray, P.K., Rani, R. and Singh, A.K. (2006) Response of young shahi litchi trees to drip irrigation. *Indian Journal of Horticulture* 63, 133–138.

Rodrigues, B.M., Souza, B.D., Nogueira, R.M. and Santos, M.G. (2010) Tolerance to water deficit in young trees of jackfruit and sugar apple. *Revista Ciência Agronômica* 41(2), 245–252. DOI: 10.1590/S1806-66902010000200011.

Roy, R.N., Finck, A., Blair, G.J. and Tandon, H.L.S. (2006) *Plant Nutrition for Food Security: A Guide for Integrated Nutrient Management*. Fertilizer and Plant Nutrition Bulletin No.16. United Nations Food and Agriculture Organization, Rome.

Ruiz-Sanchez, M.C., Domingo, R. and Castel, J.R. (2010) Deficit irrigation in fruit trees and vines in spain. *Spanish Journal of Agricultural Research* 8(Suppl. 2), 5. DOI: 10.5424/sjar/201008S2-1343.

Rupali, S., Sandeep, B. and Mahesh, C. (2014) Fertigation on fruit crops: a review. *Annals of Biology* 30, 48–53.

Saha, B.C., Sharad, A.D. and Barman, R. (2011) Everbearing jackfruit (*Artocarpus heterophyllus* Lam.). *Acta Horticulturae* 918, 785–790. DOI: 10.17660/ActaHortic.2011.918.103.

Salazar-García, S. and Lazcano-Ferrat, I. (2001) Identifying fruit mineral removal differences in four avocado cultivars. *Better Crops International* 15, 28–31.

Samaddar, H.N. and Yadav, P.S. (1982) Flowering, floral biology, pollination, fruit set and fruit growth of jack fruit. *Horticulture Journal* 1, 54–60.

Samaddar, H.N. (1985) Jackfruit. In: Bose, T.K. (ed.) *Fruits of India: Tropical and Subtropical*. Naya Prokash, Kolkata, India, pp. 487–497.

Savita, K., Ngangom, B., Thoithoi, R., Devi, M.T., Mishra, G. *et al*. (2016) Diagnosis and Recommendation Integrated System (DRIS) approach on nutritional diagnosis in fruit crops- a review. *Journal of Applied and Natural Science* 8(4), 2337–2345. DOI: 10.31018/jans.v8i4.1134.

Sharma, J., Shikhamany, S.D., Singh, R.K. and Raghupathi, H.B. (2005) Diagnosis of nutrient imbalance in Thompson seedless grape grafted on dog ridge rootstock by DRIS. *Communications in Soil Science and Plant Analysis* 36(19–20), 2823–2838. DOI: 10.1080/00103620500305991.

Sidhu, A.S. (2012) Jackfruit Improvement in the Asia-Pacific Region – A Status Report. Asia-Pacific Association of Agricultural Research Institutions (APAARI). Bangkok, Thailand, pp. 1–182.

Siebers, N., Wang, L., Funk, T., von Tucher, S., Merbach, I. *et al*. (2021) Subsoils—a sink for excess fertilizer P but a minor contribution to P plant nutrition: evidence from long-term fertilization trials. *Environmental Sciences Europe* 33(1), 60. DOI: 10.1186/s12302-021-00496-w.

Singh, S.R., Narayanaswamy, P., Banik, B.C., Shyamalamma, S. and Simon, L. (2011) Evaluation of cracking and non-cracking genotypes of jackfruit [*Artocarpus heterophyllus* Lam]. *Crop Research Hisar* 42, 157–162.

Soepadmo, E. (1992) *Artocarpus heterophyllus* Lamk. In: Verheij, E.W.M. and Coronel, R.E. (eds) *Plant Resources of Southeast Asia, No.2: Edible Fruits and Nuts*. PROSEA, Wageningen, Netherlands, pp. 86–91.

Solaimalai, A., Baskar, M., Sadasakthi, A. and Subburamu, K. (2005) Fertigation in high value crops – a review. *Agricultural Reviews* 26, 1–13.

Srivastava, A.K. and Singh, S. (2008) DRIS norms and their field validation in Nagpur mandarin. *Journal of Plant Nutrition* 31(6), 1091–1107. DOI: 10.1080/01904160802115359.

Srivastava, A.K. and Malhotra, S.K. (2017) Nutrient use efficiency in perennial fruit crops—a review. *Journal of Plant Nutrition* 40(13), 1928–1953. DOI: 10.1080/01904167.2016.1249798.

Stoll, M., Loveys, B. and Dry, P. (2000) Hormonal changes induced by partial rootzone drying of irrigated grapevine. *Journal of Experimental Botany* 51(350), 1627–1634. DOI: 10.1093/jexbot/51.350.1627.

Su, L., Bai, T., Qin, X., Yu, H., Wu, G. *et al*. (2021) Organic manure induced soil food web of microbes and nematodes drive soil organic matter under jackfruit planting. *Applied Soil Ecology* 166, 103994. DOI: 10.1016/j.apsoil.2021.103994.

Sukthumrong, A., Boonkerd, N., Khumlert, R., Feungchan, S., Luksanawimol, P. *et al*. (2000) Plant nutrient and distribution under different management in Nam Dok Mai. *Acta Horticulturae* 509, 307–314. DOI: 10.17660/ActaHortic.2000.509.35.

Sun, Y., Yang, J., Wang, H., Zu, C., Tan, L. *et al.* (2015a) Standardization of leaf sampling technique in jackfruit nutrient status diagnosis. *Agricultural Sciences* 06(2), 232–237. DOI: 10.4236/as.2015.62023.

Sun, Y., Yang, J., Wang, H., Zu, C., Wang, C. *et al.* (2015b) Diagnosis and recommendation integrated system norms for jackfruit (*Artocarpus heterophyllus* Lam.) nutrient status evaluation. *Journal of Agricultural Science* 7(11), 245–253. DOI: 10.5539/jas.v7n11p245.

Swami, S.B., Thakor, N.J., Haldankar, P.M. and Kalse, S.B. (2012) Jackfruit and its many functional components as related to human health: a review. *Comprehensive Reviews in Food Science and Food Safety* 11(0), 565–576. DOI: 10.1111/j.1541-4337.2012.00210.x.

Syvertsen, J.P. (1985) Integration of water stress in fruit trees. *HortScience* 20, 1039–1043.

Tan, C.S. (1990) Irrigation Scheduling for Fruit Crops, Ontario Agriculture Bulletin ISSN 1198-712X. pp. 1–14. Updated and republished 20 July 2022. Available at: https://www.ontario.ca/search/search-results?query=Agriculture%20fruit%20crops%20bulletins (accessed 27 September 2022).

Tan, C.S. (1995) Effect of drip and sprinkle irrigation on yield and quality of five tomato cultivars in southwestern Ontario. *Canadian Journal of Plant Science* 75(1), 225–230. DOI: 10.4141/cjps95-043.

Tandon, H.L.S. (1987) *Fertilizer Recommendations for Horticultural Crops in India – a Guidebook*. Fertilizer Development and Consultation Organisation, New Delhi, India.

Tapia, L.M., Larios, A., Abrisqueta, I., Mounzer, O., Vera, J. *et al.* (2009) Peach deficit irrigation. Fruit yield and water use efficiency analysis. *Revista Fitotecnia Mexicana* 33, 1–5.

Tarai, R.K., Dora, D.K., Swain, S.C. and Panda, A.K. (2018) Jackfruit. In: Rana, M.K. (ed.) *Vegetable Crops*. CRC Press, Boca Raton, Florida, pp. 1035–1046. DOI: 10.1201/9781315116204.

Taylor, N. and Gush, M. (2007) *The Water Use of Fruit Trees and Orchard Crops in South Africa: A Knowledge Review*. University of Pretoria, Pretoria, South Africa.

Tessy, J., Kumaran, K. and Joseph, T. (1996) Flowering, floral biology and pollination in hard and soft flaked types of jack (*Artocarpus heterophyllus* L). *Journal of Tropical Agriculture* 31, 6–9.

Therios, I.N. (2009) *Olives*. CAB International, Wallingford. DOI: 10.1079/9781845934583.0000.

Tiradhanapiboon, Y. (1989) Preliminary study on the flood tolerability in jackfruit. Kasetsart University, Bangkok: Faculty of Agriculture. Available at: https://agris.fao.org/agris-search/search.do?recordID=TH9320344 (accessed 13 July 2022).

Trajkovic, S. and Kolakovic, S. (2010) Comparison of simplified pan-based equations for estimating reference evapotranspiration. *Journal of Irrigation and Drainage Engineering* 136(2), 137–140. DOI: 10.1061/(ASCE)IR.1943-4774.0000133.

Uddin, M.M., Mannan, M.A., Ahammed, M.F. and Zaman, A.K.M. (2005) Effect of fertilizer application, hand pollination and fruit thinning on jackfruit production. *Journal of Subtropical Agricultural Research and Development* 3, 19–22.

Vozzo, J.A. (2002) *Tropical Tree Seed Manual*. Agriculture Handbook No.721. US Department of Agriculture, Washington, DC.

Wallace, A. (2008) Soil acidification from use of too much fertilizer. *Communications in Soil Science and Plant Analysis* 25(1–2), 87–92. DOI: 10.1080/00103629409369010.

Williamson, J.G. and Coston, D.C. (1990) Planting method and irrigation rate influence vegetative and reproductive growth of peach planted at high density. *Journal of the American Society for Horticultural Science* 115(2), 207–212. DOI: 10.21273/JASHS.115.2.207.

Wu, T., Feng, F. and Ye, C.-H. (2008) Studies on the anatomical structures of the leaf and flowers of jackfruit (*Artocarpus heterophyllus*). Guangxi Zhiwu/Guihaia 28, 746–749.

Xiuchong, Z., Guojian, L., Jianwu, Y., Shaoying, A. and Lixian, Y. (2001) Balanced fertilization on mango in Southern China. *Better Crops International* 15, 16–20.

Yap, A.R. (1972) Jackfruit (*Artocarpus heterophyllus* Lam., Moraceae). In: *Cultural Directions for Philippine Agricultural Crops, Vol.1, Fruits*. Bureau of Plant Industry, Department of Agriculture, Manila, Philippines, pp. 137–141.

9 Flowering, Fruit Set and Fruit Development

Kundan Kishore[1]*, Phebe Ding[2] and Ankita Sahu[3]
[1]Central Horticultural Experiment Station (ICAR-Indian Institute of Horticultural Research), Bhubaneswar, India; [2]Department of Crop Science, Faculty of Agriculture, Universiti Putra Malaysia, Selangor, Malaysia; [3]ICAR-Central Institute for Women in Agriculture, Bhubaneswar, India

9.1 Introduction

Jackfruit (*Artocarpus heterophyllus*) is one of the important tropical fruit crops blessed with high nutraceutical value. In recent years, the importance of jackfruit as a food has increased due to its multifarious uses. However, limited information is available on reproductive biology including phenology of jackfruit. Flowering in jackfruit is one of the intricate phenomena due to the differential expression of both the sex forms. Fruit set is the most critical event in ensuring reproductive success. However, it is regulated by endogenous and exogenous factors, and among exogenous factors climate and pollinators play a vital role. This chapter briefly describes floral biology, the sex ratio, flower synchronization, pollination, agamospermy and changes during fruit maturation. The information will help to increase our understanding of the reproductive phenology of jackfruit, which will in turn facilitate effective orchard management.

Reproductive phenology is the key event in ensuring the successful existence of plant species, as well as representing a critical dimension of plant functioning. The timing of reproductive events represents the interaction between plants and environments (Segrestin *et al.*, 2020). Reproductive phenology deals with the seasonal timing of recurrent biological events, which are intimately linked to the developmental phases of plant species. The timing of the switch between reproductive phases that occurs in concert with flowering is crucial for the reproductive success of plants (Caradonna *et al.*, 2014). Phenology has received much attention in the recent past as the developmental stages of plants are influenced in response to climate (Gillison, 2019). The BBCH scale provides a system of uniform numerical coding to define these stages (Hack *et al.*, 1992). This numerical scale is widely used for describing phenological stages of plants including fruit crops. A description of the reproductive phenological stages such as bud development, flowering, syncarp development and fruit maturation along with pollination and flower synchronization is essential to achieve higher yield and better fruit quality, as a number of agronomical practices (e.g. fruit drop management, irrigation, nutrient management, pest management) rely on recognition of specific phenological stages. Jackfruit (*Artocarpus heterophyllus* Lam.) is an important

*Corresponding author: kkhort12@gmail.com

DOI: 10.1079/9781800622319.0009

fruit crop of the tropics and subtropics of South-east Asia owing to its high food value and market potential. However, limited information is available about its reproductive phenology. This chapter describes the reproductive biology of jackfruit along with its phenological stages according to the BBCH scale.

9.2 Reproductive Biology

9.2.1 Floral biology

Jackfruit is a monoecious species bearing male and female inflorescences (spikes) separately on the same plant. Jackfruit has a unique character of producing specialized reproductive shoots (footstalks) from the main trunk and from older branches, thus exhibiting a cauliflory nature. Female spikes are produced only on footstalks, whereas male spikes are produced on footstalks as well as on young branches of the tree. In jackfruit, the male spikes appear before the female spikes. However, their ratios vary with tree age, germplasm and climatic conditions. The female spikes bear hundreds of small, sessile flowers with fleshy receptacles. Each flower is characterized by a perianth, stigma and unilocular ovary. The male spikes also bear hundreds of small sessile flowers with single stamens. Unlike the female spikes, the male spikes do not grow continuously and start decaying with the completion of pollen dehiscence. Jackfruit produces a multiple fruit (syncarp) with a green to yellow-brown exterior rind containing hexagona, bluntly conical carpel apices (spines). The perianth of the individual flowers becomes the fleshy pericarp and surrounds the seeds, each pericarp and seed being an individual fruit. The fleshy aril along with the seed is called the bulb, and varies in number, size, colour, texture, thickness and chemical properties. The fruit axis is the modified mature inflorescence axis and is somewhat dome-shaped, rigid and slightly fleshy (Pushpakumara, 2006). In India, ripe jackfruits are available mainly in June, while in Malaysia, the fruit can be found from June to August and from December to February.

Jackfruit produces specialized bearing and non-bearing shoots (footstalks) during the reproductive phase. The bearing footstalks produce both male and female spikes, whereas the non-bearing footstalks produce only male spikes (Fig. 9.1). The number of bearing and non-bearing footstalks varies with the germplasm. Bearing footstalks are more vigorous than non-bearing footstalks. Moreover, non-bearing footstalks appear earlier than bearing ones. The footstalks usually start producing reproductive buds 4–6 weeks after their initiation. Jackfruit is a heterogamous monoecious plant wherein male and female inflorescences (spikes) are borne separately on specialized reproductive shoots. However, male spikes are also borne on normal shoots. The two types of spike are distinguished on the basis of shape, size, colour and length of peduncle. Male spikes are elongated, light green to brown in colour and relatively smaller with long peduncle, while female spikes are dark green, larger in size, blunt in shape and supported by a short peduncle.

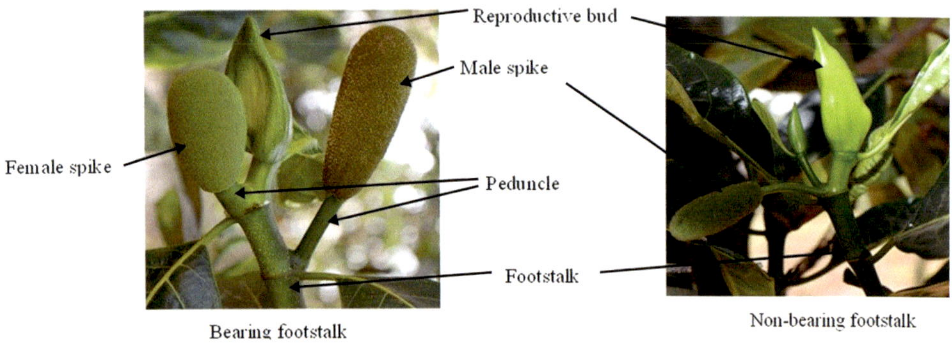

Fig. 9.1. Bearing and non-bearing footstalks of jackfruit. Photograph courtesy of Dr Kundan Kishore.

Fig. 9.2. Images of mature stamen and pollen, and of the growth stages of female spikes of jackfruit. (A) Fluorescence image of a stamen with its filament and anther. (B) Scanning electron microsope image showing the pollen sac splitting with exposure of the pollen grains (P). (C) During the early growth stage of the female inflorescence, the protruding stigma(s) can be found at the terminus of the perianth tip. (D) During the stigmatic receptive period, the tip of the stigma becomes bent. Photographs courtesy of Dr Phebe Ding.

The male spikes of jackfruit bear numerous, tiny, staminate flowers with small filament and adnate anthers (Fig. 9.2A). Usually, male spikes attain maturation 2 weeks after emergence, which is evident by pollen dehiscence. Pollen dehiscence (Fig. 9.2B) continues for a week, and is followed by gradual rotting of the spike and development of *Rhizopus* fungus. In jackfruit, the male cycle (production of male spikes) lasts 6–8 weeks; however, genotypic variation is observed. Anther dehiscence occurs mostly in the day time. However, more pollen dehiscence occurs in the afternoon (1.00–4.00 p.m.), which may be correlated with the increase in temperature. Pushpakumara (2006) reported that pollen dehiscence is greatly reduced during rainfall but during dry days pollen shedding is rapid. However, Mijin and Ding (2020) observed that anthesis in jackfruit germplasm started in the early morning (6.00 a.m.) and reached its maximum around 8.00 a.m., and there was no anthesis after 9.20 a.m.

In jackfruit, production of female spikes usually continues for 5–6 weeks; however, the duration may vary with genotype and growing conditions. The numerous, small, tightly packed, pistillate flowers are borne on fleshy receptacles (Fig. 9.2C). The flowers possess a perianth and a unilocular superior ovary with a conspicuous visible stigma at maturation (Fig. 9.2D).

However, the style is usually embedded. Female spikes develop progressively and complete their cycle of maturation within 15–20 days. The stigmas protrude out gradually during spike development. The stigmas are wet, fleshy, milky-white and spathulate (ligulate). During peak stigma receptivity, the whole spike is covered with slightly curved fleshy ligulate stigmas. The receptivity of the stigma continues for 5–7 days (Mijin and Ding, 2020).

9.2.2 Spatial variation in the occurrence of male and female spikes

Jackfruit is primarily cauliflorous in nature as the male and female spikes are borne directly from the main branches. However, male spikes are also borne axially on the terminal branches. There is a wide range of variation in the spatial arrangement of male and female spikes on a tree. Male spikes are borne primarily on the primary, secondary and tertiary branches, whereas female spikes are mainly borne on the trunk, primary branches and secondary branches (Fig. 9.3). In some genotypes, female spikes are also borne on tertiary and fourth-order branches. More than 75% of female spikes are located on the trunk and primary and secondary branches. In contrast, more than 75% of male spikes are borne on primary, secondary and tertiary branches. However, the intensity of occurrence of inflorescences varies with genotype and climatic conditions. Ullah and Haque (2008) observed a wide range of variation in the occurrence of male and female spikes in jackfruit tree.

9.2.3 Abnormalities in spikes and syncarps

Floral abnormalities are observed in jackfruit, but there is less occurrence of deformed fruits. Pushpakumara (2006) reported more than 10% inflorescence abnormalities in jackfruit cultivated in Sri Lanka. Abnormalities are associated with the occurrence of both sexual forms of the spike. Pistillate flowers may develop at the basal portion of abnormal male spikes (Fig. 9.4), whereas in abnormal female spikes, the terminal part is occupied by staminate flowers. In some cases, the basal pistillate flowers remain underdeveloped in female spikes. All the abnormal female spikes are aborted, whereas abnormal

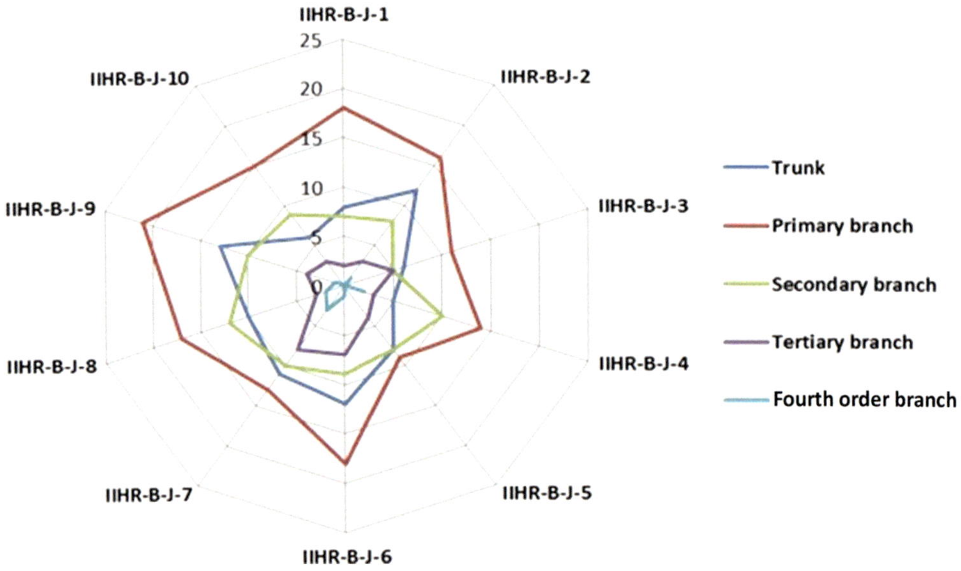

Fig. 9.3. Spatial variation of female spikes in jackfruit. The occurrence of female spikes on the trunk and branches for the jackfruit germplasm lines IIHR-B-J-1 to IIHR-B-J-10 are shown. Image courtesy of Dr Kundan Kishore.

Fig. 9.4. An abnormal male spike with basal pistillate flowers. Photograph courtesy of Dr Kundan Kishore.

male spikes remain on the tree until pollen dehiscence. Malformed, shrivelled and deformed syncarps were observed by Rymbai *et al.* (2021) in the north-eastern hill regions of India.

9.2.4 Sex ratio and flower synchronization

The relative number of male and female spikes is known as the sex ratio and varies with genotype, growing condition and time of flowering. In jackfruit, the number of male inflorescence spikes is invariably higher than the number of female spikes, i.e. the sex ratio is always more than 1, and hence the ratio is male biased. Usually, the sex ratio varies from 2 to 15. In some genotypes, fruit set is reduced (four to six fruits per tree) due to fewer female spikes, although the tree produces a significantly large number of male inflorescences. In contrast, some genotypes are highly productive as they bear higher numbers of female spikes, and in such trees the sex ratio is as low as 2. Pushpakumara (2006) reported that at the population level the mean inflorescence sex ratio was 4.5, but ranged from 0.8 to 19.2. Furthermore, he observed that in some genotypes, the number of female spikes was greater than that of the male spikes. In the eastern tropical climate of India, flowering in jackfruit continues for about 3 months (December–February), although the early phase is dominated by male spikes. In the later phase of flowering, the number of male and female spikes is substantially reduced (Fig. 9.5).

Synchrony of flowering is defined as the co-occurrence of pollen dehiscence and stigma receptivity in a tree or between trees. When flower synchronization occurs within a tree, geitonogamy is common. In contrast, synchronization between the trees facilitates allogamy. The synchrony of flowering can be assessed quantitatively in terms of the synchronization index (SI), which can be calculated with respect to both female and male phases as follows (Pushpakumara, 2006):

$$SI = N_f/N_m$$

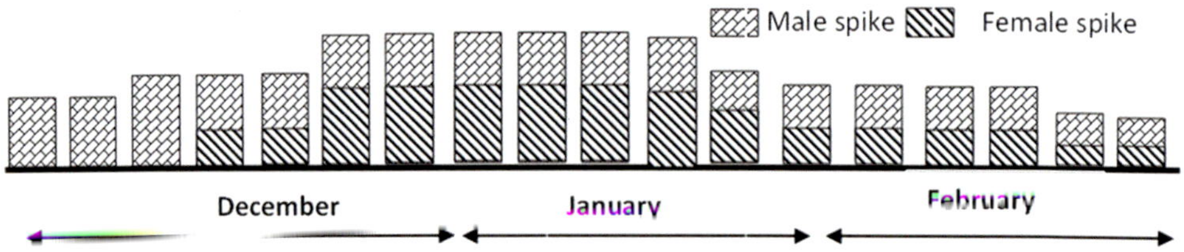

Fig. 9.5. Temporal variation in the male and female phase in jackfruit. Photograph courtesy of Dr Kundan Kishore.

where N_f is the number of days an individual produces female inflorescences synchronously and N_m is the number of days an individual produces pollen-releasing male inflorescences.

The value of SI may vary between 0 and 1. An SI value of 1 indicates perfect synchrony between the male and female phases, whereas 0 indicates no synchrony. A value between 0 and 1 indicates partial synchronization. The SI value may be assessed at the tree level or the population level.

9.3 Pollination

Pollination in jackfruit is a complex biological phenomenon owing to the spatial separation of both the form of sex and temporal variation in their maturation. Thus, although jackfruit bears male and female inflorescences on the same plant, both the forms of sex are spatially separated. Synchronization in the maturation of male and female inflorescences varies with genotype. As described above, in some genotypes, maturation is highly synchronized within the tree, which facilitates geitonogamy (pollination of pistillate flowers by staminate flowers of the same plant). In other genotypes, maturation is highly asynchronous and pollen dehiscence and stigma receptivity within a tree do not coincide. In such conditions, fruit set occurs through allogamy (cross-pollination). Lina and Protacio (2015) reported that fruit deformation in jackfruit occurs due to the unavailability of pollen for the receptive female spike or if the wind speed is too low; in such cases, assisted pollination may improve fruit set and syncarp development.

There has been debate about whether pollination in jackfruit is carried out by insects or wind. Moncur (1985) suggested that pollination in jackfruit was the outcome of an interaction between wind and insects, as pollen-laden male spikes attract insects through scent, but the pollen is subsequently transported to the female spikes by the wind. El-Sawa (1998) reported that jackfruit was primarily allogamous wherein pollination is carried out by wind-assisted insects involving various species of beetles. However, Pushpakumara (2006) considered jackfruit an autogamous fruit species as he found complete synchrony of the female flowering phase with the male phase within an individual tree, which facilitated geitonogamy. He also reported that fruit setting in isolated jackfruit trees was only possible through geitonogamy. The extended flowering duration and longer maturation period of male and female spikes in jackfruit facilitated geitonogamy and in turn reduced the level of allogamy. Pushpakumara (2011) stated that wind is the most important agent for jackfruit pollination, as they observed deposits of pollen grains on pollen traps on open-pollinated receptive female spikes. The involvement of insects in jackfruit pollination was ruled out by Pushpakumara (2011) as he did not observe the presence of insects on receptive female spikes. He detected the presence of insects only on rotting male spikes producing a distinct odour. He further observed poor pollen collection capacity and pollination efficiency of male spike-visiting insects. It is evident that a relatively greater number of male spikes, a high rate of pollen production, and substantially long periods of flowering and spike maturation make jackfruit suitable for wind pollination (Pushpakumara, 2011). In contrast, Gardner *et al.* (2018) presented evidence of entomophilous pollination in jackfruit and reported the involvement of the gall midge (*Clinidiplosis ultracrepidata*) in pollen transfer. They further reported that the gall midge is attracted equally towards mature male spikes and receptive female spikes, as both were involved in the release of volatile compounds. However, they did not rule out the possibility of wind pollination. It is evident that in jackfruit pollination is mediated by both wind and insects. The presence of small, inconspicuous flowers with a substantially large number of small pollen grains on the male spike and presence of several wet stigmas on the female spikes suggests anemophilous behaviour (Fenster *et al.*, 2004). However, the production of scent/odour during spike maturation attracts insects. Thus, a role for both wind and insects in carrying out pollination in jackfruit seems justified.

9.4 Agamospermy

Agamospermy is the phenomenon of embryo formation through asexual reproduction without gametogenesis (gamete formation) and the act of fertilization. In plants, the process of agamospermy occurs in three ways (Fig. 9.6). The first is the recurrent agamospermy wherein the formation of a diploid embryonic sac takes place from diploid nucellar cells (apospory). Formation of the embryonic sac may also take place from the unreduced megaspore mother cell (diplospory). The development of the embryo takes place without fertilization through the process of parthenogenesis. In the second case, the megaspore mother cell divides meiotically to form a haploid embryo sac, and in the absence of fertilization, sterile embryos are produced. This process is known as non-recurrent agamospermy. Lastly, adventive embryony is the process of embryo development that takes place via the nucellar cells (nucellar) or the integuments (integumental) of the ovule.

Agamospermy is quite common in gymnosperms; however, it has also been reported in jackfruit. Pushpakumara (2006) observed fruit set and seed set in jackfruits when the female spikes were covered with pollen-proof bags. Moreover, removal of stigma could not prevent fruit set. He observed more than 50% fruit set in bagged stigma, whereas in fully shaved female inflorescences, the fruit set was about 25%. He concluded that pollination is not a prerequisite for fruit set in jackfruit. The occurrence of agamospermy in jackfruit has also been substantiated with a report of below-ground syncarp development (Acedo, 1992). It is evident that fruit set in jackfruit is facilitated primarily by transfer of pollen by wind and insects, but in the absence of pollination, fruit set may also occur through agamospermy. It has also been observed that jackfruit exhibits facultative agamospermy by providing the scope for both sexual and asexual reproduction (Pushpakumara, 2006).

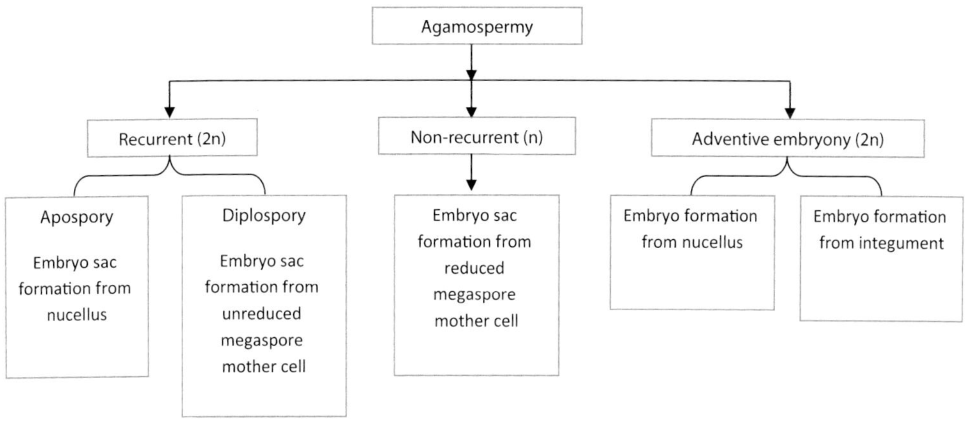

Fig. 9.6. Types of agamospermy in plants.

9.5 Fruit Set and Seed Set

Fruit set and seed set indicate successful fertilization and the start of ovary and ovule development, respectively. In jackfruit, fruit set is measured on the basis of the number of pistillate spikes and syncarps (fruits) and is expressed as a percentage. The number of fruits on a tree depends on the number of pistillate spikes and the sex ratio. The percentage of fruit set may be as high as 100%, which indicates that all female spikes have been successfully pollinated and transformed into syncarps. However, lack of pollination may also lead to dropping of the female spikes (Tessy *et al.*, 1996a, b). The final size and shape of the syncarp depends on the number of flowers pollinated and fertilized, as well as the pattern of pollination of flowers. Some varieties or genotypes of jackfruit produce cylindrical or spherical fruits without any deformation, which indicates pollination and fertilization of spikes from all sides. However, if pistillate spikes are partially pollinated, the syncarp becomes deformed with irregular shapes (APAARI, 2012). Pushpakumara (2006) observed irregular shapes of syncarps even under induced agamospermic condition.

In jackfruit, seed set is estimated on the basis of the number of seeds in each syncarp and the number of spines indicating the number of pistillate flowers. The number of pistillate flowers in a spike may vary from 1200 to 3500, whereas the number of seeds in a mature syncarp can range from 25 to 200. A wide range of variation in seed set is observed in jackfruit. Genotypes with small bulbs usually have a high seed set percentage (12%), whereas genotypes producing large bulbs exhibit a low seed set percentage (<2%). Pushpakumara (2006) also reported a low seed set under open-pollinated condition. The syncarp size (weight, length and diameter), bulb number and aril thickness of manually pollinated jackfruit was higher than naturally pollinated fruit (Mijin *et al.*, 2021). Low seed set in jackfruit might be due to the completion of resources among developing ovules. It also indicates that the strategy of plant is to ensure better development of syncarps by ensuring resources for developing fruits. Pushpakumara *et al.* (1997) stated that in *A. heterophyllus*, low fruit set was due to the result of competition rather than pollen limitation.

9.6 Phenological Growth Stages

9.6.1 Specialized reproductive shoots (footstalks)

Reproductive phenology in tree crops signifies the sequential changes in the reproductive organs in a calendar year. Over the last few decades, phenological studies gained the attention of researchers because of the sensitivity to seasonal and climatic change and the impact on crop production. Phenological events are intimately linked to specific developmental phases of plant species such as reproductive bud development, flowering, fruit development and fruit maturation. These stages are described by the BBCH numerical scale, which is a system for uniform coding of growth stages. The basic BBCH scale, represented by two digits, has primary and secondary scales, and each scale is subdivided into 10 (0–9) clearly recognizable and distinguishable developmental phases. The primary scale describes the principal stages associated with the major developmental cycle of the plant. The reproductive phenology of jackfruit is represented by five principal growth stages: development of specialized reproductive shoots (4), inflorescence development (5), flowering (6), fruit development (7) and fruit maturity (8). The secondary scale is a subsequent division of the principal stages into ten distinct developmental stages. In the extended BBCH scale, the frequency of occurrence of the growth stages is also considered (mesostage), and thus it becomes a three-digit scale. The mesostage is incorporated between the primary and secondary stages. Kishore (2018) has comprehensively described and defined the phenological growth stages of jackfruit; some of the important phenological growth stages of the development of non-bearing footstalk and bearing footstalk are described.

The development of specialized reproductive shoots is described by the code 4, with A and B defining the non-bearing and bearing footstalks, respectively (Fig. 9.7). Stage 411A indicates the early stage of non-bearing footstalk elongation. At 415A, male buds start initiating from the footstalk, and at 417, the first male spike appears. However, growth of the non-bearing footstalk continues. Stage 419A

Fig. 9.7. The phenological growth stages of bearing and non-bearing footstalks. Photographs courtesy of Dr Kundan Kishore.

indicates development of more male spikes on the non-bearing footstalk growth. In contrast, 411B describes elongation of the bearing footstalk. The start of the reproductive phase with the initiation of female and male buds is indicated by 415B. At 417B, the first female/male spike appears, but the footstalk growth continues. Stage 419B indicates the development of the female/male spike.

9.6.2 Reproductive buds

The buds of bearing footstalks enclose the spike of either form of sex as well as a developing bud, which, in turn, produces spikes along with a new bud. The phenological stages of male (5A) and female bud (5B) development are described separately for a better understanding (Fig. 9.8). Stage 511A describes elongation of male buds, whereas 513A is characterized by the start of male bud development and differentiation of the male spike. Stage 515A signifies the end of bud extension. The spathe (bract covering bud) starts separating, and the male spike is partially visible. At 517A stage, bud opening reaches the advanced stage. The male spike and new buds are clearly visible, and the spathe turns light yellow. Stage 519A indicates completion of male bud development. Similarly, 511B marks the elongation of female buds. Stage 513B describes the start of female bud development and differentiation of the female spike with a perceptible change in shape. Stage 515B signifies the end of bud extension. The spathe starts separating, the female spike is partially visible and the spathe starts to turn light yellow. At the 517B stage, the bud opening reaches the advanced stage. The female spike and fresh buds are clearly visible. Female bud development is completed at stage 519B.

9.6.3 Male and female spikes

The initial reproductive phase of jackfruit is dominated by the protandrous male phase. The male spikes are elongated, smooth and attached by a long and thin peduncle. The initial colour and size of the male spikes depends on their origin. Spikes borne on normal shoots are invariably small and light yellow, whereas those appearing on footstalks are large and light green. The male spikes are densely covered with numerous fertile and sterile staminate flowers. Each flower has a single stamen with a filament, and a bilocular anther. Pollen dehiscence signifies maturity of the male spike and the end of

Fig. 9.8. The phenological growth stages of reproductive buds of jackfruit. Photographs courtesy of Dr Kundan Kishore.

the male phase, which is followed by spike decay. This cycle is usually completed in 2 weeks.

The female spikes emerge later in the season on the bearing footstalks. They are stout, large and granular in appearance and attached by a stout peduncle. Numerous female flowers develop on the surface of the female spikes, which are borne on fleshy receptacles. The female flowers are characterized by tubular perianths, a ligulate style, a papillate fleshy stigma and a unilocular ovary. During the process of development, the stigmas emerge throughout the surface of the spike, signifying peak stigma receptivity, which is followed by fruit set and initiation of syncarp development. The female phase is completed in about 4 weeks.

According to the BBCH scale, inflorescence development is designated by the code 6, with the phenological stages of male inflorescences indicated by A and those of female inflorescences by B. Stage 611A indicates advanced male spike development characterized by the growth and differentiation of staminate flowers, while 613A indicates the end of spike development once it has attained full size (Fig. 9.9). Differentiation of the staminate flowers is complete at this stage and anthesis begins. Stage 615A is characterized by full bloom. Most of the male flowers open, followed by anther dehiscence. Stage 617A defines the start of spike fading as the spikes start to decay. Stage 619A is characterized by advanced spike decay, indicating completion of the male phase.

Stage 611B indicates advanced female spike development, characterized by spathe dehiscence and differentiation of pistillate flowers (Fig. 9.9). Stage 613B signifies the start of spike maturation and the emergence of stigmas. At stage 615B, milky-white stigmas appear on the female spikes, indicating peak stigma receptivity. Stage 617B defines the start of fruit set, as well as the fading of stigmas, and stage 619B indicates advanced fruit set, fading of stigmas, completion of the female phase and the start of syncarp development.

9.7 Syncarp Development and Maturation

The fruit of jackfruit is a syncarp (multiple fruit) composed of several fertilized ovaries (achenes). Each fertilized flower develops into a fruitlet (bulb) with a fleshy perianth (aril) enclosing a remnant of the ovary and a seed. In

Fig. 9.9. The phenological growth stages of male and female spikes. Photographs courtesy of Dr Kundan Kishore.

contrast, unfertilized pistillate flowers develop into tough, strap or string-like structures called 'rags' (perigones), which occur between the bulbs. Interestingly, the ovary does not become fleshy in *Artocarpus* spp. but develops into a dry pericarp, whereas the lower part of the perianth forms the edible fleshy pulp, which may be firm or soft in texture and yellow or pink in colour. The outer surface of the syncarp is covered with a stiff rind and spines, which are formed from the hardening of the middle and upper parts of the tubular perianth. The fleshy fruit axis is the modified inflorescence axis, which is usually non-edible (Haq, 2006). The shape and size of the syncarp and the shape, size, colour and number of bulbs (25–200) vary with different germplasms. In some cases, the syncarp is exceptionally large (>20 kg), whereas some fruits are small (<2 kg). In most cases, the fruit size varies between 3 and 8 kg. The bulb weight (10–40 g) varies significantly with different germplasms.

Fruit development follows a sigmoidal growth pattern with three distinct phases: (i) early slow growth (lag phase); (ii) rapid growth (log phase); and (iii) fruit maturation (diminishing growth phase) (Fig. 9.10). The lag phase continues for 2–3 weeks followed by rapid fruit development for 10–12 weeks. At maturation, the fruit follows a diminishing growth pattern that lasts for 5–6 weeks. The lag phase is characterized by slow development of the syncarp, perianth and seed, whereas the log phase is signified by rapid development of the syncarp and bulb. Maturation of the syncarp is characterized by physico-chemical changes in the bulb with respect to colour, firmness, sweetness and flavour. Mijin and Ding (2020) studied the anatomical changes during syncarp development and reported that the early phase of syncarp development (lag phase) is associated with cell division, whereas fruit development during the log phase is associated primarily with cell expansion. However, each stage of syncarp development is characterized by distinct cellular activities. They further observed that rapid development of the syncarp is correlated with strong sink activity exerted by the expanding cells.

The phenological growth of syncarp is divided into five distinguishable stages characterized by distinct physical changes (Fig. 9.11). According to the BBCH scale, code 7 defines fruit development, while code 8 is assigned to fruit maturation. Stage 711 indicates initial fruit growth and the syncarp attains

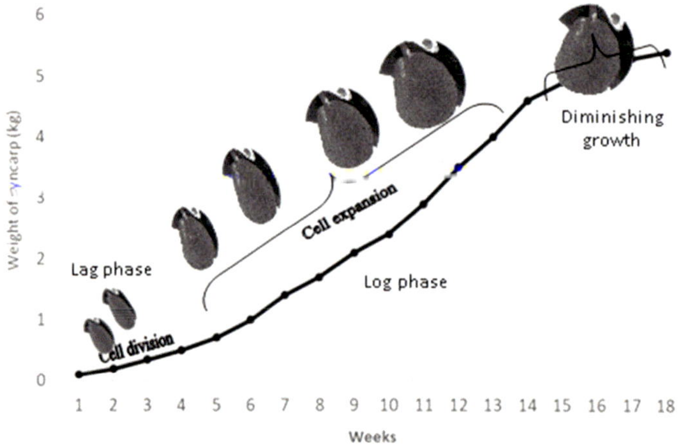

Fig. 9.10. The sigmoidal growth pattern of the jackfruit syncarp. Courtesy of Dr Kundan Kishore.

Fig. 9.11. The phenological stages of fruit during development and maturation. Photographs courtesy of Dr Kundan Kishore.

10% of its final size. The perianth and seed are less conspicuous, but development of the fruit axis continues. Stage 713 is signified by development of the syncarp (30% of final size). Fruitlets start developing along with fusion of the perianth. However, the spines are still soft. At stage 715, the fruit attains about 50% of its final size. Hardening of spines begins and development of the fruitlets continues. The fruits are harvested for culinary purposes at this stage. Stage 717 is marked by further increases in fruit size (70% of final size) with thickening of the latex and hardening of the spines. Growth of the perianth and perigone continues, and seed hardening begins. At stage 719, the fruits have almost attained their full size. However, perianth growth and hardening of the seed and perigones continues. Seed development is complete at this stage.

Fruit maturation begins after completion of fruit development, which can be judged by changes in syncarp colour, peduncle colour, perianth texture (soft and juicy), seed colour, fruit volume, cessation in latex secretion, bulb firmness, bulb sweetness and flavour. After completion of the fruit development process, maturation begins, which is completed in 5–6 weeks. The harvesting of fruits begins in June and continues up to July. However, the time of harvesting may vary with genotype. On the basis of distinguishable development, fruit maturation is divided into four phenological growth stages (Fig. 9.11). Stage 811 is characterized by the start of fruit maturation wherein the fruit colour turns light green or light yellow, flattening of spines is initiated, the interspaces between the spines increase, latex secretion is reduced, the aril changes to light yellow in colour and the seedcoat turns light brown. Stage 815 indicates advanced maturation, characterized by a change in peduncle colour (green to light yellow or light brown) and a perceptible change in the aril, which turns light yellow and fleshy. The seedcoat turns light brown, latex secretion ceases. At this stage, the fruits have attained climacteric maturation. Stage 817 indicates the perfect stage for harvesting in terms of flavour; they make a dull, hollow sound when tapped. The aril becomes fleshy and yellow, and seed maturation is completed. The overripe stage of jackfruit is indicated by stage 819. At this stage, the spines yield to light pressure, the textural consistency of the aril is lost and it becomes exceedingly soft, and in some cases, the seeds start to germinate inside the bulb.

9.8 Changes during Fruit Maturation and Ripening

In jackfruit, maturation of the fruit is associated with a sharp increase in the soluble solids fructose, glucose, fructose and β-carotene, and in flavonoid content and antioxidant activity. In contrast, bulb firmness, acid content (malic and oxalic acid) and lycopene content decline (Mijin *et al.*, 2021). Chemical changes during maturation are due to the enzyme-mediated rapid conversion of polysaccharide (starch), fatty acids and organic acids into sugar and volatile organic compounds. It has also been observed that the content of non-reducing sugars such as sucrose is greater than that of reducing sugars (glucose and fructose). The activity of sucrose synthase increases during the early stage of maturation, followed by a decline during the advanced stage. Sucrose is the major sugar species contributing to bulb sweetness, followed by fructose and glucose (Li *et al.*, 2017). The dominant acids in jackfruit are malic and citric acid. Malic acid is abundant at the early stage of ripening and marginally decreases thereafter (Ong *et al.*, 2006). Succinic and oxalic acid are found in trace amounts throughout the ripening process. A gradual reduction in the concentration of organic acids during ripening is largely due to the utilization of these compounds as respiratory substrates and synthesis of other biochemical compounds such as esters. There is a gradual reduction in the content of fatty acids during maturation, indicating that they may act as substrates for flavour compounds. The ripening process of jackfruit is characterized by a concomitant increase in flavour, which is associated with biosynthesis of volatile compounds. More than 35 organic compounds responsible for flavour have been detected in jackfruit (Maia *et al.*, 2004; Ong *et al.*, 2006). Methylbutyl acetate, propyl isovalerate, methyl butanol, isovaleric acid, methyl dodecanoate, benzaldehyde, phenylpropyl ester and others are important organic compounds associated with the flavour of jackfruit. Jackfruits can be divided into two groups on the basis of bulb texture. In one type, the bulb becomes soft and pulpy when ripe, while in the other, the flesh remains firm (*khaja*). The latter form of bulb is highly preferred and fetches a higher market price. In firm-type bulbs, the firmness is associated with reduced activity of polygalacturonase and pectin methylesterase, which in turn slow down cell-wall degradation. In contrast, in the soft form, polygalacturonase and pectin esterase activities increase by 12-fold and 40-fold, respectively, in ripe fruits (Rahman *et al.*, 1995). The chemical changes occurring during fruit maturation have been described in detail in Chapter 2 (this volume).

References

Acedo, A.L. (1992) *Jackfruit Biology, Production, Use and Philippines Research*. Monograph number 1. Forestry/Fuelwood Research and Development (F/FRED) project. Winrock International Institute for Agricultural Development, North Little Rock, Arkansas.

Asia-Pacific Association of Agricultural Research Institutions (2012) *Jackfruit Improvement in the Asia-Pacific Region – A Status Report*. APAARI, Bangkok.

Caradonna, P.J., Iler, A.M. and Inouye, D.W. (2014) Shifts in flowering phenology reshape a subalpine plant community. *Proceedings of the National Academy of Sciences of the United States of America* 111(13), 4916–4921. DOI: 10.1073/pnas.1323073111.

El-Sawa, S. (1998) Pollination and breeding of jackfruit (*Artocarpus heterophyllus* Lam.) in South Florida. MSc thesis, Florida International University, Miami, Florida.

Fenster, C.B., Armbruster, W.S., Wilson, P., Dudash, M.R. and Thomson, J.D. (2004) Pollination syndromes and floral specialization. *Annual Review of Ecology, Evolution, and Systematics* 35(1), 375–403. DOI: 10.1146/annurev.ecolsys.34.011802.132347.

Gardner, E.M., Gagné, R.J., Kendra, P.E., Montgomery, W.S., Raguso, R.A. *et al.* (2018) A flower in fruit's clothing: pollination of jackfruit (*Artocarpus heterophyllus*, Moraceae) by a new species of gall midge, *Clinodiplosis ultracrepidata* sp. nov (Diptera: Cecidomyiidae). *International Journal of Plant Sciences* 179(5), 350–367. DOI: 10.1086/697115.

Gillison, A.N. (2019) Plant functional indicators of vegetation response to climate change, past present and future: I. trends, emerging hypotheses and plant functional modality. *Flora* 254, 12–30. DOI: 10.1016/j.flora.2019.03.013.

Hack, H., Bleiholder, H., Buhr, L., Meier, U., Schnock-Fricke, U. *et al.* (1992) Einheitliche Codierung der phanologischen Entwicklungsstadien mono- und dikotyler Pflanzen-Erweiterte BBCH-Skala, Allgemein. *Nachrichtenblatt des Deutschen Pflanzenschutzdienstes* 44, 265–270.

Haq, N. (2006) *Jackfruit* (Artocarpus heterophyllus). Southampton Centre for Underutilized Crops, University of Southampton,Southampton, UK.

Kishore, K. (2018) Phenological growth stages of jackfruit (*Artocarpus heterophyllus*) according to the extended BBCH scale. *Annals of Applied Biology* 172(3), 366–374. DOI: 10.1111/aab.12427.

Li, Y., Duan, X., Liu, S., Li, Y., Zhang, X. *et al.* (2017) Changes in soluble sugar accumulation and activities of sucrose-metabolizing enzymes during fruit ripening of jackfruit. *Journal of Agricultural Science* 9(8), 155. DOI: 10.5539/jas.v9n8p155.

Lina, D.P. and Protacio, C.M. (2015) Phenology of jackfruit (*Artocarpus heterophyllus* Lam.) and effects of paclobutrazol on off-season fruit production. *Acta Horticulturae* 1088, 495–499. DOI: 10.17660/ActaHortic.2015.1088.90.

Maia, J.G.S., Andrade, E.H.A. and Zoghbi, M. das G.B. (2004) Aroma volatiles from two fruit varieties of jackfruit (*Artocarpus heterophyllus* Lam.). *Food Chemistry* 85(2), 195–197. DOI: 10.1016/S0308-8146(03)00292-9.

Mijin, S. and Ding, P. (2020) Growth development and structural changes of Malaysian jackfruit cv. Tekam Yellow syncarp. *Scientia Horticulturae* 272, 109594. DOI: 10.1016/j.scienta.2020.109594.

Mijin, S., Ding, P., Saari, N. and Ramlee, S.I. (2021) Effects of pollination techniques and harvesting stage on the physico-chemical characteristics of jackfruit. *Scientia Horticulturae* 285, 110199. DOI: 10.1016/j.scienta.2021.110199.

Moncur, M.W. (1985) Floral ontogeny of the jackfruit, *Artocarpus heterophyllus* Lam. (Moraceae). *Australian Journal of Botany* 33(5), 585. DOI: 10.1071/BT9850585.

Ong, B.T., Nazimah, S.A.H., Osman, A., Quek, S.Y., Voon, Y.Y. *et al.* (2006) Chemical and flavour changes in jackfruit (*Artocarpus heterophyllus* Lam.) cultivar J3 during ripening. *Postharvest Biology and Technology* 40(3), 279–286. DOI: 10.1016/j.postharvbio.2006.01.015.

Pushpakumara, D.K.N.G. (2006) Floral and fruit morphology and phenology in *Artocarpus heterophyllus* Lam (Moraceae). *Sri Lankan Journal of Agricultural Sciences* 43, 82–107.

Pushpakumara, D.K.N.G. (2011) Reproductive Biology. In: Valavi, S.G., Peter, K.V. and Thottappilly, G. (eds) *The Jackfruit*. Studium Press LLC, Houston, Texas, pp. 41–50.

Pushpakumara, D.K.N.G., Boshier, D.H. and Harris, S.A. (1997) Mating system in *Artocarpus heterophyllus* Lam. *Tropical Agricultural Research* 9, 1–14.

Rahman, A.K.M.M., Huq, E., Mian, A.J. and Chesson, A. (1995) Microscopic and chemical changes occurring during the ripening of two forms of jackfruit (*Artocarpus heterophyllus* L.). *Food Chemistry* 52(4), 405–410. DOI: 10.1016/0308-8146(95)93290-8.

Rymbai, H., Verma, V.K., Talang, H.D., Assumi, S.R., Devi, M.B. *et al.* (2021) Abnormality in flowers and fruits of jackfruit (*Artocarpus heterophyllus* Lam.) in Meghalaya, North eastern hills of India. *Agriculture and Environment* 2, 48–51.

Segrestin, J., Navas, M.L. and Garnier, E. (2020) Reproductive phenology as a dimension of the phenotypic space in 139 plant species from the Mediterranean. *The New Phytologist* 225(2), 740–753. DOI: 10.1111/nph.16165.

Tessy, J., Kumaran, K. and Joseph, T. (1996a) Fruit set, fruit development and fruit drop in hard and soft flaked types of jack (*Artocarpus heterophyllus* L). *Journal of Tropical Agriculture* 31, 21–24.

Tessy, J., Kumaran, K. and Joseph, T. (1996b) Flowering, floral biology and pollination in hard and soft flaked types of jack (*Artocarpus heterophyllus* L). *Journal of Tropical Agriculture* 31, 6–9.

Ullah, M.A. and Haque, M.A. (2008) Studies on fruiting, bearing habit and fruit growth of jackfruit germplasm. *Bangladesh Journal of Agricultural Research* 33(3), 391–397. DOI: 10.3329/bjar.v33i3.1598.

10 Physiological Disorders

Nor Elliza Tajidin[1]*, Azimah Hamidon[2], Munirah Mohamad[3], Hamizah Hassan[1] and Rosmah Murdad[4]

[1]Department of Crop Science, Faculty of Agriculture, Universiti Putra Malaysia, Selangor, Malaysia; [2]Department of Crop Science, Faculty of Technical and Vocational, Universiti Sultan Idris, Perak, Malaysia; [3]Horticulture Research Centre, Malaysian Agricultural Research and Development Institute, Selangor, Malaysia; [4]Faculty of Sustainable Agriculture, Universiti Malaysia Sabah Sandakan Campus, Sabah, Malaysia

10.1 Introduction

The productivity and quality of jackfruit cultivation are significantly affected by a number of physiological disorders. Disruptions caused by environmental factors such as temperature, moisture, light, aeration and nutritional imbalances affect the plant's metabolic activities, resulting in physiological disorders (Kumar and Kumar, 2016). Several physiological disorders commonly faced in jackfruit cultivation and storage include chilling injury, bronzing, fruit cracking, irregular shape of fruit, premature fruit drop and other disorders related to micronutrient deficiency. Although the symptoms of these disorders may appear disease-like, they can usually be prevented by altering growing conditions. One significant physiological disorder associated with the jackfruit is the breakdown of internal tissues in the centre pockets of the compound fruit, which could be due to mineral deficiencies in the soil. Appropriate supplementation with minerals such as calcium (Ca) and zinc (Zn) in the field could help prevent this tissue breakdown. Internal fruit breakdown leads to unmarketable production, even though the disorder is not always visible on the external part of the fruit (Yahia *et al.*, 2019).

10.2 Chilling Injury

Horticulture produce, especially fruit and vegetables, is subject to quick deterioration when exposed to high temperatures. Storing fruits and vegetables under cold conditions will preserve the quality and prolong the shelf life of these horticultural commodities. Cold storage conditions can slow down the metabolic rate, reduce spoilage by pathogens, delay senescence, and maintain the postharvest quality of fruits and vegetables (Zhang *et al.*, 2021). However, improper temperature management can lead to chilling injury, especially for tropical and subtropical fruits (Zhang *et al.*, 2021). Chilling injury, a physiological disorder that manifests as brown and black peel, black patches, epidermis pitting and water-soaked and grainy tissue in the pulp, occurs when the fruit is exposed to temperatures below 10°C in storage (Nian *et al.*, 2022).

*Corresponding author: elliza.tajidin@upm.edu.my

DOI: 10.1079/9781800622319.0010

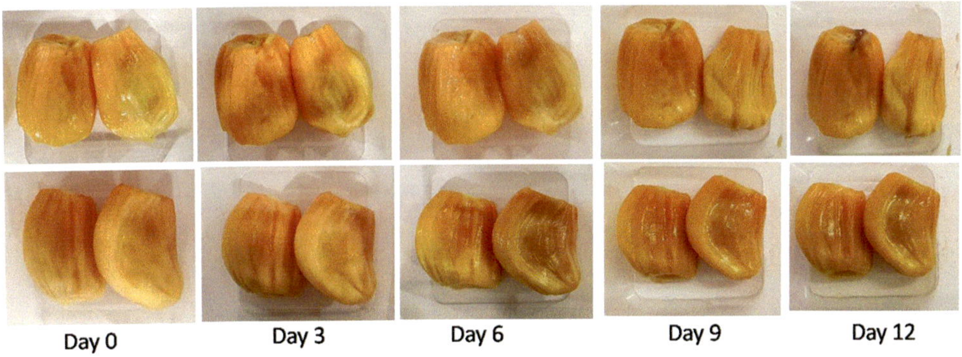

Fig. 10.1. Effects of chilling injury after 0, 3, 6, 9 and 12 days of storage at 0°C (top) and 10°C (bottom) on fresh bulb of jackfruits. Photograph courtesy of Munirah Mohamad.

Jackfruit is usually stored in cold storage in two forms: the whole jackfruit and minimally processed fresh bulbs (Fig. 10.1). These two forms are conditioned at different temperatures and relative humidity according to their physiological changes (Sutrisnol and Sugiyono, 2012). Suboptimal storage temperatures can cause spoilage of the whole jackfruit (Saxena *et al.*, 2011) as it increases its susceptibility to fruit rot caused by *Lasiodiplodia theobromae*, which tends to manifest after harvest, especially under high temperatures and relative humidity (Cho *et al.*, 2020). Kader (2001) described chilling injury symptoms in jackfruit as including dark brown discoloration of the skin, pulp browning and off-flavour development, with increased susceptibility to decay. Thus, maintaining optimum cold storage conditions to store jackfruit is critical to extending its storage life by maintaining the fruit's quality and reducing disease incidence (Singh and Sharma, 2018). The best storage temperature for jackfruits is 10–12°C, with higher and lower temperatures not exceeding an additional 2°C (Saxena *et al.*, 2011).

Minimally processed jackfruit bulbs are suitable for storage at 8–12°C (Zeng *et al.*, 2018). Bulbs stored below 4°C result in off-flavour development, reduced nutritional quality and loss of aroma, which is related to the adverse effects of chilling injury. Jackfruit bulbs stored at 8–12°C are firmer, juicier and sweeter, with a higher aroma of the flesh (Zeng *et al.*, 2018).

10.3 Irregular Fruit Shape

Irregular, deformed and shrivelled shapes of the fruit reduce the market value of jackfruit (Fig. 10.2). These arise as a result of the inability of some fruitlets to develop due to the non-uniform out-crossing of its multicarpellar flowers with asynchronous stigmatic receptivity (Lina and Protacio, 2015). Jackfruit is naturally pollinated by wind or insects. The female spikes appear on the trunk and main branches, while the male spikes are formed in the periphery of the tree. Normally, the period of anthesis does not occur simultaneously, causing irregular pollination. When there is a group of flowers that are not pollinated, the fruit will not develop fully, thus causing uneven fruits and even leading to the dropping of spikes. Studies have suggested that boron (B) can help induce the normal production of hormones, and increase the number of pollen grains as well as pollen-tube formation, and boost the stickiness of the stigma to accept pollen grains (Halder *et al.*, 2008).

An experiment was conducted in Bangladesh to determine the optimum dose of B to prevent malformed shapes and sizes of jackfruits. The results showed that treatment of 20-year-old trees with 15 g B per tree, a blanket dose of nitrogen (N), phosphorus (P), potassium (K), sulfur (S) and Zn at a ratio of 920:200:250:85:20 g year^{-1} per tree, as well as 10 kg cow dung manure, was the most effective in producing more normal and fewer deformed

Fig. 10.2. Irregular shaped (A) and normal (B) jackfruit. Photographs courtesy of the Department of Agriculture Malaysia.

fruits per tree (Halder *et al.*, 2008). Rymbai *et al.* (2021) suggested a complete package of pruning of weak, dead and diseased branches, soil application of 15 g B per tree, spraying of naphthalene acetic acid (200 p.p.m.) before flowering, addition of the fungicide Captan or Topsin-M (1%) during flower initiation, and a combined micronutrients dose of Zn (0.3%), B (0.3%) and copper (Cu; 0.3%) before and after flower initiation to minimize the incidence of fruit cracking. It is also suggested that assisted pollination, by brushing fresh pollen on to the female spikes in the morning, can effectively improve fruit set and development. As a consequence of such treatment, the number of fruitlets and seeds increased by over 30%, with the number of seeds per kg fruit increasing by 46%. The number of seedless fruitlets also more than doubled. This artificial technique is crucial, especially when the position of the female spikes is not exposed to dehiscing pollen (Lina and Protacio, 2015).

10.4 Rusty Fruit Flesh

Incidences of a physiological disorder locally called rusty fruit flesh occurs in some varieties of jackfruit where rust-coloured specks or streaks appear on the pulp of the fruit (Fig. 10.3). The symptoms of this disorder include the following: (i) the fruit pulp hardens and the typical yellowish-orange pulp changes to brownish yellow, making the fruit pulp inedible, less crunchy and bitter; (ii) rot may be visible at the base of the peduncle; (iii) brown discoloration occurs along the peduncle to the central core of the fruit; (iv) in advanced stages, brown spots also occur on the unseeded aril (perigones); and (v) in advanced stages, browning and rotting is seen on the external fruit surfaces.

In such cases, fruit quality is compromised and can deter consumers and processors. Sariah (1999) reported that this disorder was not due to disease infection but was caused by nutrient imbalances. However, this problem

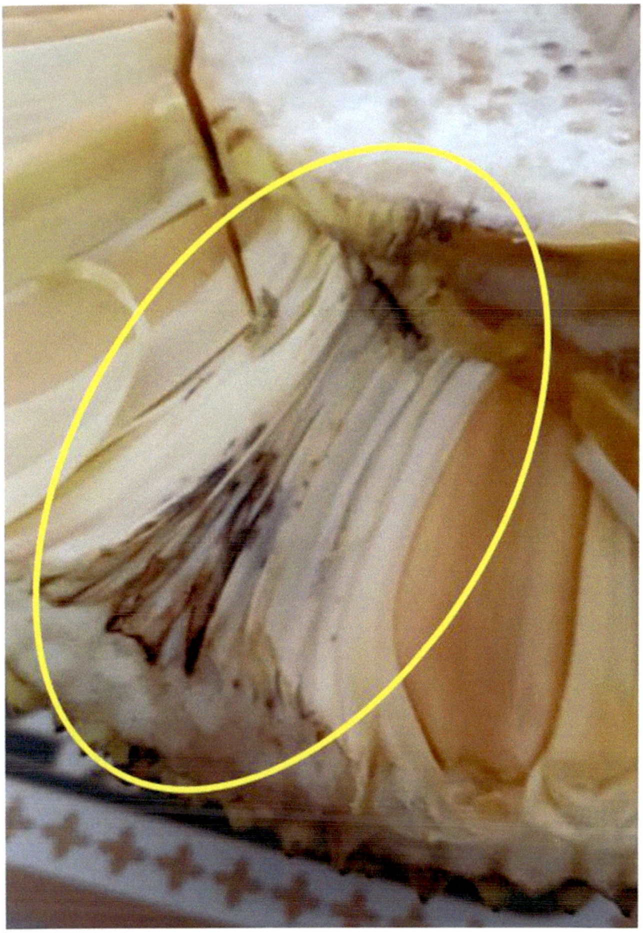

Fig. 10.3. Rust colour formation at the rags of jackfruits. Photograph courtesy of Munirah Mohamad.

has been seen to occur more often in certain varieties with high total soluble solids content such as the 'Tekam Yellow' ('J33') variety.

Researchers at the Department of Agriculture in Malaysia report that the occurrence of this condition occurred most often after or during the rainy season, when humidity was relatively high. Symptoms were more apparent in stressed plants as a result of nutritional imbalances, soil type and terrain conditions, and injury. However, healthier trees were more tolerant of this disorder. Zainudin (2013) found that the edible portion of jackfruit was highly influenced by the availability of macro- and micronutrients, particularly P, K, iron (Fe), magnesium (Mg) and Cu, which could cause occurrence of this condition. Fe was found to be the main element that affected this incidence. The condition is found in all fruits where, at initial or advanced stages, external symptoms such as dark rot spots and the presence of mycelia or spores is exhibited. Some affected fruits do not exhibit any external symptoms on the fruit surface, but initial signs of this condition are shown by brown discoloration when the fruit stem or peduncle is cut.

The incidence of this disorder varies from year to year and in some years, there may be no symptoms. This suggests that the incidence is related to the climatic conditions during fruit development. However, studies by the Department of Agriculture in Malaysia have indicated that the condition is related to infection by *Rhizopus artocarpi* and *Lasiodiplodia* spp. Initial treatment with prophylactic sprays of copper oxychloride seems to be effective in reducing disease incidence, especially during the rainy season. Cultural methods to reduce incidence include pruning the lower branches of the tree, thinning of fruits to restrict their number, removal of infected male inflorescences, disinfecting wrapping bags, avoiding injury to developing fruits and destroying diseased fruit. In general, rusty fruit flesh occurrence is believed to be caused by abiotic factors such as weather, soil type, terrain and plant nutrition, which affect plant health and may cause infection by fungal pathogens. However, further clarification is needed on this disorder.

10.5 Cracking and Splitting of Fruit

Fruit cracking and splitting is one of the preharvest disorders of jackfruit. This serious problem occurs extensively in pome and stone fruit crops (Opara *et al.*, 1996). Cracking is the physical failure of the fruit skin in the form of fractures in the cuticle or skin, which typically do not penetrate the flesh, while splitting is an extreme form of cracking that does penetrate the flesh of the fruit (Fig. 10.4) (Opara *et al.*, 1996; Rodríguez *et al.*, 2018).

Fruit cracking and splitting result from changes in fruit water relations (Milad and Shackel, 1992), fruit skin characteristics (Opara *et al.*, 1996) and mineral nutrient deficiencies (Huang *et al.*, 2008; Yahia *et al.*, 2019). Opara *et al.* (1996) found that fruit splitting is closely associated with environmental factors such as soil moisture, rainfall, relative humidity, temperature and exposure to sunlight. Fruit splitting is assumed to be due to a sudden increase in the water content of the soil, the atmospheric humidity or the temperature (Opara *et al.*, 1996). Peet (1992) reported that splitting occurs when a sudden net influx of water and solutes into the fruit coincides with other factors, leading to reduced skin elasticity and strength. An oversupply of water both during irrigation and heavy rainfall increase the soil moisture content, leading to an increase in the fruit's internal turgor pressure (Rodríguez *et al.*, 2018) and indirectly increasing the amount of fruit splitting.

Fig. 10.4. Splitting of jackfruit. Photographs courtesy of Munirah Mohamad.

Low Ca status has also been shown to be involved in the physiological process of cellular adhesion loss (Yahia *et al.*, 2019) that contributes to fruit cracking (Huang *et al.*, 2008).

Saxena *et al.* (2011) suggested that appropriate supplementation of minerals such as Ca and Zn in the field could help prevent fruit cracking. Additionally, external factors such as drought, high temperature, excessive water and high solar radiation have been determined to affect cracking resistance (Yahia *et al.*, 2019). Singh *et al.* (2011), working with 20 jackfruit genotypes, stated that the cracking trait in jackfruit is more likely to be a varietal characteristic. The investigation also revealed that cracking genotypes contained less pectin when compared with non-cracking genotypes, which may cause cracking due to the inelastic nature of the peel. However, conclusive data are still needed to derive proper explanations about this disorder, especially for jackfruit.

10.6 Conclusion

Significant challenges are faced by jackfruit farmers and producers due to physiological disorders that affect the fruit. To get high-quality fruit and good yields, nutrient deficiencies must be detected before visual symptoms are expressed, especially micronutrient deficiencies. Some symptoms of physiological disorder may appear to be disease-like or may not even be visible on the outside, such as bronzing and internal browning.

The physiological disorders of the jackfruit are caused by adverse environmental conditions such as temperature, moisture, light, aeration and nutritional imbalances, which can affect the metabolic activities of plants. These disorders have also become increasingly widespread with the diminishing use of organic manures, adoption of high-density planting strategies, the use of rootstocks for dwarfing, diseases, salt tolerance, unbalanced NPK fertilizer application and extension of horticulture activities into marginal lands. Some of these disorders could be overcome through better nutrient management, regular watering and application of plant hormones to assist in the growth and development of jackfruit. Recent advances in plant physiology and genetics may help solve these problems faced in jackfruit production. However, further studies are needed with a focus on physiological disorder management strategies and adaptation to produce higher-quality jackfruit.

References

Cho, J.L.Y., Jaffar, N.S., Wan Hussin, W.M.R.I., Abdul Hamid, M.F. and Muhsin, N.I. (2020) Maintaining postharvest quality of jackfruit during cold storage. *International Journal of Agriculture, Forestry and Plantation* 10, 282–286.

Halder, N.K., Farid, A.T.M. and Siddiky, M.A. (2008) Effect of boron for correcting the deformed shape and size of jackfruit. *Journal of Agriculture & Rural Development* 6(1), 37–42. DOI: 10.3329/jard.v6i1.1655.

Huang, X.-M., Wang, H.-C., Zhong, W.-L., Yuan, W.-Q., Lu, J.-M. *et al.* (2008) Spraying calcium is not an effective way to increase structural calcium in litchi pericarp. *Scientia Horticulturae* 117(1), 39–44. DOI: 10.1016/j.scienta.2008.03.007.

Kader, A. (2001) *Jackfruit*. Postharvest Center, University of California, Davis, California. Available at: https://postharvest.ucdavis.edu/Commodity_Resources/Fact_Sheets/Datastores/Fruit_English/?uid=28&ds=798 (accessed 14 July 2022).

Kumar, R. and Kumar, V. (2016) Physiological disorders in perennial woody tropical and subtropical fruit crops: a review. *Indian Journal of Agricultural Sciences* 86, 703–717.

Lina, D.P. and Protacio, C.M. (2015) Assisted pollination techniques for improved fruit quality development in jackfruit (*Artocarpus heterophyllus* Lam.). *Acta Horticulturae* 1088, 501–504. DOI: 10.17660/ActaHortic.2015.1088.91.

Milad, R.E. and Shackel, K.A. (1992) Water relations of fruit end cracking in French prune (*Prunus domestica* L. cv. French). *Journal of the American Society for Horticultural Science* 117(5), 824–828. DOI: 10.21273/JASHS.117.5.824.

Nian, Y., Wang, N., Li, R., Shao, Y. and Li, W. (2022) Cold shock treatment alleviates chilling injury in papaya fruit during storage by improving antioxidant capacity and related gene expression. *Scientia Horticulturae* 294, 110784. DOI: 10.1016/j.scienta.2021.110784.

Opara, L.U., Studman, C.J. and Banks, N.H. (1996). Fruit skin splitting and cracking. *Horticultural Reviews* 19, 217–262.

Peet, M.M. (1992) Fruit cracking in tomato. *HortTechnology* 2(2), 216–223. DOI: 10.21273/HORTTECH.2.2.216.

Rodríguez, P., Galindo, A., Collado-González, J., Medina, S., Corell, M. *et al.* (2018) Fruit response to water-scarcity scenarios. Water relations and biochemical changes. In: Terejo, I.F.G. and Zuazo, V.H.D. (eds) *Water Scarcity and Sustainable Agriculture in Semiarid Environment – Tools, Strategies, and Challenges for Woody Crops*. Elsevier, Cambridge, UK, pp. 349–375.

Rymbai, H., Verma, V.K., Talang, H.D., Assumi, S.R., Bilashini Devi, M. *et al.* (2021) Abnormality in flowers and fruits in jackfruit (*Artocarpus heterophyllus* Lam.) in Meghalaya, North eastern hills of India. *Agriculture and Environment* 2, 48–51.

Sariah, M. (1999) Disorder of Nangka of complex etiology. *Agro-Search* 6, 47.

Saxena, A., Bawa, A.S. and Raju, P.S. (2011) Jackfruit (Artocarpus heterophyllus Lam.). In: Yahia, E.M. (ed.) *Postharvest Biology and Technology of Tropical and Subtropical Fruits – Cocona to Mango: Vol. 3*. Elsevier, Cambridge, UK, pp. 275–299.

Singh, S.R., Narayanaswamy, P., Banik, B.C., Shyamalama, S. and Simon, L. (2011) Evaluation of cracking and non-cracking genotypes of jackfruit (*Artocarpus heterophyllus* Lam). *Crop Research* 42, 157–162.

Singh, D. and Sharma, R.R. (2018) Postharvest diseases of fruits and vegetables and their management. In: Siddiqui, M.W. (ed.) *Postharvest Disinfection of Fruits and Vegetables*. Elsevier, Cambridge, UK, pp. 1–52.

Sutrisnol, E. and Sugiyono, I.M. (2012) *Postharvest Handling of Jackfruit*. Available at: https://repository.ipb.ac.id/handle/123456789/53847 (accessed 14 July 2022).

Yahia, E.M., Carrillo-López, A. and Sañudo, A. (2019) Physiological disorders and their control. In: Yahia, E.M. (ed.) *Postharvest Technology of Perishable Horticultural Commodities*. Elsevier, Cambridge, UK, pp. 499–527.

Zainudin, M. (2013) Rusty fruit pulp incidence: physiological disorder in jackfruit. In: Mohd Roseli, A.N., Osman, N., Ying, T.F., Othman, R. and Megat Wahab, P.E. (eds) *Transactions of the Malaysian Society of Plant Physiology, Vol.22, of the 24th Malaysian Society of Plant Physiology Conference: Innovative Plant Productivity and Quality, 27–29 August 2013,Terengganu, Malaysia*. pp. 95–99.

Zeng, L.P., Fan, H.Y., Zhang, M.X., Li, X.G. and Shen, Y. (2018) Effect of different storage temperatures on biochemical quality of freshly-cut jackfruit. *Acta Agriculturae Jiangxi* 30, 28–32.

Zhang, W., Jiang, H., Cao, J. and Jiang, W. (2021) Advances in biochemical mechanisms and control technologies to treat chilling injury in postharvest fruits and vegetables. *Trends in Food Science & Technology* 113, 355–365. DOI: 10.1016/j.tifs.2021.05.009.

11 Pests

Hari S. Singh* and Gundappa Baradevanal
ICAR-Central Institute for Subtropical Horticulture, Rehnamkhera, Lucknow, Uttar Pradesh, India

11.1 Introduction

Jackfruit is a hardy plant and its fruit is the largest edible fruit in the world (Naik, 1949; Sturrock, 1959). Almost all parts of the plant have economical value but the fruits are the main useful part. The fruit pulp is rich in calories, minerals and many vitamins, particularly vitamin A, whereas the seeds are a rich source of carbohydrate and vitamin B. The fruit is also rich in phenols and tannin. The edible portion of jackfruit pulp and seed constitutes only 25–40% of the total fruit weight. The leaves of jackfruit provide a good feed for cattle and small ruminants, particularly goats.

Like many other cultivated crops, jackfruit can be infested by a number of insect pests, although most do not inflict serious loss. In a given area, jackfruit is normally damaged by two or three key pests, some secondary pests and an array of occasional pests. Butani (1979a) initially reported 39 species of insects that attack jackfruit in India. Subsequent surveys in jackfruit fields in Karnataka, Kerala and Tamil Nadu in India revealed 51 species of insect pests on jackfruit (Soumya *et al.*, 2020a). The maximum number of insect pests (46 species) was recorded in Karnataka, followed by Tamil Nadu (31 species) and Kerala (29 species). Some of the reported insect pests are given in Table 11.1 and the more important ones are described in this chapter.

11.2 Shoot and Fruit Borer: *Diaphania caesalis* (Lepidoptera: Crambidae)

Jackfruit shoot and fruit borer, *Diaphania caesalis*, is a destructive lepidopteran insect pest. Walker (1859) described it as *Glyphodes caesalis* sp. nov.; it was later shifted to the genus *Margaronia*, and then to genus *Palpita* and finally described under the genus *Diaphania* (Wang, 1963). It is considered to be one of the major insect pests of jackfruit across jackfruit-growing countries. It was reported as an important pest of jackfruit in East Pakistan (Alam, 1962) and in Bangladesh (Alam *et al.*, 1964; Karim, 1995; Khan *et al.*, 2003; Khan and Islam, 2004). In India, it has been reported in Karnataka and Maharashtra (Fletcher, 1914), Assam, Sikkim, Bihar, Uttar Pradesh, Andhra Pradesh and Tamil Nadu (Chowdhury and Majid, 1954; Prarthna *et al.*, 2014), and in other countries such as Borneo, Burma and Sri Lanka (Hampson, 1896).

The coastal hot and humid climate is most suitable for population build-up and damage by this pest. The adult female deposits eggs on flower buds and tender shoots. The egg hatching period lasts for 3–5 days. Initially, the first-instar larva is tiny and subsequently turns pinkish, with a yellowish head and prothoracic shield. Each body segment has single,

*Corresponding author: Hari.Singh1@icar.gov.in

DOI: 10.1079/9781800622319.0011

Table 11.1. Insect pests reported on jackfruit.

Common name	Species	Pest status	References
Order: Lepidoptera			
Shoot and fruit borer	*Diaphania caesalis* Walker	Major	Alam (1962); Madhava Rao (1965); Butani (1979a); Karim (1995); Tandon (1998); Khan and Islam (2004); Mridha (2006); Manjunatha *et al.* (2014)
Bark-eating caterpillar	*Indarbela tetraonis* Moore	Major	Mann and Bindra (1977); Ali *et al.* (2007); Garg and Tomar (2008)
	Inderbella quadrinotata Walker	Major	
Leaf webber	*Diaphania* (*Glyphodes*) *bivitralis* Guenée	Minor	Butani and Jotwani (1984); Muniappan *et al.* (2012)
Jackfruit leaf webber	*Perina nuda* Fabricius	Minor	Butani and Jotwani (1984); NBAIR (2013)
Castor capsule borer	*Dichocrocis punctiferalis* (Gunee)	Minor	Butani and Jotwani, 1984; Gupta (1990); NIPHM Working Group (2015)
	Batocera rufomaculata De Geer	Major	NIPHM Working Group (2015)
Order Coleoptera			
Bud weevil or leaf-eating weevil	*Ochyromera artocarpi* Marshall	Minor	Marshall (1917); Jha and Sen-Sarma (2008)
	Onychocnemis careyae Marshall	Minor	
	Teluropus ballardi Marshall	Minor	
Stem/trunk borers (longhorn beetles)	*Apriona germari* (Hope)	Minor	Chatterjee *et al.* (1969); Zhang and Shen (1980); Hua (1982); Singh and Prasad (1985); Lee (1987); Paik (1987); Singh and Singh (1987); Kim *et al.* (1994); Yoon *et al.* (1997a, b); Poland and McCullough (2006); Ahmed *et al.*, 2013a,b
	Microcolona leucosticta Meyrick	Minor	Singh (1969); Butani (1979b); Azad (2000); Haq (2006)
Order Hemiptera			
Aphids	*Greenidea artocarpi* (Westwood)	Minor	Carver (1978); Blackman and Eastop (1984)
	Toxoptera aurantii (Boyer de Fonscolombe)	Minor	
Spittle bug	*Cosmoscarta relata* Dist.	Minor	Jeyaraj and Ananthan (2009)
	Clovia lineaticollis (de Motschulsky)	Minor	

Continued

Table 11.1. Continued

Common name	Species	Pest status	References
Mealybug/coccids	*Drosicha mangiferae* (Green)	Minor	Ghosh and Ghosh (1985); Mani and Thontadarya (1987); Mani and Krishnamoorthy (1997); Srivastava (1997); Mani and Krishnamoorthy (2008); Ben-Dov *et al*. (2010)
	Planococcus lilacinus (Cockerell)	Minor	
	Nipaecoccus viridis (Newstead)	Minor	
Lesser snow scale	*Pinnaspis trachani* (Cooley)	Minor	CIE (1960); Deckle (2008); Sirisena *et al*. (2013); Ben-Dov *et al*. (2015)
Coconut scale	*Aspidiotus destructor* (Signoret)	Minor	
Mango shield scale	*Protopulvinaria mangiferae* (Green)	Minor	
Pyriform scale	*Protopulvinaria pyriformis* Cockerell	Minor	
Pink waxy scale	*Ceroplastes rubens* Maskell	Minor	
Jassid	*Apheliona indica* Dworakowska & Sohi	Minor	Soumya *et al*. (2020a)
Sharpshooter leaf hopper	*Kolla ceylonica* Melichar	Minor	
Planthopper	*Eurybrachys tomentosa* Fabricius	Minor	
True bug	*Coptosoma siamicum* Walker	Minor	
Stink bug	*Eupaleopada concinna* Westwood	Minor	
Aubergine mealy mug	*Coccidohystrix insolita* Green	Minor	
Lace bug	*Cysteochila delineata* Distant	Minor	
Spiralling whitefly	*Aleurodicus dispersus* Russel	Minor	Kumari (2011)
Order Thysanoptera			
Thrips	*Pseudodendrothrips dwivarna* Ramakrishna & Margabandhu	Minor	Ramakrishna Ayyar and Margabandhu (1931); Anandhakrishnan and Sen (1980)
Order Diptera			
Fruit flies	*Bactrocera umbrosus* (Fabricius)	Minor	Srivastava (1997); Vagalo *et al*. (1997); NIPHM Working Group (2015)
	Bactrocera dorsalis (Hendel)	Minor	

short, sharp hairs arising from the black compressed horny lumps (Fig. 11.1). The larval period lasts for 11–13 days with five instars, with a prepupal stage. Fifth-instar larvae, on maturity, pupate inside the silken cocoon, which takes 7–12 days. The colour of pupa is reddish-brown and the life cycle is completed within 26–36 days depending on the season and climatic conditions.

In another study, Soumya *et al.* (2019) reported that the life cycle of the male and female moths is completed in 30.4 ± 2.3 and 31.8 ± 2.3 days, respectively. The mean incubation period of the egg was 4.9 ± 0.6 days, the larval period 17 ± 0.6 days and the prepupal period 3.6 ± 0.4 days, with a pupal duration of 8 ± 0.4 days. The first- and last-instar larvae measured 2.6 ± 0.3 and 22.4 ± 0.4 mm in length, respectively.

Fig. 11.1. Larva of the jackfruit shoot and fruit borer, *Diaphania caesalis*. Photograph courtesy of Dr H.S. Singh.

Diet was an important factor for survival; unfed moths lived for a shorter period (6–9 days) than those with water (9–12 days), honey (12–14 days) and honey with supplements such as vitamins (14–17 days).

The larva pupates on the nearby fruit surface (Fig. 11.2) or inside it. The colour of the moth is whitish-brown and is half an inch long. The wing of the adult moth has black dots as border bands with a greyish cryptic pattern. The adult moth wingspan is 25–30 mm and the female is larger than the male (Alam, 1962; Butani, 1979b; Khan and Islam, 2004; Manjunatha *et al.*, 2014). Some of the alternative hosts of *D. indica* such as *Artocarpus chaplasha*, *A. elasticus*, *A. communis*, *A. altilis*, *A. odoratissimus* and *A. rigidus* were reported by Beeson (1941) and Martinez (1999).

This pest causes variable levels of loss ranging from 3% to 28% depending on the climate conditions. Various fruit developmental stages are attacked by this pest. In early infestation, deformation of fruits takes place and sometimes fruit drops occur in immature fruits. On maturity, the larva enters the fruit and damages the edible portion (Fig. 11.3).

Fig. 11.2. Borer damage on fruit surface. Photograph courtesy of Dr H.S. Singh.

Fig. 11.3. Borer feeding internal fruit mass. Photograph courtesy of Dr H.S. Singh.

Later, affected fruits rot as rain water enters them. The entry hole is full of a mass of excreta. Khan and Islam (2004) reported that this pest attacks the tender shoots and fruits of all developmental stages. Adult female moths deposit eggs on the apical buds and tender shoots. The larvae bore into the flower spikes and feed on the internal content. Damage at the initial stage prevents male flower anthesis, and the flowers later rot and fall from the plant. Severely infested female spikes drop before they set the fruit. The phenomenon of flower or fruit drop under severe infestation was also reported by Tandon (1998). This pest not only attacks the trees and fruits in the field but also nursery plants by inflicting damage to the tender apical buds of saplings, which retards their growth and the initiation of lateral branches (Khan *et al.*, 2003). The caterpillar damages the tips of the fruits. As estimated by Khan and Islam (2004), the average percentage of fruit infestation was 27.44%. The number of bores and amount of damage per infested fruit was 1.47 and 525.37 g, respectively. According to this study, the reduction of market price due to jackfruit borer infestation in six locations of Bangladesh ranged from 27% to 39%. Soumya *et al.* (2017) recorded the incidence of *D. caesalis* throughout the year, indicating overlapping of generations. Two distinct peaks in population were recorded during June–July and October–November. Afterwards, the insect population declined with the onset of winter and was at its minimum level in summer. They also reported the influence of abiotic factors such as maximum and minimum temperatures, maximum and minimum relative humidity, wind speed and rainfall on *D. caesalis* damage in jackfruit. The percentage of fruit infestation was found to be positively correlated with rainfall. Over this 3-year study, only rainfall was found to induce variability in total fruit infestation ranging from 77% to 97%.

A forecasting system for the prevalence of this pest was developed by Soumya *et al.* (2017). Tandon (1998) endorsed the suggestion that elimination and destruction of the affected parts including shoots and flower buds and culminating at the preliminary level of incidence can lessen infestation of this pest. Fruit bagging with polythene reduces the pest infestation considerably because it restricts the oviposition by the adults (Madhava Rao, 1965; Mridha, 2006). Karim (1995) recommended spraying with cypermethrin 10% EC at 1 ml l^{-1} or fenitrothion 50% EC at 2 ml l^{-1} at flower initiation to prevent pest infestation, but a second spray may be required 15–20 days after first, if the infestation does not subside.

As the pest exists outside the fruit during the initial infestation, it is also attacked by natural enemies. Soumya *et al.* (2020b) reported 13 species of primary parasitoids and one species of hyperparasitoid (*Tetrastichus pantnagerensis* Khan). Of these, *Apanteles stantoni* Ashmead was the most prevalent parasitoid, and parasitized nearly 35% of host larvae in Karnataka, 38.5% in Kerala and 33.1% in Tamil Nadu. The overall level of parasitization by all the parasitoids was nearly 41–45% and was similar across the studied locations.

11.3 Stem/trunk Borers: *Aprioma germari*, *Batocera rufomaculata* (Coleoptera: Cerambycidae) and *Microcolona leucosticta* (Lepidoptera: Elachistidae)

Among all the insect pests, the stem or trunk borer *Batocera rufomaculata* De Geer, also called the red-spotted longhorn beetle or bark borer, is the most destructive pest on jackfruit. It also attacks fig, mango, apple and about 50 other plant species. The borer, a beetle of almost 4–5 cm long, is an annual pest with one generation per year. It lays eggs on the main trunk of relatively older or sick trees and prefers wilt-affected, neglected or broken branches/rejuvenated trees. Various workers have reported on the biological aspects of this pest (Singh, 1969; Butani, 1979a, Butani, 1979b; Azad, 2000; Haq, 2006). The eggs are shiny white in colour, oval and 5–7 mm long. The pupae are large, measuring 50–55 mm long and yellowish-brown to dark brown. The adults are robust, dark brown beetles of 50–60 mm long with a yellowish-green pubescence. The beetles come out of the infested trunk after the onset of the monsoon season (June–July), mate and start egg laying within 1–2 days, which lasts for 20–25 days at the rate of an average of one egg day^{-1}. The eggs hatch in 7–13 days. The life cycle is long (170–190 days) and the adult survives for 60–100 days.

The wingspan of *M. leucosticta* is about 10 mm. Meyrick (1928) reported that the adult forewings are light olive and fuscous, the costal and apical areas are suffused dark grey and the base of the dorsum has a white elongated blotch. A dark grey median streak is noted from the base to the end of the cell and the hind wings are grey. There is also a white dot on the tegmen towards the apex.

Baradevanal *et al.* (2015) standardized an artificial diet to rear the larvae of *B. rufomaculata*. It mainly comprised gram powder and mango stem powder with some other nutritional components. They reported exponential development of the larvae in terms of body weight and length of the grub on this artificial diet at fortnightly intervals. Of the grubs fed on this diet, the adults emerged 10 days after pupation. Pupation was 83.3% and adult emergence was 33.3%. No specific variation was found between natural population of adults and diet-reared adults, which were robust and active with a mean body length and width of 45 and 15 mm, respectively.

The grubs of *B. rufomaculata* bore and eat below the bark; as they grow, their tunnelling becomes extensive and is invisible from the outside (Fig. 11.4). Severe damage results in yellowing, drying and dieback of terminal shoots and branches, which ultimately leads to death of the whole tree. Gurjar *et al.* (2019) reported that stem borer infestation was more severe in the months of September and October with one to three holes per tree in different germplasms of jackfruit at the ICAR-National Bureau of Plant Genetic Resources, Regional Station at Ranchi, in the eastern Indian region; however, the most damage was seen during November–December.

In Korea, *A. germari* is widely distributed and is one of the major pests of mulberry trees (Hua, 1982; Lee, 1987). A Korean study indicated that it hibernates as eggs or as larvae in the mulberry fields under conventional management practices (Yoon *et al.*, 1997a). The larvae of *A. germari* start boring into the trunk through egg-laying depressions and tunnel further (Yoon *et al.*, 1997b). The larva of *M. leucosticta* is reported to damage the fruits of jackfruit from southern India (Meyrick, 1928).

Ahmed *et al.* (2013a) found *B. rufomaculata* on jackfruit trees from June to September with a peak emergence in mid-July; the highest percentage infestation was recorded in July (7.33%) followed by June, and was lower in August (6.00%). They reported cumulative infestation of this pest over the year to levels as high as 35.33% in the month of October. They recorded lowest infestation in the month of February (0.67%) and there was no activity during the months of November–January. They further reported that the incidence of infestation by the trunk borer was affected by temperature, rainfall and relative humidity. Stepwise regression showed that the maximum temperature was the most crucial factor and had a 35.3% influence, whereas average rainfall had the lowest influence (2.1%).

These pests remain between the bark and stem wood in the early stages and then make a hole inside the stem. Pesticide sprays on the tree or stem injections are not effective due to the unpredictable translocation of insecticides within the trees (Poland and McCullough, 2006) and asynchronous larval development, both of which allow the insects to evade treatment. Although these pests seldom attack the tertiary branches, pruning of affected stem/branches and fumigation of the main stem have been recommended for pest control (Chatterjee *et al.*, 1969; Singh and Prasad, 1985; Singh and Singh, 1987). Other methods of control have been recommended by various workers such as removal and destruction of affected branches, killing the grubs with a stiff wire or closing the holes with mud (Nayar *et al.*, 1989). Ahmed *et al.* (2013a) reported that cleaning the dead bark, removing the excreta and inserting aluminium phosphide tablets inside the hole followed by sealing with Bordeaux paste was an effective method to control pest infestation up to 83.3% with the highest benefit:cost ratio in Bangladesh. Upadhyay *et al.* (2013) determined the efficacy of insecticides in trunk borer management and reported that imidacloprid 17.8% SL, thiamethoxam 25% WG and triazophos 40% SL were superior in the management of mango stem borer.

Fig. 11.4. External symptoms of trunk borer damage. Photograph courtesy of Dr H.S Singh.

11.4 Fruit Flies: *Bactrocera dorsalis* and *Bactrocera umbrosus* (Diptera: Tephritidae)

Fruit flies infest 478 varieties of fruits and vegetables, including bananas, citrus, coffee, figs, jackfruit, guava, loquat, mango, pink apple, papaya, passion fruit, peach, pear, persimmon, pineapple, Suriname cherry, tomato and many other crops (USDA, 2016). The body length of these species of fruit flies is around 8.0 mm with a wingspan of approximately 7.3 mm. The colour of different species of flies is variable, but there are prominent yellow and dark brown to black markings on the thorax (Fig. 11.5). The ovipositor is very sharp and pointed.

Fruit flies can damage crops in three ways: (i) by laying eggs in soft tissue on the fruit surface; (ii) by feeding of larvae on the internal material; and (iii) by causing pulp decay by microorganisms. The adult female lays eggs by puncturing the exocarp of the mature and ripe fruit using her ovipositor and deposits the eggs (up to 20) in small groups inside the mesocarp. A female fly

Fig. 11.5. Adult fruit flies. Photograph courtesy of Dr H.S. Singh.

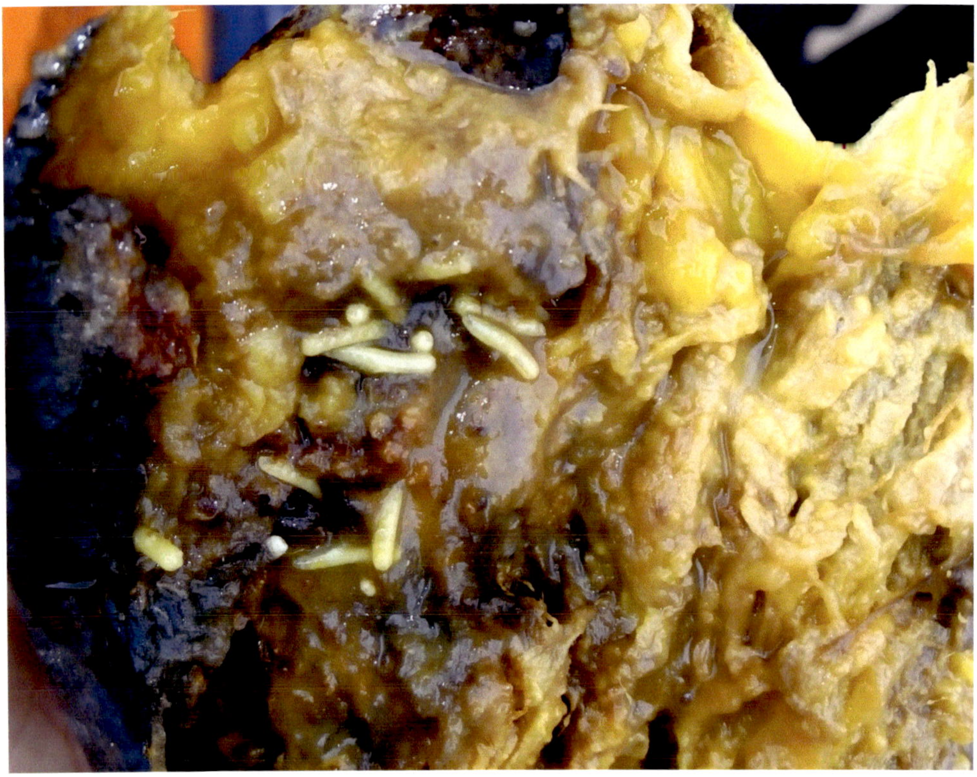

Fig. 11.6. Fruit fly-infested pulp. Photograph courtesy of Dr H.S. Singh.

can lay 400 or more eggs in her lifetime. In 2–4 days, the eggs hatch inside the fruit. A variety of bacteria may be introduced within the fruit when the female lays her eggs. After hatching, the larvae eat the pulp of the fruit; the damage is usually not visible from the outside. The larvae decompose the pulp (Fig. 11.6) and bacteria cause the fruit to decay internally, but this varies with the type of fruit attacked (Steiner, 1957). Affected fruits develop the appearance of water-soaked patches on the peel and these affected fruits fall prematurely from the tree. The tunnels made by the feeding larvae provide bacteria and fungi with an entry route, causing fruit rot.

Fully mature larvae leave the fruit, fall to the ground and burrow into the soil. In the soil, they form hard, barrel-shaped pupae that turn from light to dark brown in colour. Adult flies hatch from the pupae within a week in summer, but this takes several months in winter. The adult fly searches for food such as protein and sugar needed for reproduction and maturation to lay eggs. Emergence of the fruit fly begins in early summer and the maximum population appears in May–July, which coincides with fruit maturation and rainfall. Its life cycle is 10–27 days in summer, depending on the weather conditions (Srivastava, 1997; Carroll *et al.*, 2002). It takes a week from emergence of the adults to sexual maturity. The adult fly rests in non-host shade trees (e.g. fruit trees, ornamental trees and shrubs) during the day and feeds on nectar found in the host tree or in nearby plants.

B. umbrosus is very common in Malaysia and southern Thailand, and its host range is limited to jackfruit. The percentage loss is estimated to be between 30% and 70% (Vagalo *et al.*, 1997; Carroll *et al.*, 2002). Liquido (1990) points out that the rotten and fallen fruits left on land and fields are the main breeding grounds and reservoirs for the fruit fly population. They reported that the number of adults in the orchard was significantly related to the level of fruit infestations on the ground and on the tree. This information further supports the importance of removing fallen fruit for fruit fly population management in orchards. Thus, the cleanliness of the field is of the utmost importance and destroying all fallen and affected fruits is an effective measure, because this practice reduces the stress of reinfestation. After harvest, the ground should be ploughed.

van den Bosch *et al.* (1951) reported that one predator and 13 parasites were specific for *B. dorsalis*. These parasites lay eggs in fruit flies or in their eggs, feed inside them and are released in the pupal stage. These parasites are effective against specific species, mainly on Oriental (*B. dorsalis*) and Mediterranean fruit flies. *Opius longicaudatus* is a parasite of the second- and third-instar fruit fly larvae; *O. vandenboschi* is a parasite of first-instar larva and *O. oophilus* is an egg–larval parasite (van den Bosch and Haramoto, 1953).

Singh *et al.* (2008) developed an integrated pest management schedule of the orchard fruit fly with a combination of a male annihilation technique and a bait application technique. They reported the control and cost-effectiveness of the male annihilation technique ($50 \times 50 \times 12$ mm plywood blocks soaked in ethyl alcohol:methyl eugenol:malathion (6:4:1), soaked for 48 h and hung in trees at 10 ha^{-1} in bottle traps in two rounds) and the bait application technique (2 g carbaryl and 100 g jaggery in 1 g water applied in spots at weekly intervals), supplemented with field hygiene. This module used a 11.11 times lower pesticide content and was 1.32 times cheaper than pesticide spray but gave almost double the protection (93–97%), as well as double the cost:benefit ratio (1:21.76). If fruit fly infection is high, deltamethrin (0.0014%) plus azadirachtin (3000 p.p.m.) at 2 ml l^{-1} can be applied 3 weeks before harvesting.

11.5 Bark-Eating Caterpillars: *Indarbela tetraonis* and *Inderbella quadrinotata* (Lepidoptera: Metarbelidae)

About 14 species of bark-eating caterpillar within three genera have been reported from India (Mathew and Rugmini, 1997). The genus *Indarbela* is the largest with 12 species, of which five – *I. campbelli* Hampson, *I. minima* Hampson, *I. watsoni* Hampson, *I. tetraonis* Moore and *I. quadrinotata* Walker – have been reported from India. Of these species, *I. quadrinotata* is reported to be very common and is widely distributed.

I. tetraonis and *I. quadrinotata* are the most damaging for a large number of fruit trees. Ali *et al.* (2007) reported that these are polyphagous pests that inflict damage to many fruit crops, street trees and decorative perennial trees. They have been reported to invade forest trees and fruit crops such as guava, mango, citrus, orange, jamun, lychee, amaranth, jackfruit and pomegranate (Garg and Tomar, 2008). They are widely dispersed in India and are more common in Haryana, Punjab, Uttar Pradesh, Rajasthan and Madhya Pradesh. The adults are light brown with dark brown spots and are 35–40 mm long with prolonged wings. The female deposits its eggs in groups on the bark of the tree. The fully grown caterpillar is brown and 35–45 mm long. The larvae feed from April to December. There is only one generation per year so the life cycle is annual. The males are shorter than females with white short and thick brown stripes on the forewings. The hind wings are yellow. The female moths are light brown and have dark red spots on their forewings. The females have hairs at the end of the abdomen. The males range in length from 22 to 25 mm, and the females from 32 to 34 mm. Eggs are laid in groups of 15–20 directly on the stalk or bark of the tree. Freshly laid eggs are pale and oval in shape and after a while, turn creamy white.

Satyanarayana and Arunakumara (2016) determined the biology and management of *I. tetraonis* and reported that the eggs were 0.67–0.73 mm long and 0.42–0.46 mm wide, and hatching took place in 7–11 days. Fully grown larvae were found to be as long as 37–43 mm and 6–8 mm wide with a growth period of 274–298 days. The pupae were 11–15 mm long and 4–6 mm width with a growth period of 21–26 days. Male adults had a wing size of 11–13 mm long by 20–24 mm wide, whereas female had a wing size of 12–14 mm long and 25–27 mm wide.

The fully grown larvae are about 3.5–5 cm long, with dark chitinized patches on the smooth skin segments. The larvae usually enter the trunk or branches by making holes at the crotches or angles to a depth of 15–25 cm. This tunnel forms a shelter for the larvae during the day, and which they use to pupate in the next stage. They leave the hole at night and eat the bark of the tree around the hole. Large, dark brown, webby masses, comprising chewed wood particles and faecal matter, are conspicuously seen plastered loosely on the tree trunks (Fig. 11.7) or main branches, especially near the forks (Mann and Bindra, 1977). Small trees can be easily killed by this pest. Sometimes, the bark is eaten in large quantities, causing a block in the flow of sap, even in large trees, and the trees may show drying of the twigs. Usually, one caterpillar is found inside a tunnel, but occasionally, there may be 15–30 larvae in heavily infected trees. Pupation takes place in a tunnel of wood; the pupae have a series of teeth or hooks on the abdomen through which the larvae exit the tunnel and release the exuviae. The pupal period lasts about 3 weeks. Old, shady and neglected orchards are prone to attack by this pest.

Keeping the ground clean and preventing tree congestion has been found to reduce pest infestation. This pest is highly prevalent on deciduous trees, so it is important to prune and to paste the cut ends with a pesticide. The gallery can be inspected for a bore hole, which is commonly found on branch crotches (Fig. 11.8). The larva can be killed by inserting a thin, iron spoke into the borehole. Egg clusters can be collected and destroyed from the bark in May–June. After removing the infested and dead bark and galleries, 0.05% dichlorvos can be sprayed on the trunk. During November–January, in the event of a severe attack, the entire web or galleries are removed and 10 ml 0.05% dichlorvos 76% EC is added to the tunnel, which kills the larvae effectively. Pesticides can be injected directly or into boreholes and sealed with soil (Srivastava, 1997).

Two types of entomogenous fungi – *Aspergillus candidus* Link. and *Beauveria bassiana* (Bals.) – have reportedly caused the death of these insects in the field. *A. candidus* was recorded on *Indarbela* spp. in pomegranate up to a level of infection of about 5–11% (Singh and Singh, 1982). Laboratory experiments using this pathogen have shown 100% pest mortality, demonstrating its potential as a biocontrol agent against this pest (Fasih and Srivastava, 1988). About 41.6% and 65% parasitism of *B. bassiana* was recorded on *I. quadrinotata* on guava plants in Lucknow during the months of March and July, respectively. In another study, Arshad and Hafiz (1983) recorded 33.8–45% field mortality and 21.4–33.6% mortality in laboratory experiments. Narendran and Sureshan (1988) isolated the insect *Podagrionella indarbelae* (Chalcidoidea: Torymydiae) from the eggs of *I. tetraonis* in a cashew plantation.

Sidhu and Poon (1983) reported satisfactory control of *I. disciplaga* Swinhoe in Malaysia by injecting chlorpyrifos into pest tunnels and pasting the holes with insecticide-treated soil. Verma (1985) also tried a similar procedure and reported phorate (10% at 1 g per hole) gave 100% mortality in 7 days. Singh and Dhamdhere (1989) used six insecticides in cotton plugs soaked in water; of these, 0.05% demeton methyl, dichlorvos and quinalphos provided 57.5%, 62.03% and 50.67% control, respectively. Petrol and paraffin also provided 95.5% and 73.5% control, respectively, while aluminium phosphide (one tablet per hole) provided 60.6% control of the insect. Patil *et al.* (1990) used dichlorvos and gasoline by inserting cotton swabs embedded in 10 ml of these materials and reported 100% mortality with dichlorvos and 96.7% with gasoline. Mote and Tambe (1990) injected ten insecticides into pomegranate borer holes in Maharashtra and reported fenvalerate (0.01%) and quinalphos (0.05%) as the most effective pesticides. Shevale (1991) also tested the effectiveness of eight insecticides on pomegranate and reported dichlorvos at 0.08% to be most effective, followed by monocrotophos (0.08%) and fenvalerate (0.04%). Sandhu *et al.* (1987) experimented with various treatment application methods and reported that topical application works on trees such as oranges with a rough bark where 90–95% control was recorded but was less effective on trees such as guava, loquat and jujube. Satyanarayana and Arunakumara (2016) reported that chlorantraniliprole 18.5% SC gave 100% pest control and high yields in guava in experiments carried out over 2 years.

Fig. 11.7. Bark-eating caterpillar damage on tree trunks. Photograph courtesy of Dr H.S. Singh.

Fig. 11.8. Internal gallery on tree trunk caused by trunk borers. Photograph courtesy of Dr H.S. Singh.

11.6 Leaf Webbers: Tussock Moth, *Perina nuda* F. (Lepidoptera: Erebidae) and *Glyphodes bivitralis* Guenée (Lepidoptera: Crambidae)

The tussock moth is also known as the transparent wing moth, clear wing tussock moth or banyan tussock moth. This pest attacks palms, figs and mangoes, as well as jackfruit. It is a sporadic but a serious pest of banyan (*Ficus* spp.) and is a major defoliator of forest and shade trees in South-east Asia (Su *et al.*, 1983), causing heavy damage in certain years. The tussock moth is commonly found in many species of mango, *Ficus* and jackfruit in India (NBAIR, 2013). Cheanban *et al.* (2017) reported its nocturnal copulation. Females lay eggs in groups, averaging 46 eggs per group. The number of eggs laid by each female was found to vary from 168 to 387. The incubation period of the eggs was up to 5.5 days. They continued the growth and development process with seven larval instars with the developmental periods of the first- to seventh-instar larval as 2.28, 3.08, 3.02, 3.42, 3.67, 3.34 and 6.10 days, respectively. The larvae shed their hair and used it as a protective weapon. Hair saete were carried across the pupal stage. In another study, Muniappan *et al.* (2012) found that females laid 50–400 faded-red eggs in clusters or rows on leaves, and the eggs later changed to red. The larvae had six moults in the course of 16–20 days. The larvae pupated at the top of the leaves. The adults emerged 5–9 days after pupa formation and the whole life cycle was completed in 27–39 days.

P. nuda is more common in the wet season. The larvae fold or web the leaves collectively and feed inside the folds. In extreme case, it results in reasonably sizeable defoliation.

The leaf webber *Glyphodes bivitralis* Guenée 1854 is local to South-east Asia, Queensland (Australia) and Hawaii. The larvae feed on *Erythrina speciosa*, *Ficus variolosa*, *Ficus elastica* and *Ficus microcarpa*. The larva is greenish with black spots on the thorax and on the remaining abdominal segments. The larvae feed on the rims of the leaves and live inside by twisting them with silky threads. The mature larvae change to brown with uniform black markings. The adults have a wingspan of 30 mm. The forewings are brown with white spots and the hind wings are white with a vast brown margin. Removal and destruction of twisted leaves and spraying with neem oil at 1500 p.p.m. at 3 ml l^{-1} water with a spreader sticker has been recommended. Hand collection can be carried out to destroy the caterpillar in the early stage of the infestation, preventing a pest population increase.

11.7 Scale Insects: *Planococcus lilacinus*, *Aspidiotus destructor* Sign. (Diaspididae), *Pinnaspis strachani*, *Protopulvinaria mangiferae*, *Protopulvinaria pyriformis* and *Ceroplastes rubens* (Coccidae: Hemiptera)

Coccids are found all over India. They attack many tropical and subtropical fruits, and are often grouped near the tips of twigs, on fruit twigs and on the fruit. Hard scale or armoured scale insects are small, about 3 mm long. They can be white, yellow, brown, red or purple. The armoured body varies greatly with host and environment (Sirisena *et al.*, 2013). Adults are oval or rounded, with tight bumps and sometimes a nipple or dimple in the middle. They may or may not produce honeydew, but make an oyster shell, round shell or shell wax in a circular pattern around the body. As the shells grow, they become attached to the host plant, and remain attached until the shells are lifted up.

The eggs of soft scales are covered or found under the mother scales. Crawlers (nymphs) are formed from the eggs, which are flattened and look like dust on the plant surface. Adult females are ovate or rounded, smooth, bulging without legs and wingless. Males have small yellow feathers with soft scales. Soft scale insects secrete honeydew, which attracts ants. The bodies of soft scale insects are not left on the plant when lifted up. The soft cover they secrete is inseparable from the body. Soft scales usually move between branches and leaves during their life cycle. Several species of scale insects attack fruit crops in India. Of these, the five most serious are: *Aspidiotus destructor* Sign., *Pulvinaria polygonata* Cock., *Parlatoria pergandii* Comst., *Parlatoria cinerea* Hadden and *Lepidosaphes gloverii* Pack. *A. destructor* occurs in the mango-growing tracts of Uttar Pradesh and the Punjab; *Parlatoria cinerea* and *Pulvinaria polygonata* in Uttar Pradesh, and *Parlatoria mangiferae* in Orissa. *Aspidiotus dictyospermi* Morgan, *Diaspis mangiferae* Green, and *Lecanium bicruciatum* Green occur in South India. *Lepidosaphes gloverii* occurs in Uttar Pradesh only. *Ceroplastes rubens* is about 3.5–4.5 mm long; the waxy cover of the adult female is pinkish to red, convex, with two conspicuous pairs of white bands that extend dorsally from the spiracular area. The nymphs are pink (Deckle, 2008).

Planococcus lilacinus is commonly known as the Oriental cocoa mealybug. It is a polyphagous insect, attacking more than 30 families of tropical and subtropical fruit trees. *A. destructor* is commonly known as coconut scale, and is also polyphagous, infesting 60 plant families (Ben-Dov *et al.*, 2015). *C. rubens* is a red wax scale that infects more than 50 hosts (CIE, 1960). The scales tend to congregate in huge numbers and remove plant sap using their sucking mouth parts. In severe infections, the affected areas become dry and wither. The leaves dry up and turn yellow, and the twigs and branches die. Both adults and nymphs suck sap from the shoots, leaves, fruit and fruit stalks. They weaken the plants by puncturing the tissues and feeding on the sap, and some species produce honeydew, which promotes the development of saprophytic fungi, resulting in a thick, black layer of sooty mould that significantly reduces photosynthesis and causes premature leaf fall. Fruits heavily infested with scale are unmarketable. The production of honeydew attracts ants, which probably provide protection against parasites and predators.

Affected twigs can be pruned and burnt, especially in summer to control minor infestations. Nearly 40 species of predators and parasites have been reported to attack these coccids. In cases of severe infestation, spraying with dimethoate 30% EC at 2 ml l^{-1} or buprofezin at 1 ml l^{-1} is recommended, but sprays are effective only at the crawler stage of the scales as insecticidal-based control is difficult at other stages of life cycle. The adults are strongly attached to the plant and remain so even after their death, which can give a false impression about the presence of the insect. Chemical applications should only be used if the parasites make the crop economically unviable. The use of pesticides causes large-scale destruction of natural enemies and may lead to pest resurgence.

11.8 Thrips: *Pseudodendrothrips dwivarna* (Thysanoptera: Thripidae)

Thrips are tiny, small, winged and widely distributed insects and are polyphagous. The species was first reported from southern India (Ramakrishna Ayyar and Margabandhu, 1931). Thrips can be important pests but are often less serious than other pests. Thrips can easily be carried long distances on air currents from infected plants. Most adult thrips are small (less than 1 mm long) and have long, narrow wings with two long margins at the edges. The adults have a reddish-brown head and thorax and yellow abdomen. The adult hind tarsi are elongated, and the forewings are narrow with only a faint upper vein (Anandhakrishnan and Sen, 1980). Adults have nine segmented antenna and their third and fourth segments with fork-shaped sensory cones. The immature nymphs are long or slender and compact and have no wings. Most thrips vary in colour from opaque white or yellow to brown or black. Females lay their long, guarded eggs in the leaves, in other plant parts or in the open where the nymphs hatch. Later, the larvae dramatically alter their appearance and behaviour and there is no real pupal stage. Yellow pre-pupae and pupae enter the soil or litter, or settle inside plant cracks. Thrips can have several generations in a year (up to about eight). When the weather is hot, the life cycle from egg to adult is completed in 2 weeks. They attack the leaves, inflorescences and fruits.

The nymphs and adults rasp the ventral surface of the tender leaves and flower stalks and suck the oozing sap. They can change the colour of the leaf, flower and fruit, creating a scar and deforming parts of plants or veins. Affected leaves have a silvery shine and bear small spots of faecal matter. Where numerous thrips occur together, the leaves remain small, and the tips of the leaves turn brown and folded. The twigs of infested shoots are much shorter than those of uninfested ones. Infested inflorescences and young fruits show discoloured tissues at the feeding site, which subsequently turn brown. Populations of thrips are low during the winter, increasing in early spring and reaching high levels during the summer.

Effective control of thrips through the use of pesticides is difficult in part because of their mobility, diet, and the stages of eggs and pupae. Yellow sticky traps may be installed to capture flying adult thrips. In cases of severe incidence, spraying with spinosad 45% SC at 1 ml 5 l^{-1} water is recommended.

11.9 Giant Mealybug: *Drosicha mangiferae* (Monophlebidae), and Spherical Mealybug: *Nipaecoccus viridis* (Pseudococcidae: Hemiptera)

The giant mealybug *Drosicha mangiferae* (Green) is a polyphagous insect, recorded to feed on 70 plant species (Tandon and Lal, 1978; Narula, 2003; Bhau *et al.*, 2017). Tandon and Verghese (1985) reported that *D. mangiferae* is harmful to fruit crops including mango. It is a serious, dilapidating, polyphagous, dimorphic and notorious insect pest of fruit trees of the Indian subcontinent, which is often distributed in the interior of the Gangetic Plains. Other fruits crops such as jackfruit, forest trees, ornamental plants and weeds serve as alternative hosts to this bug. During peak season of infestation, this pest has been reported on different forest trees, with large numbers of crawlers and females seen on *Dalbergia sissoo*, *Bombax ceiba*, *Ficus religiosa* and *Populus alata* trees (Khan, 2001). Mealybugs cause damage to plants principally due to sucking of cell sap from tender leaves, stems, inflorescence and growing fruits, which often causes fruit drop and reduces its market value. Honeydew excretion from mealybugs promotes growth of sooty mould, causing blackening and malformation of growing leaves, stems and fruits.

The body of *D. mangiferae* is usually covered with a waxy powder. Older-instar nymphs and older female bugs are flat, oval and white (Fig. 11.9). The adult male has wings and is small, whereas the female is large and has no wings.

After mating, the female crawls down the tree in April–May and enters cracks in the ground to lay white eggs in large numbers in sacks. One female lays up to 400–500 eggs in the soil. The eggs remain in the diapause stage in contact with the soil until favourable conditions return in November–December. Depending on crop phenology and climate, sometimes emergence is extended to January. Shortly after hatching, the pink-brown minute nymphs crawl to the tree where they develop into adults through three successive nymphal stages. The highest numbers of nymphs are seen between March and April. The time of migration plays a critical role in management decisions.

The duration of the first instar lasts 45–71 days, the second instar 18–38 days and the third instar 15–26 days for females and 5–10 days for males. The total duration of female development is 77–135 days and 67–119 days for males (Srivastava, 1997).

The circular mealybug *Nipaecoccus viridis* was first described in India as *Dactylopius viridis* and was then reported several times (Nechols, 1985). It originated in Asia and has become widespread in tropical and subtropical areas. As a polyphagous insect, it is known to invade more than 35 families of plants including Euphorbiaceae, Fabaceae, Myrtaceae, Leguminoeae, Vitaceae, Rubiaceae, Malvaceae, Tiliaceae, Verbenaceae, Anacardiaceae, Liliaceae and Solanaceae (Ben-Dov *et al.*, 2010). Ghosh and Ghosh (1985) reported the nature of the damage caused by this mealybug on host plants as an arrestant of linear growth of stems and petioles, and causing a reduction and crumbling of leaves. *N. viridis* damage has been recorded on horticultural crops such as grapes (Mani and Thontadarya, 1987), jackfruit (Mani and Krishnamoorthy, 1997; Ben-Dov *et al.*, 2010) and pummelo (Mani and Krishnamoorthy, 2008).

N. viridis sucks sap from stems and petioles, which affects their growth. It also leads to the reduction and crumbling of leaves (Ghosh and Ghosh, 1985).

Repeated long-term control is needed because of the bug's habit of hiding and covering its body. Chemical treatment is not recommended until control becomes unsatisfactory using other means. Biological and physical control methods have been found to be effective in managing mealybug populations, but application of systemic insecticide is needed during heavy infestations. The presence of a wide range of alternative hosts allows persistence of mealybugs and makes them a greater threat in agricultural fields. Alteration of environmental factors may affect their life cycle, thereby affecting the time of infestation. Their life cycle is well known, and the seasonal dynamics of the mealybug help in population

Fig. 11.9. Giant mealybugs. Photograph courtesy of Dr H.S. Singh.

control. Thus, spatial and temporal variations in the life cycle of this insect provide an opportunity to apply a range of cultural, biological and chemical control measures alone or in combination.

Orchard ploughing in November–December destroys many eggs and exposes them to predators. Wrapping the trunk of the tree with a 25 cm wide polythene sheet (400 gauge) (Fig. 11.10) and placing it at a height of 30 cm from the ground during the second week of December has been recommended. Grease is applied to the bottom of the band to prevent the nymphs crawling up. Application of the fungus *Beauveria bassiana* (2 g l^{-1}) or 5% neem seed kernal extract can also be used around the trunk of the tree to kill nymphs. Mounding the soil around the tree trunk and application of 1.5%

Fig. 11.10. A polythene band on a tree trunk used to control mealybug. Photograph courtesy of Dr H.S. Singh.

chlorpyrifos dust at 250 g per tree in the second week of December has also been found to be effective. When nymphs climb the tree, spraying of carbosulfan (0.05%) or dimethoate (0.06%) can be done (Srivastava, 1997).

Mani and Krishnamoorthy (1997) reported that *N. viridis* was suppressed by the encyrtid parasitoid *Anagyrus dactylopii* (Howard) and drosophilid invader *Cacoxenus perspicax* (Knab) within a month.

11.10 Spittlebugs: *Cosmoscarta relata* and *Clovia lineaticollis* (Hemiptera: Cercopidae)

Spittlebugs are known for the frothy spittle mass they produce over their body, which protects them from their enemies. The bugs are 15 mm long, with a reddish-brown head and bright frog-like eyes and are known as froghoppers. The presence of the spittle is an easy way to identify spittlebugs. These sticky lumps are about 2 cm in size, and the nymphs are found inside. They have soft, compact bodies up to about 6 mm long. Initially, they are green and later turn orange to yellow-brown or grey. The nymphs have large red eyes on the sides of their heads. Adult spittle bugs are about 6 mm long but are not usually seen. They have enlarged hind legs for jumping and are similar to leafhoppers but are fatter.

Spittlebugs can survive the winter in leaf litter. The nymphs emerge in late April or early May and begin feeding under the plants. They continue to climb, looking for soft leaves and flowers. Nymphs secrete a liquid that is released as foam at feeding time. They mature in 5–8 weeks. As adults, they move to nearby grasslands, pastures or areas with a wide range of weeds. The females return in September and October to lay their eggs in plant litter or on leaves and tree branches or stems. The total life cycle takes about 2 months and there is only one generation per year.

Spittlebugs are a pest of regional significance and are common in southern India. Infestation of spittlebugs on jackfruit has been reported, for example on the Kolli Hills in Tamil Nadu (Kareem *et al.*, 1971). The adult lays eggs inside the tender shoots and the nymphs pierce the soft twig stems and leaves of the plant and absorb plant sap juices. They also feed on a variety of plants such as ornamental grasses, roses, chrysanthemum, clover, strawberries, herbs and many other orchard plants. In many cases, spittlebug feeding does not harm the plants, but if there are large numbers of spittlebugs, their feeding can cause the leaves to lose their shape, leading to drying of the younger shoots, yellowing and bending of the leaves, and weakening of the tree. The highest levels of infection occur in March during flowering.

Spittlebug foam and nymphs need be checked regularly on the underside of young leaves. Ordinarily, control of spittlebugs is not needed because they occur in non-damaging numbers and are only found on the plants for a short time causing little damage. However, orchard cleaning, proper tree training and pruning may be carried out in order to get rid of this pest if necessary. The foamy mass can be collected and destroyed with a brush and affected shoots can be removed and destroyed. A strong blast of water spray can be applied to remove nymphs from the plants. Insecticides are normally not effective against this insect as nymphs are protected from the pesticide droplets by the large volume of their saliva. However, spraying with quinalphos (0.05%) or neem oil (0.03%) may be useful. The adults can be controlled by dust formulations of insecticides (Jeyaraj and Ananthan, 2009). In field experiments in Salem District, Tamil Nadu, India, spraying jackfruits with 0.07% monocrotophos gave 89% control of *Cosmoscarta relata* (Venkatesan *et al.*, 1992). Kareem *et al.* (1971) reported that pyrethrum (0.1%), malathion (0.1%), endrin (0.04%) and naphthalene powder all gave complete control of *C. relata*.

11.11 Bud Weevils or Brown Weevils: *Ochyromera artocarpi*, *Onychocnemis careyae* and *Teluropus ballardi* (Coleoptera: Curculionidae)

Bud weevil, *Onychocnemis careyae* is a reddish-brown insect with a dark head and shiny short white setae. It is reported to feed on the leaves of the jackfruit (Marshall, 1917) and is specific to jackfruit. It has a brown elytral suture and the patch extends from the base to the centre. The pronotum is coarsely coated and the legs are pierced. *Teluropus ballardi* (Marshall) is another species of weevil that has been reported on the leaves of jackfruit in India (Marshall, 1917; Jha and Sen-Sarma, 2008).

The small whitish grubs of *O. artocarpi* bore into tender flower buds and fruits and cause premature fruit drop. The adult weevils nibble on leaves (Butani, 1979b). Ayyar (1940) reported *O. artocarpi* in the north Malabar area of Kerala and Mysore in Karnataka as a pest on jackfruit, while Subramanian (1958) reported its outbreak in the Coimbatore region.

Soumya *et al.* (2020b) conducted a study on biology of *O. artocarpi* and reported that the gravid females of *O. artocarpi* made small holes in small jackfruits and laid eggs in them. After hatching, the larvae scrape the soft tissue in the outer part of the fruit. The first-instar larvae were found feeding in the pericarp region of the fruit. Fully grown caterpillars caused very deep damage to the endocarp of the fruit. The freshly emerged adult weevils fed in the pericarp region. The adult emerged from the dried fallen fruits by making small circular holes. Male and female weevils completed their life cycle in 25.18 ± 1.09 days and 26.14 ± 1.12 days, respectively, with five larval instars. The incubation period was 3.90 ± 0.20 days. The mean growth period for caterpillars, pre-pupae and pupae was 11.84 days ± 1.44, 3.10 ± 0.07 and 7.30 ± 0.50 days, respectively. The older larval instars are compact and C-shaped. Subramanian (1958) also reported that *O. artocarpi* completes its life cycle with five larval stages. The larvae complete their growth within 14–15 days, and the first and last (fifth) larvae were 1.27 ± 0.05 and 4.98 ± 0.61 mm long, respectively.

Small, white-coloured grubs of *O. artocarpi* pierce the bud, flower, tender sprouts and fruit and feeds on them. Severe infection causes premature fruit fall. Soumya *et al.* (2020b) reported that the insect population fluctuates during different months of the year. A declining population of jackfruit bud weevil was strongly associated with weather conditions. The first infestation was seen in December, which coincides with the onset of the fruit set. This increased from January to its peak in March, and completely subsided in April. The population level was associated with evaporation ($r = 0.70$) and was closely related to rainfall ($r = 0.63$). Adults were sometimes found nibbling on leaves and feeding on the outer portion of immature fruits during March. The incidence was similar during all 3 years studied (November–May), with a peak density in March.

The egg and grub stages are found inside the soil, which is why pesticides are often not effective enough to control them in the egg and adult stages. However, soil treatment with insecticides may control them effectively and can prevent further infestation. The adult weevils leave the tree canopy where pesticides are used. Manual collection and destruction of insects along with affected flowers, buds and fallen fruit can check the further spread of weevil damage. Application of insecticides at night is required when the weevils remain active and may come in contact with the insecticide. Another option for weevil prevention is covering 8–10 cm of the trunk with an adhesive to prevent the adults from crawling up the tree trunk. A preparation of the fungus *Beauveria bassiana* may be applied as a soil drench and may kill the larvae and emerged adult stages of the weevil. Insect parasitic nematodes can also be used to contain the juvenile stages of weevils multiplying in the soil.

11.12 Aphids: *Greenidia artocarpi* and *Toxoptera aurantii* (Hemiptera: Aphididae)

The aphid *Greenidia artocarpi* has been reported from Meghalaya, Karnataka and Tamil Nadu in India. It is light brownish to green and feeds on the lower surface of tender leaves in January–February and August–October. It is medium sized and relatively long in the body. The cornices are dark brown, very long and thin.

Toxoptera aurantii or black citrus aphid is very important in agriculture, partly because of its polyphagy but mainly because of its ability to spread various viruses (Blackman and Eastop, 2000). It is found in tropical and subtropical areas on nearly 120 plant species including *Annona*, *Ficus*, citrus, coffee, sandalwood and tamarind. It usually feeds on new shoots, flowers and emerging buds (Carver, 1978). Black citrus aphid reproduces parthenogenetically, allowing rapid population growth. The aphids are usually attended by ants, which provide protection from natural enemies. Depending on temperature, its life cycle is completed within 6 days at 25°C, but takes 20 days below 15°C (Blackman and Eastop, 1984). The adults are dark brown, glittery black or red, alate or without wings, and with a black and white stripe on the horns. The cauda is finger-shaped, dull and slightly bumpy. It has a stridulatory device consisting of rims on the abdomen and hooks on the rear tibia.

The nymphs and adults suck sap from newly emerged tender branches, leaves and soft fruits. This leads to weakening of the trees and restricts growth. The aphids produce honeydew on the outer surface of the fruit and leaves, which promotes the growth of a sooty fungus that inhibits photosynthesis, weakens the plant and makes the fruit unattractive. Its feeding often causes plants to become deformed, and the leaves curl and wither and, in some cases, galls form on the leaves (Metcalf, 1962).

In most cases, black citrus aphid is a minor jackfruit insect pest and is not usually a problem other than on small trees or perennials, as their numbers decrease as the leaves harden and their natural enemies become more effective at controlling them. Natural enemies often control aphids over a short period of time, and pesticide application is rarely needed. The main parasites are braconids and chalcidoids (two families of parasitoid wasp). Lacewings, many coccinellids (e.g. lady beetles) and syrphid predators, parasites and fungal infections often keep aphid numbers below harmful levels. A moderate number of aphids (about 40% of growth flushes infested) can be considered beneficial on mature trees because the aphids and their honeydew provide a good food source for natural enemies of other pests early in the season when other hosts are not available (UC IPM, 2020).

11.13 Spiralling Whitefly: *Aleurodicus dispersus* (Aleyrodidae: Hemiptera)

The spiralling whitefly is a small sap-sucking insect. The adults have a body length of about 2 mm, and their wings are whitish, sometimes with pale or black spots on the front. It is a polyphagous insect with a circular pattern of oviposition under the leaves (Russell, 1965). The species is native to Central America, the Caribbean region and islands of the Pacific. Palaniswami *et al.* (1995) reported it from Kerala in India in 1993, and later Reddy and Chandurkar (1999) from other parts of peninsular India. Srinivasa (2000) has compiled a comprehensive list of whiteflies covering 481 plants under 295 genera from 90 families of vegetables, fruits and ornamental trees. Botha *et al.* (2000) reported its key hosts such as citrus, avocados, guavas, bananas, coconuts, beans, cassava and stone fruits.

An older female usually lays her eggs in the normal way just under the lower leaves, but when the population is large, eggs can also be seen on the upper surface of leaves and on fruit and non-vegetative material. Kumari (2011) reported that the eggs are round and yellow, 0.3 mm long and placed separately at right angles in the veins of leaves, and are associated with irregular deposits of white wax. At a temperature of 20–39°C, growth from egg to adult takes 34–38 days (eggs 9–11 days, first instar 6–7 days, second instar 4–5 days, third instar 5–7 days and fourth instar (pseudo-pupa) 10–11 days). Under laboratory conditions, an adult can live for 39 days.

The insect has three larval instars and a fourth pseudo-pupa form. The life cycle is generally completed within 21–48 days. The adults are clear on appearance but develop a white powdery covering over the next few hours, and measure 4 mm in size. The nymphs and adults suck sap from the leaves, leading to excessive sugar loss and reduced flowering and fertility.

As with other aleyrodids, the damage is caused by the immature phases, which suck the plant sap. Their feeding reduces the plant's vitality and allows it to be attacked by other insects. The larvae also produce honeydew, which promotes the growth of a dark brown fungus on the leaves, reducing the plant's ability to photosynthesize. The appearance of plants is greatly affected by the presence of these insects, as both eggs and larvae are covered with a visible white substance.

Affected leaves dry out and fall to the ground. The first stage of the larva can move but the later stages settle down with a soft disc-shaped body that is light in colour. The final stage of maturation is the pupa, which is about 1 mm long. The sedentary larvae have characteristic waxy tufts and the pupa has glass-like flecks of wax along the sides of the body.

Whitefly can be controlled by collecting and destroying fallen leaves, installing yellow sticky traps (20 ha^{-1}) to attract adults and spraying with fish oil soap in the morning (1 kg l^{-1} water). In India, neem oil, rosin soap and cleaning soap have been found to be effective without a large amount of pesticides (Srinivasan *et al.*, 2002). Formulations based on botanicals such as neem and pongamia may serve as effective management methods for low populations of whiteflies (Baradevanal *et al.*, 2013). It has been reported that this pest can successfully be controlled using triazophos 40% EC (0.06%), dimethoate 3%0 EC (0.05%) and profenofos (Kumari, 2011). Neem oil and cotton seed oil (0.01%) have also been found to cause significant deaths of various instars of the insect. Several exotic parasitoids have been found to be very effective, including the two aphelinid parasitoids *Encarsia haitiensis* Dozier and *E. meritoria* Gahan, which is highly promising and is reported to reduce the number of pesticides required (Kumari, 2011).

11.14 Conclusion

Organized plantations of jackfruit and manual intervention therein for commercial exploitation is not very prevalent and so this crop has had less exposure to pesticides. As a less- or undisturbed ecosystem, jackfruit trees and groves probably have a more diverse fauna, which may help to maintain checks on pest populations. Perhaps this is the reason that, despite the recorded incidence of 37 insect pests on jackfruit (six Lepidoptera, six Coleoptera, 21 Hemiptera, one Thysanoptera and three Diptera), only three – the shoot and fruit borer *Diaphania caesalis* Walker, the trunk borer *Batocera rufomaculata* De Geer and the bark-eating caterpillar *Indarbela* spp. – have been reported to be major pests causing considerable damage to jackfruit. Although the literature describes around 46 insect pests on jackfruit, most from Karnataka State in India, the majority are of minor importance. These minor pests sometimes cause sporadic damage but rarely warrant intervention. They are naturally controlled by their natural enemies, and there is little need for treatment of jackfruit with pesticides. Detailed studies on the three major insect pests have been carried out and a management protocol has been developed.

References

Ahmed, K.U., Rahman, M.M., Alam, M.Z., Hossain, M.M. and Miah, M.G. (2013a) Effect of seasonal variations on jackfruit trunk borer (*Batocera rufomaculata* de Geer) infestation. *Pakistan Journal of Biological Sciences* 16(7), 339–344.

Ahmed, K.U., Rahman, M.M., Alam, M.Z., Hossain, M.M. and Miah, M.G. (2013b) Evaluation of some control methods against the jackfruit trunk borer, Batocera rufomaculata de Geer (Cerambycidae: Coleoptera). *Bangladesh Journal of Zoology* 41, 181 187.

Alam, M.Z. (1962) *Insect and Mite Pests of Fruit and Fruit Trees in East Pakistan and Their Control.* Department of Agriculture, East Pakistan, Dacca.

Alam, M.Z., Ahmad, A., Alam, S. and Islam, M.A. (1964) *A Review of Research, Division of Entomology (1947–1964).* Department of Agriculture, East Pakistan, Dacca.

Ali, M.S., Azam, F. and Chaturvedi, O.P. (2007) Occurrence, host spectrum and host specificity of bark eating caterpillar, *Inderbela quadrinotata* Wlk. in relation to trees of Bihar. *Journal of Tropical Forestry* 23, 59–64.

Anandhakrishnan, T.N., and Sen, S. (1980) Taxonomy of Indian Thysanoptera. Hand Book Series no.1, Zoological Survey of India. Government of India, India.

Arshad, M. and Hafiz, I.A. (1983) Field performance of *Beauveria bassiana* (Bals.) Vuvill. on *Indarbela quadrinotata* Wlk., Pseudarbelidae: Lepidoptera. *Bulletin of Zoology* 1, 11–16.

Ayyar, T.V.R (1940) *Handbook of Economic Entomology for South India.* Government Press, Madras, India.

Azad, A.K. (2000) Genetic diversity of jackfruit in Bangladesh and development of propagation methods. PhD thesis, University of Southampton, Southampton, UK.

Baradevanal, G., Jayanthi, P.D.K. and Verghese, A. (2013) Management of spiraling whitefly, *Aleurodicus dispersus* (Russel) in guava, *Psidium guajava* L. *Pest Management in Horticultural Ecosystems* 19(1), 102–105.

Baradevanal, G., Balaji, R., Srivastava, K. and Singh, S. (2015) Rearing of mango stem borer, *Batocera rufomaculata* De Geer (Coleoptera: Cerambycidae) on artificial diet. *Pest Management in Horticultural Ecosystems* 21, 219–220.

Beeson, C.F.C. (1941) *Ecology and Control of the Forest Insects of India and the Neighbouring Countries*. C.F.C. Beeson, Vasant Press, Dehra Dun, India.

Ben-Dov, Y., Miller, D.R. and Gibson, G.A.P. (2010) *Nipaecoccus viridis* (Newstead, 1894). Available at: http://scalenet.info/catalogue/Nipaecoccus viridis/ (accessed 15 July 2022).

Ben-Dov, Y., Miller, D.R. and Gibson, G.A.P. (2015) *Aspidiotus destructor* Signoret. Available at: http://scalenet.info/catalogue/Aspidiotus destructor/ (accessed 15 July 2022).

Bhau, B., Shankar, U. and Abrol, D.P. (2017) Studies on host range and biology of mango mealybug (*Drosicha mangiferae*) in Jammu Region. *International Journal of Current Microbiology and Applied Sciences* 6, 230–240.

Blackman, R.L. and Eastop, V.F. (1984) *Toxoptera aurantii* (Boyer de Fonscolombe), black citrus aphid. In: *Aphids on the World's Crops: An Identification and Information Guide*. Wiley, Chichester, UK, pp. 364–365.

Blackman, R.L. and Eastop, V.F. (2000) *Aphids on the World's Trees – An Identification and Information Guide*. CAB International, Wallingford, UK.

Botha, J., Hardie, D. and Power, G. (2000) *Spiraling Whitefly* Aleurodicus dispersus: *Exotic Threat to Western Australia*. HortGuard™ Initiative AGWEST. Available at: https://ausveg.com.au/app/data/technical-insights/docs/TL38.pdf (accessed 15 July 2022).

Butani, D.K. (1979a) *Insects and Fruits*. Periodical Expert Book Agency, Vivek Vihar, Delhi.

Butani, D.K. (1979b) Insect pests of fruit crops and their control: jackfruit. *Pesticides* 12, 36–44.

Butani, D.K. and Jotwani, M.G. (1984) *Insects in Vegetables*. Periodical Expert Book Agency, Vivek Vihar, Delhi.

Carroll, L.E., White, I.M., Freidberg, A., Norrbom, A.L., Dallwitz, M.J. *et al*. (2002) Pest fruit flies of the world– larvae. *Bactrocera latifrons* (Hendel). Available at: www.delta-intkey.com/ffl/www/bac_lati.htm (accessed 15 July 2022).

Carver, M. (1978) The black citrus aphids, *Toxoptera citricidus* (Kirkaldy) and *T. auranti* (Boyer de Fonscolombe) (Homptera: Aphidiae). *Journal of the Australian Entomological Society* 17, 263–270.

Chatterjee, P.N., Singh, P. and Prasad, G. (1969) Insect pests of fast-growing tree species II. Effect of various fumigants on poplar stem borers, *Apriona cinerea* Chevrolat (Lamiidae) in New Forest Plantations. *Indian Forester* 95, 369–371.

Cheanban, S., Bumroongsook, S. and Tigvattananont, S. (2017) *Perina nuda* F. (Lepidoptera: Lymantriidae): an important leaf eating caterpillar of fig trees. *International Journal of Agricultural Technology* 13, 485–492.

Chowdhury, S. and Majid, S. (1954) *HandBook of Plant Protection*. Department of Agriculture, Assam, India.

Commonwealth Institute of Entomology (1960) Distribution Maps of Pests, Series A (Agricultural), Map no.118. CIE, UK.

Deckle, G.W. (2008) *Red Wax Scale,* Ceroplastes Rubens *Maskell (Insecta: Hemiptera: Coccidae*). EENY 237. University of Florida, Gainesville, Florida.

Fasih, M. and Srivastava, R.P. (1988) Natural occurrence of *Beauveria bassiana*, an entomogenous fungus on bark eating caterpillar, *Indarbela* spp. *Indian Journal of Plant Pathology* 6, 11–16.

Fletcher, T.B. (1914) *Some South Indian Insects*. Suprintendent, Government Press, Madras, India.

Garg, V.K., and Tomar, D.S. (2008) Preferential feeding of bark eating caterpillar, *Indarbela quadrinotata* (Walker) on horticultural and forest plant species. *Indian Forester* 134, 579–580.

Ghosh, A.B. and Ghosh, S.K. (1985) Effect of infestation of *Nipaecoccus vastator* (Maskell) on host plants. *Indian Agriculturist* 29, 141–147.

Gupta, S.L. (1990) Key for the identity of some major lepidopterous pests of vegetables in India. *Bulletin of Entomology* 31, 69–84.

Gurjar, S., Choudhary, J., Choudhary, S., Dhakar, M., Seema, H. *et al*. (2019) Stem borer (*Batocera rufomaculata* De Geer): a threat to jackfruit cultivation in Hill and Plateau Region of India. In: *Conference Proceedings: National Conference on Integrated Plant Health Management in Fruit Crops, September 3–4, 2019*, p. 169.

Hampson, G.F. (1896) *The Fauna of British India Including Ceylon and Burma*. Moths, Vol. I. Taylor& Francis, London.

Haq, N. (2006) *Jackfruit* (Artocarpus heterophyllus). Southampton Centre for Underutilized Crops, University of Southampton, Southampton, UK.

Hua, L.-Z. (1982) *A Check List of The Longicorn Beetles of China (Coleoptera: Cerambycidae*). Zhongshan University, Guangzhou, China.

Jeyaraj, J. and Ananthan, M. (2009) Spittle bug menace in jackfruit, science and technology. *The Hindu Daily*, 9 February, 2009.

Jha, L.K. and Sen-Sarma, P.K. (2008) *Forest Entomology*. APH Publishing Company, New Delhi.

Kareem, A.A., Nachiappan, R.M., Subramaniam, T.R. and Govindarajan, V. (1971) Severe occurrence of a spittle bug, *Cosmoscarta relata*, Dist. (Cercopidae: Homoptera) on jack at Kolli hills in Tamil Nadu and its control. *South Indian Horticulture* 19, 87–89.

Karim, M.A. (1995) Insect pests of fruits and their control in Bangladesh. In: *Fruit Production Manual*. Horticulture Research and Development Project in collaboration with Department of Corporation, Dhaka, p. 113.

Khan, M.A. (2001) Control of insect pests of mango. In: *Proceedings of the International Mango Workshop, 27 February–1 March 2001*. Directorate of Agriculture Multan, p. 224.

Khan, M.A.M. and Islam, K.S. (2004) Nature and extent of damage of jackfruit borer, *Diaphania caesalis* Walker in Bangladesh. *Journal of Biological Sciences* 4, 327–330.

Khan, M.A.M., Islam, K.S. and Haque, M.A. (2003) Biology of jackfruit borer, *Diaphania caesalis* Walker in Bangladesh. *Bangladesh Journal of Environment Science* 9, 417–421.

Kim, S.K., Park, I.G., Park, K.Y. and Kim, K.J. (1994) Seasonal change and biological features of *Apriona germari* Hope as a pest on a fig. In: *Korean Journal of Applied Entomology,* Abstract. p. 30.

Kumari, N.V. (2011) Spiralling white fly, *Aleurodicus disperses* Russel: new major pest of mulberry. *Indian Streams Research Journal* 1, 1–5.

Lee, S.M. (1987) *Cerambycidae in Korea*. The National Institute of Science, Taejon, South Korea (in Korean).

Liquido, N.J. (1990) Sanitation as a cultural method of suppressing melon fly population in papaya orchards. In: *Proceedings of the 26th Annual Hawaii Papaya Industry Association Conference*, College of Tropical Agriculture & Human Resources, University of Hawaii.

Madhava Rao, V.N. (1965) The Jackfruit in India. *Farming Bulletin* (New Series), no.34. Indian Council of Agricultural Research, New Delhi.

Mani, M. and Thontadarya, T.S. (1987) Record of mealybug species on grapevine in Karnataka. *Current Science* 56, 1192.

Mani, M. and Krishnamoorthy, A. (1997) Suppression of spherical mealybug, *Nipaecoccus viridis* (Newstead) (Homoptera: Pseudococcidae) on jack fruit. *Entomon* 22, 161–163.

Mani, M. and Krishnamoorthy, A. (2008) Biological suppression of the mealybugs *Planococcus citri* (Risso), *Ferrisia virgata* (Cockerell) and *Nipaecoccus viridis* (Newstead) on pummelo with *Cryptolaemus montrouzieri* Mulsant in India. *Journal of Biological Control* 22, 169–172.

Manjunatha, R., Kumar, R.M. and Halikatti, G. (2014) Biology of jackfruit shoot and fruit borer, *Diaphania caesalis* (Walker) (Pyralidae: Lepidoptera). *Trends in Biosciences* 7, 3705–3707.

Mann, G.S. and Bindra, O.S. (1977) Field screening of different jujube cultivars against the bark eating caterpillar. *Indian Journal of Horticulture* 34, 313–315.

Marshall, G.A.K. (1917) On new species of Indian Curculionidae. Part III. *Annals and Magazine of Natural History* 19, 188–198.

Martinez, M.A. (1999) Biology of jackfruit fruit borers. *Philippine Journal of Crop Science* 24(Suppl. 1), 27.

Mathew, G. and Rugmini, P. (1997) Management of bark caterpillar *Indarbela quadrinotata* on *Albizia falcataria* and its control. *KFRI Research Report* 122, 24.

Metcalf, R.L. (1962) *Destructive and Useful Insects Their Habits and Control*. McGraw-Hill, New York/San Francisco/Toronto/London.

Meyrick, E. (1928) Exotic microlepidopterans. *Bulletin of the British Museum (Natural History)* 3, 391.

Mote, U.N. and Tambe, A.B. (1990) Chemical control of bark eating caterpillar *Indarbela quadrinotata* (Walker) in pomegranate. *Plant Protection Bulletin* 42, 7–8.

Mridha, S.H. (2006) Biology, pest status and management of jackfruit borer, *Diaphania caesalis* Walker. Sher-E-Bangla Agricultural University, Sher-E-Bangla Nagar, Dhaka.

Muniappan, R., Shepard, B.M., Carner, G.R. and Ooi, P.A.C. (2012) *Arthropod Pests of Horticultural Crops in Tropical Asia*. CAB International, UK. DOI: 10.1079/9781845939519.0000.

Naik, K.C. (1949) *South Indian Fruits and Their Culture*. P. Varadachery and Co, Madras, India.

Narendran, T.C. and Sureshan, P.M. (1988) A contribution to our knowledge of Torymidae of India (Hymenoptera: Chalcidoidea). *Bollettino del Laboratorio di Entomologia Agraria "Filippo Silvestri" Portici* 45, 37–47.

Narula, S. (2003) Time to tackle bugs that attack mango trees. *The Tribune Spectrum, 12 January 2003*. Available at: www.tribuneindia.com/2003/20030112/spectrum/main2.htm (accessed 15 July 2022).

National Bureau of Agricultural Insect Resources (2013) *Insects in Indian Agro-Ecosystems*. Ministry of Environment and Forests, India.

Nayar, K.K., Ananthakrishnan, T.N. and David, B.V. (1989) *General and Applied Entomology*. Tata McGraw-Hill, New Delhi.

Nechols, J.R. (1985) Biological control of the spherical mealybug Guam and in the Northern Marianas Islands: a classic example of fortuitous biological control. In: *Proceedings of the 1st International Symposium on Biological Control of Arthropods,* January 14–18, pp. 324–329.

NIPHM Working Group (2015) The AESA Based IPM – Jack Fruit. National Institute of Plant Health Management (NIPHM), India.

Paik, W.H. (1987) *The New Pestology*. Hyangmunsa, Seoul, pp. 367–368.

Palaniswami, M.S., Pillai, M.S., Nair, K.S. and Mohandas, C. (1995) A new cassava pest in India. *Cassava Newsletter* 19, 6–7.

Patil, P.V., Patil, V.K. and Deshpande, A.D. (1990) Incidence and control of bark eating caterpillar, *Indarbela* sp. on guava trees. *Journal of Maharashtra Agricultural University* 15, 230.

Poland, T.M. and McCullough, D.G. (2006) Emerald ash borer: invasion of the urban forest and the threat to North America's ash resource. *Journal of Forestry* 104, 118–124.

Prarthna, R., Debanand, D., Baruah, K. and Dutta, S.K. (2014) New record of *Glyphodes caesalis* Walker (Lepidoptera: Pyralidae) as a pest of jackfruit from Jorhat (Assam). *Insect Environment* 19, 247.

Ramakrishna Ayyar, T.V. and Margabandhu, V. (1931) Notes on Indian Thysanoptera with brief description of new species. *Journal of the Bombay Natural History Society* 34, 1029–1040.

Reddy, G.R.S. and Chandurkar, P.S. (1999) First record of spiralling whitefly *Aleurodicus dispersus* (Russell) on guava plants at Hyderabad. *Plant Protection Bulletin (Faridabad)* 51, 24.

Russell, L.M. (1965) A new species of *Aleurodicus* Douglas and two close relatives (Homoptera: Aleyrodidae). *The Florida Entomologist* 48(1), 47. DOI: 10.2307/3493523.

Sandhu, G.S., Sohi, A.S. and Batra, R.C. (1987) Comparison of different methods for the management of the bark-eating caterpillar *Indarbela quadrinotata* (Metarbelidae: Lepidoptera) infesting fruit and avenue trees in Punjab. *Punjab Horticultural Journal* 27, 255–259.

Satyanarayana, C. and Arunakumara, K.T. (2016) Biology and management of guava bark eating caterpillar (*Indarbela tetraonis* Moore). *Agricultural Science Digest - A Research Journal* 36, 197–201. DOI: 10.18805/asd.v0iof.11153.

Shevale, B.S. (1991) Control of bark eating caterpillar, *Indarbela quadrinotata* (Walker) in pomegranate through insecticidal sprays. *Plant Protection Bulletin* 43, 31–32.

Sidhu, M. and Poon, C.K. (1983) Efficacy of insecticide-soaked cotton plugs against caterpillars of *Indarbela* sp. (Lepidoptera: Metarbelidae) infesting guava trees. *Pesticides* 22, 19–20.

Singh, H.S., Verghese, A., Stonehouse, J.M., Mumford, J.D., George, S. *et al*. (2008) Developing bait and lure-based integrated pest management module for mango fruit fly (*Bactrocera dorsalis*) management in Orissa. *Indian Journal of Agricultural Sciences* 78, 609–613.

Singh, P. and Prasad, G. (1985) Poplar stem borer, *Apriona cinere*a Chevrolat (Coleoptera: Cerambycidae) and its biology and control. *Indian Forester* 111, 517–524.

Singh, P. and Singh, S. (1987) Pest and pathogen management in agroforestry systems. In: Khosla, P.K. and Khurana, D.K. (eds) *Agroforestry for Rural Needs, Vol. 1*. Indian Society of Tree Scientists, Solan, India, pp. 153–177.

Singh, R. (1969) *Fruits*. National Book Trust, New Delhi, India.

Singh, R. and Singh, J. (1982) New record of *Aspergillus candidus* Link – a potential entomogenous fungus on *Indarbela* spp. *Science and Culture* 48, 282–283.

Singh, U.C. and Dhamdhere, S.V. (1989) Field screening of some guava varieties against the bark eating caterpillar, *Indarbela quadrinotata* (Walker). *Indian Journal of Entomology* 51, 216–217.

Sirisena, U.G.A.I., Watson, G.W., Hemachandra, K.S. and Wijayagunasekara, H.N.P. (2013) Mealybugs (Hemiptera: Pseudococcidae) species on economically important fruit crops in Sri Lanka. *Tropical Agricultural Research* 25, 69–82.

Soumya, K., Krishnamoorthy, A., Patil, P., Rangarajan, V. and Venkatesha, M.G. (2017) Forecasting the incidence of jackfruit shoot and fruit borer *Diaphania caesalis* Walker (Pyralidae: Lepidoptera) in jackfruit (*Artocarpus heterophyllus* Lam.) ecosystems. *Journal of Entomology and Zoology Studies* 5, 483–487.

Soumya, K., Krishnamoorthy, A. and Venkatesha, M.G. (2019) Biology and seasonal incidence of the jack shoot and fruit borer, *Diaphania caesalis* (Walker) (Lepidoptera: Crambidae). *International Journal of Tropical Insect Science* 39, 235–241.

Soumya, K., Krishnamoorthy, A. and Venkatesha, M.G. (2020a) Insect pest complex of jackfruit in South India. *International Journal of Tropical Insect Science* 40(3), 513–525. DOI: 10.1007/s42690-020-00099-7.

Soumya, K., Krishnamoorthy, A. and Venkatesha, M. (2020b) Biology and incidence of the jackfruit bud weevil *ochyromera artocarpi* marshall (coleoptera: curculionidae). *Indian Journal of Agricultural Research* Article ID: A-5653. DOI: 10.18805/IJARe.A-5653.

Srinivasa, M.V. (2000) Host plants of the spiraling whitefly *Aleurodicus dispersus* Russell (Hemiptera: Aleyrodidae). *Pest Management in Horticultural Ecosystems* 6, 79–105.

Srinivasan, R., Poorani, J. and Bhumannavar, B. S (2002) Spiralling whitefly, *Aleurodicus dispersus*, in India. *Biocontrol News and Information* 23, 55–62.

Srivastava, R.P. (1997) Fruit flies. In: *Mango Quarantine Pests and Export Rules*. International Book Distributing Co, Lucknow, India.

Steiner, L.F. (1957) Field evaluation of oriental fruit fly insecticides in Hawaii. *Journal of Economic Entomology* 50, 16–24.

Su, X., Tai, K.C., Shih, M.P. and Hsuan, P.J. (1983) Two nuclear polyhedrosis viruses of two forest insect pests. *Scientia Silvae Sinicae* 19, 114–115.

Subramanian (1958) The occurrence and life cycle of jack weevil (*Ochyromera artocarpi* Marshall) at Coimbatore. *Current Science* 27, 214.

Tandon, P.L. (1998) Management of insect pests in tropical fruit crops. In: Arora, R.K. and Ramanatha Rao, V. (eds) *Proceeding of the IPGRI-ICAR-UIFANET. Regional Training Course on the Conservation and Use of Germplasm of Tropical Fruits in Asia, 18–31 May 1997*. Indian Institute of Horticultural Research, Bangalore, India, pp. 237–244.

Tandon, P.L. and Lal, B. (1978) The mango coccid, *Rastrococcus iceryoides* Green (Homoptera: Coccidae) and its natural enemies. *Current Science* 13, 46–48.

Tandon, P.L. and Verghese, A. (1985) *World List of Insect, Mite and Other Pests of Mango*. Technical Document no.5, Indian Institute of Horticultural Research, Bengaluru, India.

University of California Statewide Integrated Pest Management Program (2020) *Pest Management Guidelines: Citrus*. UC ANR Publication 344. University of California. Los Angeles, California. Available at: https://anrcatalog.ucanr.edu/Details.aspx?itemNo=3441

Upadhyay, S., Chaudhary, B. and Sapkota, B. (2013) Integrated management of mango stem borer (*Batocera rufomaculata* Dejan) in Nepal. *Global Journal of Biology, Agriculture and Health Sciences* 2, 132–135.

US Department of Agriculture (2016) *A Review of Recorded Host Plants of Oriental Fruit Fly, Bactrocera Dorsalis (Hendel) (Diptera: Tephritidae)*. Version 2.1. A Product of the USDA Compendium of Fruit Fly Host Information (CoFFHI).

Vagalo, M., Hollingsworth, R. and Tsatsia, F. (1997) Fruit fly fauna in Solomon Islands. *ACIAR Proceedings* 76, 81–86.

van den Bosch, R. and Haramoto, F.H. (1953) Competition among parasites of the Oriental fruit fly. *Proceedings of the Hawaiian Entomological Society* 15, 201–206.

van den Bosch, R., Bess, H.A. and Haramoto, F.H. (1951) Status of Oriental fruit fly parasites in Hawaii. *Journal of Economic Entomology* 44, 753–759.

Venkatesan, S., Letchoumanane, S., Thamburaj, S. and Prakasam, V. (1992) Chemical control of spittle bug, *Cosmoscarta relata* D. on jack in Kolli hills. *South Indian Horticulture* 40, 295–296.

Verma, T.D. (1985) Incidence and chemical control of bark-eating caterpillar, *Indarbela quadrinotata* Walker on plum trees. *Indian Journal of Agricultural Sciences* 55, 131–132.

Walker, F. (1859) *Catalogue Lepidoptera*. British Museum, London 17, 499.

Wang, P.-Y. (1963) Notes on Chinese pyralid moths of genus *Diapania* (Hubner) Peking. *Acta Entomologica Sinica* 12, 358–365.

Yoon, H.J., Mah, Y.J., Park, J.G., Lee, S.B. and Yang, S.Y. (1997a) The mode of hibernation of mulberry longicorn beetle, *Apriona germari* Hope in Korea. *Journal of Sericulture Science Japan* 66, 128–131.

Yoon, H.J., Park, J.G., Mah, Y.J., Lee, S.B. and Yang, S.Y. (1997b) Ecological characteristics of mulberry longicorn beetle, *Apriona germari* Hope, at the hibernation stage in mulberry fields. *Korean Journal of Applied Entomology* 36, 67–72.

Zhang, S. and Shen, Y.W. (1980) In: Xiao, G. (ed.) *Forest Insects of China*. China Forestry Publishing House, Beijing, pp 461–463.

12 Diseases

Van Tri Mai[1]*, Dinh Thi Yen Phuong[2] and David Guest[3]
[1]Southern Horticultural Research Institute, Tien Giang, Vietnam; [2]School of Biotechnology, Tan Tao University, Long An, Vietnam; [3]School of Life and Environmental Sciences, University of Sydney, Sydney, Australia

12.1 Introduction

Commercial cultivation of jackfruit (*Artocarpus heterophyllus*) has expanded in the last 10 years in the Asia-Pacific Region, with many newly established large-scale plantings (APAARI, 2012). However, the widespread and intensified planting of a limited variety of genotypes risks increasing the impact of jackfruit diseases. Most of the commercial jackfruit varieties with good quality and high yield tend to be susceptible to diseases (Mai, 2018a). Jackfruit diseases can be caused by fungi, fungal-like organisms, oomycetes and bacteria. In this chapter, we review the major diseases that limit commercial jackfruit production.

12.2 Diseases Caused by Fungi and Fungus-Like Organisms

12.2.1 *Phytophthora* disease

Phytophthora spp. are oomycetes. *Phytophthora* infection is one of the most serious jackfruit diseases in commercial jackfruit-growing regions (Borines *et al.*, 2014; Mai and Nguyen, 2014). The disease has been officially reported in the Philippines, Vietnam and Bangladesh (Borines *et al.*, 2014; Mai *et al.*, 2015; Jenny *et al.*, 2020).

Symptoms

The disease causes symptoms such as trunk cankers and gummosis, root rot, collar rot, tree decline, leaf blight and fruit rot (Mai, 2018a). These symptoms have been also observed in *Phytophthora palmivora*-infested durian and cocoa (Appiah *et al.*, 2004; Guest *et al.*, 2004). Early symptoms of trunk canker and gummosis, root rot and collar rot are very difficult to visually detect.

TRUNK CANKER. The visible symptom of trunk canker (Fig. 12.1A) on mature trees is a wet lesion on the bark (Fig. 12.1B), usually found at the base of the trunk and on parts close to the branch unions. Bark splitting can be found later in the areas of infected bark and tree sap oozes from the cracks. The lesions expand around the circumference of the trunk, slowly girdling the tree. Decline may occur over several years before tree death (Fig. 12.1C). On young trees, the canker appears as water-soaked lesions on the bark found mainly near the base of the trunk as a symptom of crown rot or collar rot (Fig. 12.1D). At an advanced stage, the infected bark becomes soft and when stripped away reveals dull and

*Corresponding author: maivantri2017@gmail.com

 Jackfruit : Botany, Production and Uses (ed. S. Mitra)
DOI: 10.1079/9781800622319.0012

Fig. 12.1. Symptoms of *Phytophthora* disease on mature jackfruit trees. (A) Trunk canker. (B) Bark and internal stem discoloration. (C) Leaf yellowing and defoliation. (D) Collar rot. (E) Shoot dieback on young trees. (F) Leaf wilting and dieback symptoms on the canopy of a diseased young tree. Photographs courtesy of V.T. Mai.

cream to reddish-brown discoloured internal tissues. In severe cases it may lead to leaf blight and shoot dieback (Fig. 12.1E).

Advanced infections progress to total girdling of the trunk, with wilting and dropping of the leaves, and eventually tree death (Fig. 12.1F)

Root rot. On mature trees, the first visual symptoms are those of water stress, chlorosis and stunting. New leaves are often small and a pale green to light yellow colour. Young trees are very susceptible to root rot, especially on the variety 'Sieu Som' (Mai, 2018a). The leaves of the diseased young trees lose their fresh and shiny appearance, and become yellow and drooping, with wilting of the leaves over the entire tree or only on one or several branches. The infected root tissue is soft, water-soaked and discoloured to dark brown. In advanced root rot, few new roots are produced, resulting in a lack of secondary and tertiary roots and a lack of healthy root tips (Drenth and Guest, 2004a).

Fig. 12.2. Symptoms of leaf blight (A) and fruit rot (B) of jackfruit caused by *Phytophthora palmivora*. Photographs courtesy of V.T. Mai.

LEAF BLIGHT. The symptom can occur on individual leaves or on all the leaves of a shoot in extreme cases (Fig. 12.1E). The first visible symptoms appear as small flecks on leaves that expand within 3–5 days to produce large greyish-green to dark brown, water-soaked lesions that become chlorotic and necrotic (Fig. 12.2A) (Mai *et al.*, 2019). The lesions are often surrounded by a halo of dark green tissue. A layer of white velvety fungal growth may be seen at the edge of the lesions under high-humidity conditions, mainly on the underside of the leaf.

FRUIT ROT. The first sign of fruit rot is a water-soaked, light brown lesion. Under favourable conditions, the lesions expand rapidly and can completely rot an entire fruit after 1 week (Mai, 2018a). Under conditions of high humidity, a greyish-white mycelial growth can be found at the advancing margin of the lesions (Fig. 12.2B). A strong odour coming from the advanced lesions is another characteristic of the disease. *Phytophthora* fruit rot should be distinguished from other rots, including *Lasiodiplodia theobromae*, which is reported to cause fruit rot in Taiwan (Ni *et al.*, 2008). *Lasiodiplodia* disease symptoms appears as a dark brown to black rot on mature fruits rather than the brown, water-soaked lesion caused by *P. palmivora*.

Causal agent

Based on morphological, pathological and molecular characterization, *P. palmivora* (Butler) has been confirmed as the cause of jackfruit decline in the Philippines (Borines *et al.*, 2014) and Vietnam (Mai *et al.*, 2015). *P. palmivora* is one of the most destructive plant pathogens in tropical regions. The pathogen causes severe diseases on many crops including fruit crops, including root rot and fruit rot of citrus; bud rot in coconut and oil palm; root rot, trunk canker and fruit rot in cocoa and durian; stem and fruit rot of pawpaw; fruit rot of breadfruit; and fruit rot of longan (Drenth and Guest, 2004a).

Phytophthora disease has been reported in jackfruit-growing countries across Asia, including India, the Philippines (Tsao *et al.*, 1994; Borines *et al.*, 2014), Vietnam (Mai *et al.*, 2015) and Bangladesh (Jenny *et al.*, 2020). Very probably, the disease has been present but unnoticed on many jackfruit-growing regions because of the limited economic significance of jackfruit, which, until recently, has been cultivated at a small household level for domestic consumption or for local markets (Mai *et al.*, 2015).

Phytophthora disease poses a serious threat for commercial jackfruit cultivation, with a significant impact both on yields and the long-term viability of plantations (Mai *et al.*, 2015).

Epidemiology

P. palmivora has a wide host range, including some very important economic crops (Erwin and Ribeiro, 1996). Some related species in the genus *Artocarpus*, such as breadfruit (*A. altilis*), breadnut (*A. camansi*) and cempedak (*A. integer*) have been reported as hosts of the pathogen (Chee, 1969; Gerlach and Salevao, 1984; Erwin and Ribeiro, 1996; Mai, 2018a).

P. palmivora has four different spore forms – sporangia, zoospores, chlamydospores and oospores – which are able to directly, or indirectly, infect the plant hosts (Erwin and Ribeiro, 1996). There is no evidence that oospores are involved in the disease cycle of

P. palmivora in tropical Asia, but the pathogen is extremely efficient in terms of spread and infection, even at low numbers. Sporangia are produced during damp weather in the lesion on the host trunk, leaves, fruits or root system. Aerially dispersed or soil-borne sporangia could infect the host directly by germinating on the host surface or indirectly by differentiating to produce zoospores. The zoospores are then encysted or dispersed before germination. Zoospores swim in water or in wet soil to infect host roots and stems, and can reach fruits and leaves by rainsplash (Erwin and Ribeiro, 1996). Sporangia and zoospores of *P. palmivora* have short-term survival and dispersal, while the thick cell-walled chlamydospores are adapted for long-term survival (Drenth and Guest, 2004a). Chlamydospores are produced from hyphae and then germinate to form sporangia, thereby continuing the cycle of spread and infection (Erwin and Ribeiro, 1996).

The primary inoculum is mainly from the soil and infected plant material. The pathogen is disseminated through wind-blown rain, rain splash and insect, animal and human activity in the canopy of trees, where the symptoms appear later. The secondary inoculum spreads rapidly by wind, wind-blown rain and rainsplash, and by contact and vector activity under humid weather conditions (Drenth and Guest, 2004b).

P. palmivora needs high humidity to induce sporulation and germination of sporangiospores and zoospores for infection (Erwin and Ribeiro, 1996). The majority of jackfruit-growing areas are located in a tropical monsoon climate. Seasonal heavy rainfall and high levels of humidity are favourable for development of the pathogen. Field surveys in the south-east region of Vietnam from 2013 to 2014 revealed that *Phytophthora* disease occurs mainly in the rainy season (May–October). The number of infected plants increased rapidly from May, peaked in October and then decreased rapidly to April (end of the dry season) (Mai, 2018a). A higher incidence was found in poorly drained, densely planted orchards.

Management

Controlling *Phytophthora* disease after it has become established is not easy. Integrated management approaches that include cultural, biological and chemical methods are recommended (Erwin and Ribeiro, 1996; Drenth and Guest, 2004b).

Management of the disease aims to avoid the favourable conditions that promote *P. palmivora* and includes good soil drainage, prevention of flooding, planting raised beds or mounds, appropriate tree spacing, and careful pruning allowing good ventilation, and stringent sanitation (Drenth and Guest, 2004a; Mai, 2018a). Additionally, management practices that promote soil and tree health, such as correcting soil pH, applying organic manure, regular mowing of weeds in orchards to the proper height, and optimal soil, nutrient and water management are effective (Mai, 2018a). Cultivation of resistant varieties is a most effective control measure. In Vietnam, 'Sieu Som' (a common variety) is very susceptible while 'La Lon' (also called 'La Bang') is more tolerant to the disease (Mai *et al.*, 2016). 'La Lon' should be used to replace the susceptible variety 'Sieu Som' in high disease pressure regions (Mai, 2018a). Unfortunately, 'La Lon' is not as tasty for consumers as 'Sieu Som' (Mai *et al.*, 2016). However, the 'La Lon' variety is recommended as a rootstock for 'Sieu Som' and other preferred varieties (Mai *et al.*, 2016).

Biological control is an environmentally friendly control method. *Trichoderma* spp. have been found to inhibit the growth of *P. palmivora* mycelia by more than 60% *in vitro* and to efficiently control *Phytophthora* disease on jackfruit under field conditions (Promwee *et al.*, 2017; Mai, 2018a). *Trichoderma harzianum* was found to prevent the increase of disease incidence and severity in field experiments. *T. harzianum* strain SR18 was recommended for field application twice (May and July) as a soil drench and three times (June, August and October) as a foliar spray during rainy season (May–October) in southern Vietnam (Mai, 2018a).

Chemicals are used to help supplement the other management measures. Bordeaux mix, mancozeb, metalaxyl (or combination of mancozeb and metalaxyl), fosetyl aluminium, and potassium phosphite were found to be effective to prevent the incidence and severity of *Phytophthora* disease on jackfruit (Mai, 2018a). Scheduled alternating applications (monthly or quarterly) of potassium phosphite and Bordeaux mix in heavy rainy months are recommended for diseased jackfruit groves.

A package of measures was developed and recommended for the management of *Phytophthora* disease of jackfruit in Vietnam (Mai, 2018a). These measures include:

- using the resistant variety 'La Lon';
- making drainage trenches between trees or rows to create better drainage conditions;
- annually applying chicken manure (12 t ha^{-1}) or cow manure (16 t ha^{-1}),
- applying *T. harzianum* strain SR18 twice (May and July) as a soil drench and three times (June, August and October) as a foliar spray during rainy season (May–October); and
- applying potassium phosphate (200 g l^{-1} applied at 6–9 l ha^{-1}) twice (May and August) as a foliar spray and twice (July and September) as a soil drench (Mai, 2018a).

12.2.2 Anthracnose leaf spot

Anthracnose affects the leaves, twigs and fruit of jackfruit both pre- and postharvest. Postharvest infection causes major losses of jackfruit in terms of yield and quality (Basak *et al.*, 1990; Srivasta and Mehra, 2004).

Symptoms

Anthracnose develops on jackfruit leaves from small, dark brown, water-soaked lesions that expand and fuse to induce leaf blight in favourable conditions. The expanded lesions have a grey centre surrounded by dark brown margins with concentric rings of acervuli, which serve as the diagnostic sign of anthracnose. Anthracnose causes defoliation on diseased twigs and young shoot blight (Abraham *et al.*, 1988). Anthracnose infects immature jackfruit through wounds (Trujillo, 1971; Basak, 1995). As the fruits start to ripen, soft, round, dark brown lesions develop on the fruit rinds and cause black-brown flesh rots underneath that downgrade the fruit quality and taste (Basak, 1995). Under favourable conditions, anthracnose causes serious damage to jackfruits during postharvest storage.

Causal agent

Colletotrichum gloeosporioides is the fungal pathogen that causes anthracnose in almost all jackfruit-growing regions and reduces the growth, productivity and quality of jackfruit (Sangchote *et al.*, 2003). Another *Colletotrichum* species, *Colletotrichum acutatum* induces immature jackfruit fall (McKenzie and Jackson, 1990).

Epidemiology

C. gloeosporioides produces numerous conidia on lesions on the infected canopy and ripe fruits. The primary inoculum is spread by rain, water splash and strong winds, and is favoured by a high relative humidity and temperature (Sangchote *et al.*, 2003).

Hot and humid conditions encourage anthracnose spread and development in the fields and during postharvest storage.

Management

Appropriate planting density and fertilization, insect control and avoiding wounds, as well as farm sanitation, are good cultural practices to manage anthracnose (Sangchote *et al.*, 2003). Proper pruning removes infected parts, opens up the canopy and maintains an adequate distance between the canopy and the ground, which helps to reduce anthracnose incidence and severity (Sirayoi, 1993; Srivasta and Mehra, 2004). Removing and burying infected fruit is highly recommended to reduce the spread of anthracnose. Fruit bagging also helps to protect jackfruits from insects and anthracnose. Fresh fruit should be washed and stored in a well-ventilated warehouse or fruit-processing facility. Alternatively, spraying with mancozeb and difenoconazole in the wet season or at the early stage of the disease efficiently manages anthracnose (Sirayoi, 1993)

12.2.3 Pink disease

Pink disease is common in jackfruit-growing regions with high rainfall and wet weather. Pink disease occurs on a wide range of woody plants such as fruit crops, plantation crops and forest trees in humid tropical areas. Infection

Fig. 12.3. Symptoms of pink disease. (A) Pink mycelial growth on branches. (B) Trunk dieback on a diseased tree. Photographs courtesy of V.T. Mai.

causes the death of jackfruit twigs and shoots, leading to a sparse canopy as well as sunburn and cracking on young trunks and twigs. Pink disease has been studied in eucalyptus (Ferreira and Alfenas, 1977), cocoa (Singh, 1989), citrus (Pradhanang, 1994) and rubber trees (Shamsuri *et al.*, 1997).

Symptoms

The initial signs appear on branches, particularly on the crook between branches or with the trunks, as a white thready patch of *Erythricium salmonicolor* mycelium (Lim and Khoo, 1985). The mycelium penetrates the host bark and colonizes the tree, blocking vascular transport (Subramaniam and Ramaswamy, 1987). The typical symptom of pink disease is a salmon-pink mycelial mass turning into a greyish-white encrusted hyphal sheath encircling infected branches (Fig. 12.3A) (Ferreira and Alfenas, 1977). The branches eventually crack, wilt, dry and die within weeks or months. Gumming can be associated with split branches (Fig. 12.3B) (Sharma *et al.*, 1984).

Causal agent

Erythricium salmonicolor, a fungal pathogen, is the causal agent of pink disease, named due to its creamy white-pink hymenium in the infected spots of jackfruit bark. *E. salmonicolor* has a number of synonyms including *Botryobasidium salmonicolor*, *Corticium salmonicolor* and *Pellicularia salmonicolor* (Sangchote *et al.*, 2003).

Epidemiology

Pink disease is widespread in the subtropical and tropical jackfruit-growing areas. A high incidence of pink disease is found in high-density, closed-canopy jackfruit farms during the productive stages after periods of high rainfall (Schneider-Christians *et al.*, 1986; Singh, 1989). The hymenium produces hyaline, ellipsoid basidiospores borne on the sterigma of basidia. The conidia and basidiospores are spread by rain, water splash and wind. The pathogen survives inside infected host tissue/debris and produces the secondary inoculum under favourable conditions (Schneider-Christians *et al.*, 1986).

Management

Management of this disease relies on cultural measures to reduce favourable conditions for the disease and early detection to remove affected plant parts from the groves. If the removal of diseased parts is impractical, control depends on treatment with fungicides. After removing and destroying the infected parts, wounds should be treated with copper-based fungicides. An open canopy and appropriate spacings are recommended as preventative measures. Fungicides such as Bordeaux mix and copper-based fungicides, tridemorph, triadimefon, flusilazole and oxycarboxin have been recommended for spraying the canopy or treating wounds (Ram *et al.*, 1982; Singh, 1989; Pradhanang, 1994; Mai *et al.*, 2019).

12.2.4 *Rhizopus* rot

Rhizopus rot is a common disease of jackfruit flowers and young fruits. However, the pathogen can affect the fruit at all development stages. The disease also commonly occurs on the stem ends of fruit or on wounds causing stem-end rot or fruit rot postharvest. On mature fruits, infection commonly occurs through wounds caused by strong winds, insects and animals, cracks on fruits and pruning activity (Fig. 12.4D) (Almeida and de Landim, 1980; Mai *et al.*, 2019). Normally, *Rhizopus* disease has little economic impact in the field. However, postharvest losses caused by *Rhizopus* can be high if sanitation is poor and the temperatures and humidity during storage are high.

Symptoms

Soft, watery brown lesions are the initial symptoms, developing on female inflorescences and also on young fruits. The lesions are then covered by white mycelia with a thin, powdery layer of black sporangia (Fig. 12.4A) (Nelson, 2005). Gradually, the black mass of mycelia and sporangia completely engulfs the diseased flowers or young fruits, turning to dry, black flowers or young fruits remaining on the trees or dropping to the ground (Fig. 12.4B,C).

Causal agent

Three species of the saprophytic fungus *Rhizopus* – *R. artocarpi*, *R. oryzae* and *R. stolonifer* – are reported to affect jackfruit (McMillan, 1975; Almeida and de Landim, 1980; Singh and Singh, 1989; Nelson, 2005). *R. stolonifer* is a common and widely distributed species on jackfruit (Ghosh *et al.*, 2015; García-Estrada *et al.*, 2019; Mai *et al.*, 2019). Sporangia are produced on diseased fruits and plant debris. *Rhizopus* spp. have a wide host range and are prevalent in the tropics and subtropics. *Rhizopus* rot is considered to be widespread in all jackfruit-growing areas in the tropics including India, Vietnam and Bangladesh (Nelson, 2005; Mai *et al.*, 2019).

Epidemiology

Spores released from the sporangia are dispersed by rain, wind and insects (Nelson, 2005). The pathogen survives saprophytically or as thick-walled spores on diseased fruits, in the soil and on plant debris. The spores can withstand a long period under dry conditions before initiating new infections. In the field, warm, humid conditions after rain favour the disease.

When in contact with a moist fruit surface, the spores germinate and the mycelia penetrate and colonize the fruit tissues. A black spore mass is formed on the lesion surface, which serves as a secondary inoculum to start a new infection cycle (Nelson, 2005).

Management

No jackfruit varieties have been reported to be significantly resistant to *Rhizopus* rot (Nelson, 2005; Mai *et al.*, 2019). In most instances, chemical control is not cost-effective and cultural control is recommended for management. Adequate plant spacing, appropriate pruning, good drainage and weed management reduce canopy humidity and improve ventilation (Nelson, 2005; Mai *et al.*, 2019). Early removal of diseased flowers and fruits, and removal and destruction of infested flowers and fruits as well as organic debris under the canopy are also recommended. Harvesting of the fruits should be avoided during rain, as well as direct contact of the fruits with the soil, and care should be taken that the fruits are not damaged during harvest.

Fig. 12.4. Rhizopus rot causes a soft, watery, brown spot in the female inflorescence that becomes a black furry mould (A), which remains attached to the branch (B) and then drops to the ground (C). Symptoms on a mature fruit are shown in (D). Photographs courtesy of V.T. Mai.

Sanitation and good ventilation during harvest and storage are highly recommended to reduce *Rhizopus* rot in storage (Nelson, 2005). Some antagonistic bacteria including *Burkholderia cenocepacia* VBC7, *Pseudomonas poae* VBK1 and *Lactococcus lactis* subsp. *lactis* are claimed to inhibit *Rhizopus* rot on jackfruit (Ghosh *et al.*, 2015).

When chemical control is required, spraying with copper-based fungicides or registered products is recommended to protect the fruits (Nelson, 2005). Some chemical fungicides have been tested for their antifungal activity (Parashar *et al.*, 2018). Application of tebuconazole and mancozeb before fruit set and after fruit set also showed good control (Mai *et al.*, 2019).

Damaged and rotten fruits and fruits that fall to the ground are removed before storage. Clean water is used for fruit washing and the fruits are dried thoroughly before packing. The fruits are never stored in a hot, poorly ventilated storehouse.

12.2.5 *Lasiodiplodia* fruit rot

Lasiodiplodia fruit rot usually infects mature and injured fruits, in contrast to *Rhizopus* fruit rot, which occurs mainly on female inflorescences and young fruits. The disease has also been reported to cause fruit rot in storage (Ni *et al.*, 2008).

Symptoms

The symptoms are usually characterized by flat, grey or black spots occurring on the fruit rind. The earliest symptom is fruit rot, with thin, yellow-brown lesions appearing on the rind. Under high humidity, these lesions can rapidly expand to 5–15 cm in diameter within several days with further fruit ripening. The expanded lesions become circular to oval shaped, water-soaked, soft and dark brown or blackish-grey with a reddish-brown margin (Adikaram *et al.*, 2020). The lesions progress quickly from the fruit surface to the inner tissues (rags and pulps) about 5–7 cm deep (Ganesan *et al.*, 2020). The centre of advanced lesions has a greyish mycelial mass and is relatively more tender than the margin (Taylaran *et al.*, 2021). The diseased fruits are unfit for consumption.

Causal agent

The cause of the disease was identified as the fungus *Lasiodiplodia theobromae* (Pat.) Grifon and Maubl. (synonyms *Botryodiplodia theobromae*, *Diplodia theobromae*). *Lasiodiplodia* fruit rot was reported on jackfruit in Taiwan by Ni *et al.* (2008), in India by Ganesan *et al.* (2020), in Bangladesh by Adikaram *et al.* (2020) and in the Philippines by Taylaran *et al.* (2021).

L. theobromae is a plant pathogen with a very wide host range. It is a common pathogen of many tropical tree fruits. The fungus causes rotting and dieback in most species it infects.

Epidemiology

L. theobromae can survive in soil, as an endophyte in living asymptomatic stems, on dead twigs, on mummified fruit and on organic debris (Sangchote *et al.*, 2003). It infects through wounds, and causes symptoms on mature fruits at harvest and ripening. The disease is dispersed by wind and rain splash, and thus heavy rains with strong winds are favourable conditions for spread of the disease.

Management

Cultural control methods reduce the disease. Efforts should be made to prevent or protect wounds (Sangchote *et al.*, 2003). Pruning should be avoided during wet weather and pruning wounds minimized; pruning wounds that do occur should be protect with copper-based fungicides. Diseased fruits should be removed to prevent the spread of the pathogen, and good hygiene and sanitation practised (Mai, 2018b). Careful harvesting and handling of fruit avoids wounds and contact with the soil and contaminated material.

12.2.6 *Phomopsis* fruit rot and gummosis

Phomopsis sp. causes leaf spots, fruit rot and stem gummosis on jackfruit. Leaf spot is considered a minor problem on jackfruit (Sangchote *et al.*, 2003; Elevitch and Manner, 2006).

Symptoms

The fungus causes fruit rot with small, brown spots that expand, become soft and dark brown, and penetrate the mesocarp (Sangchote *et al.*, 2003). The fungus can penetrate the fruit surface directly or through style remnants, and sporulation occurs on advanced lesions on infected fruit (Sangchote *et al.*, 2003).

On stems and branches, symptoms appear as small splits on the bark with a brown gummy exudation (Rahman and Afroz, 2016). The inner bark tissues and sapwood appear water soaked and discoloured brown. The diseased plant becomes weak and ultimately fruit yield is reduced.

Causal agent

The ascomycete fungus *Phomopsis artocarpi* is considered the causal agent of the *Phomopsis* fruit rot and gummosis in jackfruit. The fungus has been reported in India, Bangladesh and Thailand (Sangchote *et al.*, 2003; Rahman and Afroz, 2016; Khan *et al.*, 2021).

Epidemiology

Very little is known about the epidemiology of the disease. Spores of *Phomopsis* need moist conditions to germinate and infect. Conditions of heavy rain and poor orchard hygiene increase disease incidence.

Management

Hygiene during harvest and handling should be implemented carefully to avoid damage to the trunk and fruit. The fruit can be sprayed with fungicides when mature (Sangchote *et al.*, 2003). Removal and burial of diseased trunk tissues and fruits reduces the amount of inoculum. Insects that cause wounds on fruit should be controlled.

Bordeaux paste and coal tar significantly reduce the area of infection (Rahman and Afroz, 2016). Bordeaux paste is considered the most promising treatment in controlling gummosis of jackfruit. Infected tissues on the trunk should be removed and the wound painted with Bordeaux paste or warm coal tar.

12.2.7 *Phyllosticta* leaf spot

Symptoms

Leaf spot is found commonly on jackfruit in tropical regions. The symptom is characterized by sporadic, round or irregular dark brown to brick red necrotic spots on the upper side of leaves (Chowdhury *et al.*, 2015). However, the lesion can also be seen from the underside. The lesion gradually enlarges and turns greyish-white with dark brown boundaries.

Causal agent

The causal agent is the fungus *Phyllosticta artocarpina* (Awasthi *et al.*, 2005; Mazed, 2015). *Phyllosticta* leaf spot is a minor disease on jackfruit with a low economic impact.

Epidemiology

The disease occurs commonly on nursery plants and on newly planted trees in the field. High planting density, poor drainage and poor ventilation are favourable conditions for this disease.

The disease occurs mainly in wet and warm weather conditions (Chowdhury and Hossain, 2012). Under favourable conditions, the spot expands gradually. Increased disease severity and severe defoliation can occur rapidly in conditions of extended heavy rainfall and high humidity (Chowdhury and Hossain, 2012). The fungus can readily be observed on the lesion under such conditions. The fungus also attacks leaf petioles and young shoots, and in cases of severe infection, the disease results in bare shoots or dieback symptoms.

Management

Good ventilation should be created by using an appropriate planting density, pruning and weed management. A balanced irrigation and fertilization regime should be applied. Copper-based fungicides can be used as a preventative measure. Sprays of fungicides such as mancozeb

Fig. 12.5. Algal leaf spot. Photograph courtesy of V.T. Mai

(0.3%) and chlorothalonil (0.2%) are recommended (Rawal and Saxana, 1997).

12.2.8 Algal leaf spot

Algal leaf spot or red rust commonly occurs on the foliage of many species of horticultural and plantation crops, including jackfruit trees (Wolf, 1930). The damage can reduce the photosynthetic area of leaves, and results in defoliation in severe cases.

Symptoms

A raised, flattened, orange to brown or rust-coloured, velvety, circular area on the upper leaf surface is typically a sign of algal leaf spot (Fig. 12.5) (Nelson, 2008). As they develop, the spots turn from grey to orange and then to dark brown with reddish-brown fruiting bodies. Underneath the spots, the leaf tissue dies, causing yellowing and leaf fall in severe cases (Nelson, 2008; Mai, 2018b).

Causal agent

The causal agent of algal leaf spot of jackfruit is *Cephaleuros virescens*, a green alga of the family Trentepohliaceae.

Epidemiology

Algal leaf spot is widely distributed in tropical and subtropical regions (Joubert and Rijkenberg, 1971). *C. virescens* has a relatively wide host range including fruit crops. The infection is mostly dependent on frequent wetting of leaves, high humidity in the plant canopy and tree health. Unhealthy, poorly maintained, crowded groves tend to have more diseased leaves. The spores of *C. virescens* are continuously produced in rainy seasons with long periods of wet weather and are dispersed by wind, rain and water splash (Nelson, 2008). Disease damage becomes severe in poorly managed groves, particularly in association with damage caused by other pests (Mai, 2018b).

Management

Algal leaf spot rarely affects well-maintained healthy jackfruit fields. A balanced nutrient and irrigation regime and appropriate growing density combined with frequent pruning and efficient pest management are good practices for management of algal leaf spot (Mai, 2018b). Application of copper-based fungicides is only considered necessary in severe cases (Nelson, 2008).

12.3 Diseases Caused by Bacteria

12.3.1 Bronzing disease

Jackfruit bronzing is an emerging disease, causing serious losses of jackfruit production in Malaysia, the Philippines, Mexico and Vietnam (Cangao, 2012; Gapasin *et al.*, 2014; Hernández-Morales *et al.*, 2017; Mai, 2018b). Infected mature fruits are unmarketable.

Symptoms

The pulp, rags and cores of infected fruit have discoloured specks, spots and streaks that vary in colour from yellow-orange to red-brown or dark brown depending on the variety (Fig. 12.6) (Gapasin *et al.*, 2014; Hernández-Morales *et al.*, 2017; Mai *et al.*, 2019). Infected young fruits will not mature fully, and are smaller in size, misshapen and dull green in colour (Fig. 12.7). These fruits have smaller and thinner pulps, and smaller seeds (Mai *et al.*, 2019). Bronzing disease is much worse in susceptible varieties, and the proportion of diseased fruits can reach 80% (Gapasin *et al.*, 2014; Mai *et al.*, 2019).

Causal agent

The causal agent is *Pantoea stewartii* subsp. *stewartii*, a Gram-negative gammaproteobacterium (Jeger *et al.*, 2018; CABI, 2020). *P. stewartii* subsp. *stewartia* is indigenous to the Americas, and was introduced to other parts of the world mainly through maize seeds (CABI/EPPO, 2009; Jeger *et al.*, 2018). *P. stewartii* subsp. *stewartii* has a wide host range (Roper, 2011; EPPO, 2016). Maize is considered the primary host for *P. stewartii* subsp. *stewartii*, which is widely known as Stewart's bacterial wilt, as well as leaf blight of maize (EPPO, 2016). Jackfruit is listed as a host on which the bacterium causes jackfruit bronzing disease (EPPO, 2019), which has been reported in the Philippines, Malaysia, Mexico (Gapasin *et al.*, 2014; Hernández-Morales *et al.*, 2017; Zulperi *et al.*, 2017) and Vietnam (Hua *et al.*, 2020). The strains isolated from jackfruit were able to infect maize, producing the symptoms of bacterial wilt or Stewart's bacterial wilt (Gapasin *et al.*, 2014). Bronzing disease is commonly reported in some jackfruit-growing areas in Asia and America (Cangao, 2012; Gapasin *et al.*, 2014; Hernández-Morales *et al.*, 2017; Zulperi *et al.*, 2017; Mai *et al.*, 2019; Hua *et al.*, 2020).

Fig. 12.6. Symptoms of bronzing disease on pulps and rags (A, B), and the core (C) of a mature fruit of jackfruit. Photographs courtesy of V.T. Mai.

Fig. 12.7. Symptoms of bronzing disease on rags (A) and core (B) and rind (C) of a diseased young jackfruit. Photograph courtesy of V.T. Mai.

Epidemiology

P. stewartii subsp. *stewartii* has wide host range (Roper, 2011; EPPO, 2016; Jeger *et al.*, 2018; Abidin *et al.*, 2020). The disease occurs mainly in the wet season and decreases over the dry season. A higher disease incidence was observed in high-density, poorly pruned plantings, with unbalanced application of nutrients and excessive application of nitrogen fertilizers, and poorly drained orchards (Mai *et al.*, 2019). Fruit set and young fruits are very susceptible to infection. If conditions of heavy and prolonged rain occur in this stage, the disease incidence is high.

Management

To deal with this disease, prevention is key and integrated management needs to be implemented. Some management measures have been recommended by Mai *et al.* (2019), including the following:

- Use disease-free seedlings for planting new orchards, and mixed cropping with non-host plants (avoiding mixed cropping with maize).
- Use an appropriate planting density, and plant on raised mounds or beds, in permeable, well-drained soils with good drainage, and carry out appropriate pruning, trimming low branches to reduce favourable conditions for the disease.
- Take care to improve tree health, by application of organic manures and soil organic amendments, balanced fertilization and an irrigation regime.
- Remove fallen leaves and fruits on the ground in diseased orchards.
- Cycle three crop to avoid flowering and fruit harvest in the rainy season, especially in months with heavy rainfall.
- Thin the fruit to remove excess fruit combining with removing and destroying potentially diseased young fruits (unusually and irregularly shaped fruits, deformed fruits, smaller fruits; fruits with an uneven appearance on the surface of rind, and fruits with brown discoloration or rot along the peduncle).
- Use disease-free bags to wrap fruits, and control insects that cause wounds on fruit.
- Covering selected female inflorescences and young fruits until they are 45 days old to prevent rainwater from contacting the fruits during the rainy season, and avoid injury to the developing fruits (Le *et al.*, 2017).
- Using less susceptible varieties for commercial planting. In Vietnam, the variety 'La Lon' is less susceptible to the disease, while 'Sieu Som' is highly susceptible.

Chemical control of bronzing disease on susceptible varieties and in conducive environment conditions requires periodic pesticide sprays. The timing and frequency of applications are very important for reducing the disease incidence. Four critical sprays when female inflorescences appear, before blooming and after the fruit-set stage (two applications, 10–14 days apart) are required. Copper-based fungicides can be used including copper oxychloride and copper hydroxide. Potassium phosphite 400 (6 l ha^{-1}) applied four to eight times a year also helps to reduce the disease incidence.

12.3.2 Bacterial trunk canker

Bacterial trunk canker, also locally called gummosis stem canker or gummosis trunk-splitting disease, is an emerging disease on jackfruit. This devastating disease has been causing heavy damage on commercial orchards and is one of the main limiting factors for jackfruit production in Vietnam. In severe cases, estimated yield loss can reach up to 50%, and shorten the life of the orchard from 15–20 years to only 8–10 years. The disease causes a number of symptoms including canker, gummosis and splitting on the trunk and larger branches (Mai *et al.*, 2019; Hua *et al.*, 2020).

Symptoms

Infections usually occur at ground level, just above the site of the graft, but can also appear on the trunk or the scaffold branches in highly susceptible varieties (Mai *et al.*, 2019).

The visible early signs of bacterial trunk canker are a drop of white sap formed on the bark (Fig. 12.8A). The tree sap continues oozing and from the lesion (Fig. 12.8B), producing a wet stain around and below the lesion, which becomes blackish-grey to white and powdery as it dries (Mai *et al.*, 2019). A longitudinal crack across the lesions can be seen from the outside and inside of the bark (Fig. 12.8C) (Mai *et al.*, 2019). When the bark is stripped away, the inner bark tissues and sapwood appear dull, water soaked and discoloured, changing from cream to light brown (Fig. 12.8D) (Vareket *et al.*, 2005; Hua *et al.*, 2020). Longitudinal cracks can also be seen on the trunk wood tissue (Mai *et al.*, 2019). The wood lesions are often a vertical ellipse shape or irregular in shape but are well defined (Fig. 12.8E). Expanding lesions severely restrict water and nutrient flow, leading to yellowing leaves and premature leaf drop. The lesions usually expand vertically upwards, but also horizontally (Fig. 12.8D,E) (Mai *et al.*, 2019). When the expanded lesion girdles the trunk or branch, progressive death of twigs and branches occurs starting at the tips and can be seen on the crown (Fig. 12.8F).

The symptoms of bacterial trunk canker are similar to those of *Phytophthora* trunk canker. The distinguishing feature is the drop of white sap and resinous substance oozing from the lesion and the longitudinal crack across the lesions (Mai *et al.*, 2019).

Causal agent

Pectobacterium carotovora (synonym *Erwinia carotovora*), a Gram-negative, rod-shaped bacterium, was identified as the causal agent of bacterial trunk canker (Hua *et al.*, 2020). Bacterial trunk canker has been found on jackfruit fields in Vietnam, Thailand and Indonesia (Vareket *et al.*, 2005; Hua *et al.*, 2020).

In Malaysia, *P. carotovora* has been recorded on chempedak (Agrolink, 1999), causing bacterial dieback with symptoms of leaf yellowing and a gummy exudate from the stems and branches.

Epidemiology

P. carotovora infects more than 100 crops belonging to the families Cruciferae, Solanaceae, Cucurbitaceae and others. This phytopathogenic bacterium is responsible for soft rot, wilt and blackleg in potato, tomato and Chinese cabbage. Several bacteriophages have been developed and used to control soft rot disease, but not in jackfruit (Ravensdale *et al.*, 2007; Kassa, 2021).

Insect wounding allows *P. carotovorum* entry into jackfruit tissue leading to infection. Using its four to eight peritrichous flagella, *P. carotovora* moves through the xylem causing brown lesions 3 weeks after infection. On vegetable crops, *P. carotovora* is prevalent in the plant

Fig. 12.8. Symptoms of bacterial trunk canker. (A) White sap on the bark. (B) Sap oozing from the lesion. (C) Splitting of the trunk. (D) Dull, water-soaked and discoloured inner bark tissues. (E) Lesions on the sapwood. (F) Dieback of the canopy. Photographs courtesy of V.T. Mai.

litter, and can be carried on dust, rain and water splash, insects, contaminated tools and contaminated hands (Roper, 2011). High relative humidity during the wet season is favourable for the disease (Roper, 2011).

In Vietnam, the variety 'Sieu Som' is a very susceptible to bacterial trunk canker, while the 'La Lon' variety is more tolerant of the disease (Mai *et al.*, 2019).

Management

Once a tree has become infected, the disease is difficult to control. Preventative measures are the most important strategy for management. The measures used for management of bronzing disease listed earlier in the chapter (Mai *et al.*, 2019) are also helpful for the prevention of bacterial trunk canker. Application of whitewash (ratio of 3:1–2 of water:hydrated lime) to the trunk and to the main branches after harvesting and at the beginning and end of the wet season, or a lime wash (1% hydrated lime) four to eight times a year is recommended. Alternatively, Bordeaux mixture, copper-based fungicides or coal tar applied to the infected tissue was found to provide 80–90% control (Rahman and Afroz, 2016). Oxolinic acid and bronopol were reported to reduce the population density of the bacterium on applied trees (Le *et al.*, 2019).

Controlling insects that wound the bark, careful management of handling and hygiene, and avoiding injury to trees also reduces new infections (Roper, 2011; Mai *et al.*, 2019). Reducing the relative humidity by improving canopy ventilation, avoiding irrigation that wets

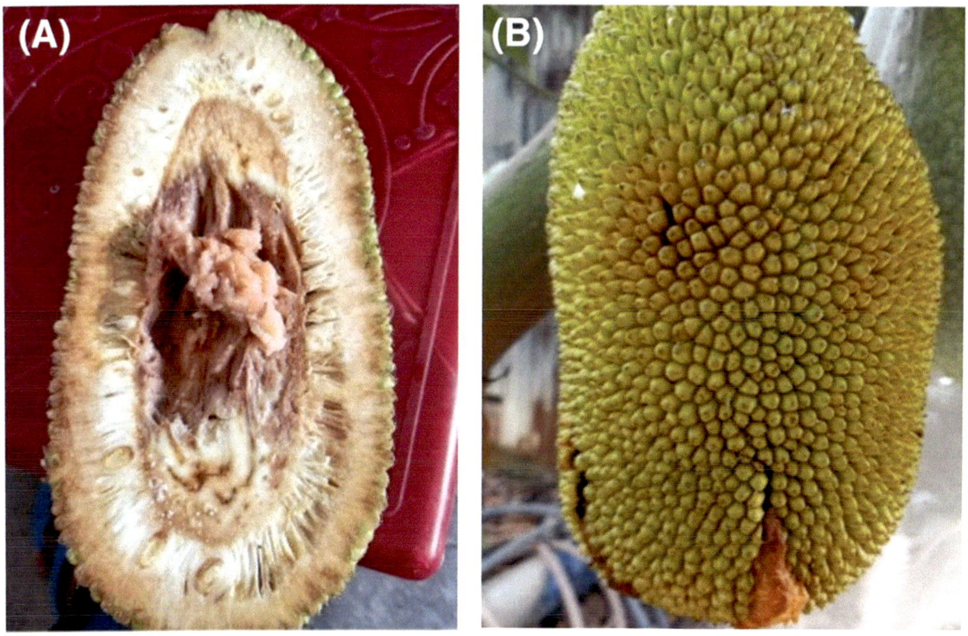

Fig. 12.9. Symptoms of bacterial fruit rot. (A) Damage to the internal fruit. (B) Damage to the rind. Photographs courtesy of N.P. Tuyen.

the bark, and using organic soil amendments and balanced fertilization are recommended (Mai *et al.*, 2019).

12.3.3 Bacterial fruit soft rot

Bacterial fruit soft rot is an emerging disease on jackfruit and has been causing damage in commercial orchards. In severe cases, the estimated yield loss caused by bacterial fruit soft rot can be up to 50% of developing fruits on trees.

Symptoms

The symptoms of this disease are internal water-soaked brown rot of pulps, rags, seeds and the core in both young and mature fruits (Fig. 12.9A), discoloured green, yellowish-green to reddish-brown spots on the rind (Fig. 12.9B) and brown on the outer surfaces of the fruit stalks (Jaffar *et al.*, 2018; Mai *et al.*, 2019).

Causal agent

Dickeya fangzhongdai was reported as the causal agent of bacterial fruit soft rot (Jaffar *et al.*, 2018). This Gram-positive bacterium is a new member of the genus *Dickeya* and the only one that infects woody plants.

The bacterium has been reported in mainland China, Malaysia, Taiwan, Scotland, Slovenia, St Lucia and the USA, and its known host range is primarily orchids (*Phalaenopsis* spp.), onions (*Allium fistulosum*, *Allium cepa*), ornamental plants (*Aglaonema* spp.,), pear (*Pyrus pyrifolia*) and jackfruit (*Artocarpus heterophyllus*) (Tian *et al.*, 2016; Alič *et al.*, 2017, 2018; Jaffar *et al.*, 2018; Zhang *et al.*, 2018; CABI, 2019; Ma *et al.*, 2020). It is likely that the disease has been present but unnoticed on many commercial jackfruit-growing regions in tropical Asia.

Epidemiology

The bacterium is most often associated with hot, wet and humid weather. These conditions are commonly found in high-density commercial orchards in the rainy season and also in the nursery. The spread of *Dickeya* spp. over long distances results from latent infections in planting materials, and over shorter distances via water. Spread to other parts of the same plant or to other plants is primarily by the movement of rain water through washing or spattering or from irrigation. Workers may also carry and spread the pathogens locally by handling plants with unsanitary cultural practices and over long distances by transporting infected planting materials or plant parts (Agrios, 2005).

The young fruit is more susceptible to the disease. Some varieties are more susceptible to the pathogen including 'Sieu Som' (Mai *et al.*, 2019).

Management

Prevention is key to management of bacterial fruit soft rot. The preventative measures for bronzing disease listed earlier in the chapter by Mai *et al.* (2019) are also helpful in reducing bacterial fruit soft rot.

12.4 Conclusion

Phytophthora trunk canker, bronzing disease, bacterial trunk canker and bacterial fruit rot could become the most important economic diseases on commercial jackfruit farms. Controlling these diseases is not easy once they have become established. The use of integrated management methods remains the best practice to limit loss caused by these diseases, as well as to reduce the amount of pesticides used. Starting with disease-free planting materials, having a well-drained soil, using an appropriate planting density and proper pruning will mitigate the conducive conditions for the pathogens. Using organic manures with application of balanced fertilization and proper irrigation regimes, adjusting soil pH and proper weed management will improve not only soil health but also tree health. In addition to the use of good sanitation, regular monitoring of the appearance and development of disease helps early detection and treatment, and minimizes the impact of diseases. Choosing resistant varieties or ones offering a good level of disease resistance is ecologically and environmentally advantageous. However, resistant varieties with good yield and quality are not yet widely available.

Chemical pesticides should be used as a last resort. Priority should be given to choosing pesticides that are toxic only to the target pathogen. Safe and effective pesticide use also requires selecting the appropriate product, optimal timing of sprays, appropriate rates of application and proper application methods.

References

Abidin, N., Ismail, S.I., Vadamalai, G., Yusof, M.T., Hakiman, M. *et al.* (2020) Genetic diversity of *Pantoea stewartii* subspecies *stewartii* causing jackfruit-bronzing disease in Malaysia. *PloS One* 15(6), e0234350. DOI: 10.1371/journal.pone.0234350.

Abraham, M., Padmakumary, G. and Nair, M.C. (1988) Twig blight (die-back) of *Artocarpus incisa*. *Indian Phytopathology* 41, 629–630.

Adikaram, N., Manawadu, L., Jayasinghe, L. and Yakandawala, D. (2020) First report of *Lasiodiplodia theobromae* rot in ripe jack fruit (*Artocarpus heterophyllus* Lam.) in Sri Lanka. *Indian Phytopathology* 73(3), 583–585. DOI: 10.1007/s42360-020-00243-w.

Agrios, G.N. (2005) *Plant Pathology*, 5th edn. Elsevier Academic Press, Burlington, Massachsetts.

Agrolink (1999) Department of Agriculture Malaysia. Available at: http://agrolink.moa.my/doa/english/croptech/che_pes.html

Alič, N.T., Tušek-Žnidarič, M., Peterka, M., Ravnikar, M. and Dreo, T. (2017) Putative new species of the genus *Dickeya* as major soft rot pathogens in *Phalaenopsis* orchid production. *Plant Pathology* 66, 1357–1368. DOI: 10.1111/ppa.12677.

Alič, Š., Van Gijsegem, F., Pédron, J., Ravnikar, M. and Dreo, T. (2018) Diversity within the novel *Dickeya fangzhongdai* sp., isolated from infected orchids, water and pears. *Plant Pathology* 67(7), 1612–1620. DOI: 10.1111/ppa.12866.

Almeida, R.T. and de Landim, C.M.U. (1980) *Rhizopus stolonifer* (Ehrenb. ex Fr.) Vuill., causal agent of soft rot of jackfruit (*Artocarpus heterophyllus* Lam.) and breadfruit (*Artocarpus altilis* Fosberg.) in Ceara State. *Fitossanidade* 4, 23–24.

Appiah, A.A., Opoku, I.Y. and Akrofi, A.Y. (2004) Natural occurrence and distribution of stem cankers caused by *Phytophthora megakarya* and *Phytophthora palmivora* on cocoa. *European Journal of Plant Pathology* 110(10), 983–990. DOI: 10.1007/s10658-004-0811-6.

Asia-Pacific Association of Agricultural Research Institutions (2012) *Jackfruit Improvement in the Asia-Pacific Region – A Status Report*. APAARI, Bangkok.

Awasthi, D.P., Sarkar, S., Mishra, N.K. and Kaisar, S.A.K.M. (2005) Disease situation of some major fruit crops in new alluvial plains of west Bengal. *Environment and Ecology* 235(Special Issue 3), 497–499.

Basak, A.B. (1995) Fruit rot disease of jackfruit caused by *Colletotrichum gloeosporioides* Penz in Chittagong. *Bangladesh Journal of Botany* 24, 197–199.

Basak, A.B., Mridha, M.A.U. and Uddin, M.J. (1990) Studies on the occurrence and severity of leaf spot disease of jack fruit trees caused by *Colletotrichum gloeosporioides* Penz. in Chittagong (Bangladesh). *Chittagong University Studies* 14, 11–14.

Borines, L.M., Palermo, V.G., Guadalquiver, G.A., Dwyer, C., Drenth, A. *et al.* (2014) Jackfruit decline caused by *Phytophthora palmivora* (Butler). *Australasian Plant Pathology* 43(2), 123–129. DOI: 10.1007/s13313-013-0241-z.

CABI (2019) *Dickeya fangzhongdai*. In: *Invasive Species Compendium*. CAB International, Wallingford, UK. Available at: www.cabi.org/ISC (accessed 17 July 2022).

CABI (2020) *Pantoea stewartii* (bacterial wilt of maize). In: *Invasive Species Compendium*. CAB International, Wallingford, UK. Available at: www.cabi.org/ISC (accessed 17 July 2022).

CABI/EPPO (2009) *Pantoea stewartii* (distribution map). Map 41, Vol. 5. CAB International, Wallingford, UK. Available at: www.cabi.org/ISC/abstract/20093074276 (accessed 17 July 2022).

Cangao, C.A. (2012) Rust-like symptoms and rot on jackfruit: a combination of disease and abiotic factors? *Tropical Fruit Net Newsletter* 2012, 16–17.

Chee, K.H. (1969) Hosts of *Phytophthora palmivora*. *Review of Applied Mycology* 48, 337–344.

Chowdhury, M.S.M. and Hossain, I. (2012) Effects of temperature, rainfall and relative humidity on leaf spot of jackfruit seedling and its eco-friendly management. *The Agriculturists* 9(1–2), 126–136. DOI: 10.3329/agric.v9i1-2.9487.

Chowdhury, M.S.M., Mazed, H.E.M.K., Irin, I.J., Rahman, M.H. and Moonmoon, J.F. (2015) Study on seedling diseases of jackfruit (*Artocarpus heterophyllus* L.) in Bangladesh. *International Journal of Applied Research* 1, 24–29.

Drenth, A. and Guest, D.I. (2004a) Phytophthora in the tropics. In: Drenth, A. and Guest, D.I. (eds) *Diversity and Management of Phytophthora in Southeast Asia*. ACIAR Monograph 114. Australian Centre for International Agricultural Research, Canberra, pp. 30–41.

Drenth, A. and Guest, D.I. (2004b) Principles of *Phytophthora* diseases management. In: Drenth, A. and Guest, D.I. (eds) *Diversity and Management of Phytophthora in Southeast Asia. ACIAR Monograph 114*. Australian Centre for International Agricultural Research, Canberra, pp. 154–160.

Elevitch, C.R. and Manner, H.I. (2006) *Artocarpus heterophyllus*, Jackfruit. Version I.IV. In: Elevitch, C.R. (ed.) *Species Profiles for Pacific Islands Agroforestry*. Permanent Agriculture Resources, Holualoa, Hawaii. Available at: https://agroforestry.org/free-publications/traditional-tree-profiles (accessed 17 July 2022).

EPPO (2016) EPPO standard PM 7/60(2) *Pantoea stewartii* subsp. *stewartii*. *Bulletin OEPP/EPPO Bulletin* 46, 226–236. DOI: 10.1111/epp.12303.

EPPO (2019) EPPO Global Database. Available at: https://gd.eppo.int (accessed 17 July 2022).

Erwin, D.C. and Ribeiro, O.K. (1996) *Phytophthora Diseases Worldwide*. APS Press, St Paul, Minnesota.

Ferreira, F.A. and Alfenas, A.C. (1977) Pink disease of eucalyptus caused by *Corticium salmonicolor* Berk and Br. in Brazil. *Fitopatologia* 2, 109–115.

Ganesan, S.S., Panda, M. and Kishore, K. (2020) A report on occurrence of *Lasiodiplodia* fruit rot infecting jackfruit in Odisha state. *Pest Management in Horticultural Ecosystems* 26(1), 160. DOI: 10.5958/0974-4541.2020.00026.0.

Gapasin, R.M., Garcia, R.P., Christine, T., Cruz, C.S. and Borines, L.M. (2014) Fruit bronzing, a new disease affecting jackfruit caused by *Pantoea stewartii* (Smith) Mergaert et al. *Annals of Tropical Research* 36, 17–31. DOI: 10.32945/atr3612.2014.

García-Estrada, R.S., López-Martínez, J.C., Medina-López, R., Márquez-Zequera, I. and Cruz-Lachica, I. (2019) First report of *Rhizopus stolonifer* causing fruit rot in jackfruit (*Artocarpus heterophyllus*) in Mexico. *Plant Disease* 103, 2957.

Gerlach, W.W.P. and Salevao, F. (1984) Fruit rot of breadfruit, *Artocarpus altilis*, caused by *Phytophthora palmivora* in western Samoa. *Alafua Agricultural Bulletin* 9, 21–26.

Ghosh, R., Barman, S., Mukhopadhyay, A. and Mandal, N.C. (2015) Biological control of fruit-rot of jackfruit by rhizobacteria and food grade lactic acid bacteria. *Biological Control* 83, 29–36. DOI: 10.1016/j.biocontrol.2014.12.020.

Guest, D.I., Chau, N.M., Sangchote, S., Vawdrey, L. and Diczbalis, Y. (2004) Integrated management of *Phytophthora* diseases of durian: recommendations and benefit–cost analysis. In: Drenth, A. and Guest, D.I. (eds) *Diversity and Management of Phytophthora in Southeast Asia*. ACIAR Monograph 114. Australian Centre for International Agricultural Research Monograph 114, Canberra, pp. 202–226.

Hernández-Morales, A., Pérez-Casillas, J.M., Soria-Guerra, R.E., Velázquez Fernández, J.B. and Arvizu-Gómez, J.L. (2017) First report of *Pantoea stewartii* subsp. *stewartii* causing jackfruit bronzing disease in Mexico. *Plant Pathology* 99, 799–818.

Hua, T.H., Nguyen, H.C.Q., Doan, T.K.T., Le, T.T., Mai, V.T. *et al.* (2020) Identification of bacterial agent causing gummosis stem canker on jackfruit (*Artocarpus heterophyllus* Lam). In: *Proceedings of 29th National Conference of the Vietnamese Phytological Society, 23–25 October 2020, Hanoi, Vietnam*, vol. 29, pp. 71–80.

Jaffar, N.S., Osman, M.S. and Koyube, M.N.K. (2018) New bacterial fruit rot disease of jackfruit (*Artocarpus heterophyllus*) caused by *Dickeya fangzhongdai* in Malaysia. *Malaysian Journal of Microbiology* 15, 214–219.

Jeger, M., Bragard, C., Candresse, T., Chatzivassiliou, E., Dehnen-Schmutz, K. *et al.* (2018) Pest categorisation of *Pantoea stewartii* subsp. *stewartii*. *EFSA Journal* 16(7), 5356. DOI: 10.2903/j.efsa.2018.5356.

Jenny, F., Kader, M.A. and Bhuiyan, M.A.B. (2020) First report of jackfruit decline caused by *Phytophthora* spp. in Bangladesh. *Fundamental and Applied Agriculture* 5, 1. DOI: 10.5455/faa.2194.

Joubert, J.J. and Rijkenberg, F.H.J. (1971) Parasitic green algae. *Annual Review of Phytopathology* 9(1), 45–64. DOI: 10.1146/annurev.py.09.090171.000401.

Kassa, T. (2021) Bacteriophages against pathogenic bacteria and possibilities for future application in Africa. *Infection and Drug Resistance* 6, 17–31. DOI: 10.2147/IDR.S284331.

Khan, A.U., Choudhury, M.A.R., Maleque, M.A., Dash, C.K., Talucder, M.S.A. *et al.* (2021) Management of insect pests and diseases of jackfruit *(Artocarpus heterophyllus* L.) in agroforestry system: a review. *Acta Entomology and Zoology* 2(1), 37–46. DOI: 10.33545/27080013.2021.v2.i1a.29.

Le, T.T., Le, T.T.H., Nguyen, V.C.H., Thai, N.O., Pham, V.H. *et al.* (2019) Diagnosis of pathogen and efficacy of some bactericides on gummosis trunk canker in jackfruit cv. Changai. *Science and Technology Journal of Agriculture and Rural Development* 18, 17–23.

Le, V.B., Nguyen, T.N., Truong, H.N. and Nguyen, X.L. (2017) Study on the cause of black rag disease on jackfruit (*Artocarpus heterophyllus* Lam.) c.v. Chiang Rai and suggesting the initial solutions for its control. In: *Proceedings of the 2nd National Conference on Plant Physiology – Plant Physiology Applied in High-Tech Agriculture, 8–9 December 2017*, vol. 29, pp. 71–80.

Lim, T.K. and Khoo, K.C. (1985) *Diseases and Disorders of Mango in Malaysia*. Tropical Press, Kuala Lumpur.

Ma, X., Bonasera, J.M., Asselin, J.A.E., Beer, S.V. and Swingle, B. (2020) First report of *Dickeya fangzhongdai* causing soft rot of onion in New York state. *Plant Disease* 104(4), 1251. DOI: 10.1094/PDIS-09-19-1940-PDN.

Mai, V.T. (2018a) Study on the causal agent and control measures of trunk canker on jackfruit in the South-eastern region. PhD thesis, Vietnam Academy of Agricultural Sciences, Hanoi, Vietnam (in Vietnamese).

Mai, V.T. (2018b) Diseases of jackfruit. In: Vu, T.M., Nguyen, V.T., Bui, C.T. and Pham, V.K. (eds) *Diseases of Plants in Vietnam*. VNUA Publisher, Hanoi (in Vietnamese), pp. 289–297.

Mai, V.T. and Nguyen, V.H. (2014) Jackfruit production in Vietnam. In: *Compendium of the International Symposium on Jackfruit and Breadfruit of the Tropics, 15 May 2014, Bangalore, India*, pp. 69–75.

Mai, V.T., Nguyen, V.H., Nguyen, M.C., Pane, A., Faedda, R. *et al.* (2015) Decline of jackfruit (*Artocarpus heterophyllus*) incited by *Phytophthora palmivora* in Vietnam. *Phytopathologia Mediterranea* 54, 275–280.

Mai, V.T., Nguyen, T.N.V., Nguyen, V.H. and Nguyen, V.T. (2016) Rapid screening for the varietal resistance to *Phytophthora palmivora* causing stem canker on jackfruit in the south-eastern Vietnam. *Journal of Vietnam Agricultural Science and Technology* 5, 11–15.

Mai, V.T., Nguyen, T.V. and Nguyen, T.N.V. (2019) *Jackfruit Diseases and Their Management*. Southeast Horticultural Research Center, Vietnam (in Vietnamese).

Mazed, H.E.M. (2015) Study on seedling diseases of jackfruit (*Artocarpus heterophyllus* L.) in Bangladesh. *International Journal of Applied Research* 1, 24–29.

McKenzie, E.H.C. and Jackson, G.V.H. (1990) *The Fungi, Bacteria and Pathogenic Algae of the Federated States of Micronesia*. SPC Technical Paper 19. South Pacific Commission, Noumea, New Caledonia.

McMillan, R.T. (1975) *Rhizopus artocarpi* rot of jackfruit (*Artocarpus heterophyllus*). *Proceedings of the Florida State Horticultural Society* 87, 392–393.

Nelson, S.C. (2005) *Rhizopus Rot of Jackfruit*. Plant Disease 29. College of Tropical Agriculture and Human Resources. University of Hawaii at Manoa, Honolulu, Hawaii.

Nelson, S.C. (2008) *Cephaleuros Species, the Plant-Parasitic Green Algae*. Plant Disease 43. College of Tropical Agriculture and Human Resources. University of Hawaii at Manoa, Honolulu, Hawaii.

Ni, H.F., Chen, R.S., Chang, S.F. and Yang, H.R. (2008) First report of *Lasiodiplodia* fruit rot of jackfruit in Taiwan. *Plant Disease* 92(7), 1137. DOI: 10.1094/PDIS-92-7-1137A.

Parashar, R., Rizvi, G. and Zargar, S.A. (2018) Efficacy of chemical fungitoxicans in fruit rot of jackfruit in jhansi vicinity. *Journal of Pharmacognosy and Phytochemistry* 2(Special Issue), 228–230.

Pradhanang, P.M. (1994) Pink disease (*Corticium salmonicolor* Berks & Broome) control in mandarin orange (*Citrus reticulata* Blanco) by application of bordeaux mixture. *Crop Protection* 13(7), 550–552. DOI: 10.1016/0261-2194(94)90109-0.

Promwee, A., Yenjit, P., Issarakraisila, M., Intana, W. and Chamswarng, C. (2017) Efficacy of indigenous *Trichoderma harzianum* in controlling *Phytophthora* leaf fall (*Phytophthora palmivora*) in Thai rubber trees. *Journal of Plant Diseases and Protection* 124(1), 41–50. DOI: 10.1007/s41348-016-0051-y.

Rahman, M.A. and Afroz, M. (2016) Survey on the diseases of jackfruit and some aspects of control measures for gummosis disease in Bangladesh. *Eco-friendly Agriculture Journal* 9, 10–14.

Ram, A., Bastos, S.T.G. and de Figueiredo, J.M. (1982) Chemical control of *Corticium salmonicolor* (Berk.) & Br., causal agent of pink disease of cocoa. *Revista Theobroma* 12, 241–247.

Ravensdale, M., Blom, T.J., Gracia-Garza, J.A., Svircev, A.M. and Smith, R.J. (2007) Bacteriophages and the control of *Erwinia carotovora* subsp. *carotovora*. *Canadian Journal of Plant Pathology* 29, 121–130.

Rawal, R.D. and Saxana, A.K. (1997) Diseases of dryland horticulture and their management. In: *Silver Jubilee National Symposium on Arid Horticulture, Chaudhary Charan Singh Haryana Agricultural University, Hisar, India*, pp. 5–7.

Roper, M.C. (2011) *Pantoea stewartii* subsp. *stewartii*: lessons learned from a xylem-dwelling pathogen of sweet corn. *Molecular Plant Pathology* 12(7), 628–637. DOI: 10.1111/j.1364-3703.2010.00698.x.

Sangchote, S., Wright, J. and Johnson, G. (2003) Diseases of breadfruit, jackfruit and related crops. In: Ploetz, R.C. (ed.) *Diseases of Tropical Crops*. CAB International, Wallingford, UK, pp. 135–144.

Schneider-Christians, J., Fliege, F. and Schlösser, E. (1986) On the release, survival and importance of basidiospores of *Corticium salmonicolor,* the pathogen causing pink disease. *Journal of Plant Diseases and Protection* 93, 397–403.

Shamsuri, M.H., Omar, M. and Napi, D. (1997) Studies on epidemiology of pink disease and the effect of temperature on mycelial growth of *Corticium salmonicolor* of Hevea rubber. *Journal of Natural Rubber Research* 12, 58–66.

Sharma, J.K., Mohanan, C. and Florence, E.J.M. (1984) Outbreak of pink disease caused by *Corticium salmonicolor* in *Eucalyptus grandis*. *Tropical Pest Management* 30, 253–255.

Singh, G. (1989) Evaluation of fungicides against vascular streak dieback, white thread blight and pink disease of cocoa. *Journal of Plant Protection in the Tropics* 6, 19–24.

Singh, N.I. and Singh, K.U. (1989) Efficacy of certain fungicides against *Rhizopus* rot of jackfruit. *Indian Phytopathology* 42, 465–466.

Sirayoi, A. (1993) *Diseases of Fruit Crops, Spices and Control*. Takansak Foundation, Kasetsart University, Bankok.

Srivasta, M.P. and Mehra, R. (2004) Diseases of minor tropical and sub-tropical fruits and their management. In: *Diseases of Fruits and Vegetables: Vol.II*. Springer, Dordrecht, Netherlands, pp. 559–632.

Subramaniam, S.V. and Ramaswamy, V. (1987) Histopathological observations on pink disease of eucalyptus. *Current Science* 56, 1042–1044.

Taylaran, A.D., Bagsic-Posada, I., Cueva, F.D. and Balendres, M.A. (2021) First report of *Lasiodiplodia theobromae* causing jackfruit (*Artocarpus heterophyllus*) brown rot in the Philippines. *Indian Phytopathology* 74(4), 1151–1153. DOI: 10.1007/s42360-021-00392-6.

Tian, Y., Zhao, Y., Yuan, X., Yi, J., Fan, J. *et al.* (2016) *Dickeya fangzhongdai* sp. nov., a plant-pathogenic bacterium isolated from pear trees (*Pyrus pyrifolia*). *International Journal of Systematic and Evolutionary Microbiology* 66(8), 2831–2835. DOI: 10.1099/ijsem.0.001060.

Trujillo, E.E. (1971) *The Breadfruit Diseases of the Pacific Basin*. Information Document 27. South Pacific Commission, Noumea, New Caledonia.

Tsao, P.H., Gruber, L.C., Portales, L.A., Gochangco, A.M., Luzaran, P.B. *et al.* (1994) Some new records of *Phytophthora* crown and root rots in the Philippines and in world literature. *Phytophathology* 84, 871.

Vareket, R., Thaveechai, N., Hongprayoon, C., Kositcharoenkul, N., Chuenchitt, S. *et al.* (2005) Identification and serological detection of bacterial organism causing die-back disease of jackfruit and champedak. In: *Proceedings of the 43rd Kasetsart University Annual Conference, 1–4 February 2005, Kasetsart University, Thailand*, pp. 254–261.

Wolf, F.A. (1930) A parasitic alga, *Cephaleuros virescens* Kunze, on citrus and certain other plants. *Journal of the Elisha Mitchell Scientific Society* 45, 187–205.

Zhang, J., Hu, J., Shen, H., Zhang, Y., Sun, D. *et al.* (2018) Genomic analysis of the *Phalaenopsis* pathogen *Dickeya* sp. PA1, representing the emerging species *Dickeya fangzhongdai*. *BMC Genomics* 19(1), 782. DOI: 10.1186/s12864-018-5154-3.

Zulperi, D., Manaf, N., Ismail, S.I., Karam, D.S. and Yusof, M.T. (2017) First report of *Pantoea stewartii* subsp. *stewartia* causing fruit bronzing of jackfruit (*Artocarpus heterophyllus*), a new emerging disease in Peninsular Malaysia. *Plant Disease* 101, 831–831.

13 Postharvest Management

Sisir Mitra[1]*, Jigzcel Divine F. Basoy[2], Josephine G. Tangonan[2] and Pradyot K. Pathak[3]

[1]Former Professor, Faculty of Horticulture, Bidhan Chandra Krishi Viswavidyalaya, Mohanpur, West Bengal, India; [2]College of Human Ecology and Food Sciences, University of Southern Mindanao, Kabacan, Cotabato, The Philippines; [3]KVK, West Bengal University of Animal and Fishery Science, Kolkata, West Bengal, India

13.1 Introduction

Jackfruit is known for its unique quality attributes and is popular as a vegetable with a high market value in South Asian countries. Jackfruits are eaten unripe at 25–50% of full size as a vegetable, or ripe as a fruit. When used as a vegetable, it is peeled, sliced, boiled and then mixed with other food. The leaves of the jackfruit tree are useful for curing fever, boils and skin diseases.

Jackfruit is the national fruit of Bangladesh. It is also called 'jak fruit', 'jak' or Ceylon Jak. In the Philippines, it is named 'langka' in general, but 'nangka' among other tribes and 'badak' in Muslim communities. The ripe fruit contains well-flavoured yellow sweet bulbs and seeds (embedded in the bulb). The edible bulbs of ripe jackfruit are consumed fresh or processed into various food products. The seeds comprise 10–15% of the total weight of the fruit and are high in carbohydrates and protein.

There are several processed products made of jackfruit that can currently be bought in the supermarket. It comes in various forms: fresh (minimally processed), canned and ready-to-eat. The jackfruit bulb is usually preserved in sweet, heavy syrup like other canned fruits. These processed jackfruits are usually used in making baked goods and other desserts. Some are also served salted or seasoned with a savoury barbeque or teriyaki sauce. Some people use these as a meat substitute. In rural areas of the Philippines, young, immature jackfruit is also a favourite food fare. It is usually added as extender in some protein dishes such as 'dinuguan' (a dish made of meat sundries or edible internal organs and pork blood), or as the main ingredient for the famous 'KBL' (Kadyos (pigeon pea), Baboy (pork), at Langka), a most prized delicacy of the Ilonggo tribe in the Visayas islands in the Philippines. The seeds of the matured jackfruit are also boiled and, with added salt, are eaten as a snack.

Fresh jackfruit seeds cannot be kept for a long time, so processing them into jackfruit seed flour is a promising alternative that can be used in some food products. The pulp of ripe jackfruit can be eaten fresh, made into various local delicacies including chutney, jam and jelly or can be preserved as candies and dried leather. The fruit is not easy to eat when fresh, due to the need to separate the edible bulbs from the rind (Jagadeesh *et al.*, 2007). The by-products of jackfruit – the shaft, skin, latex and seeds – are widely used in the production of various

*Corresponding author: sisirm55@gmail.com

DOI: 10.1079/9781800622319.0013

value-added products such as edible cutlery, edible films, leather and baked products such as cookies, waffles and cones. Nowadays, jackfruit is processed into diversified products that are widely available to keep up with the changing tastes and preferences of consumers. Processing and value addition of jackfruit have been described in detail in Chapter 3 (this volume).

13.2 Ripening

Ripeness brings out the best flavour and texture of jackfruit, and even an appetizing smell. When ripe, jackfruit is a delicious fruit. In its mature stage, it is composed of an outer rind, inner rind, rags, bulbs/arils, seeds and core. The most prized parts are the bulbs/arils. It is the bulb that gives the sweet, delicious taste of jackfruit. Jackfruit is a climacteric fruit with a short shelf life. Climacteric fruits reach a certain developmental stage and once this stage has been attained, continue to develop to full physiological maturity, even when removed from the plant. Jackfruit respiratory behaviour follows three distinct phases of respiration (the preclimacteric, climacteric and postclimacteric stages). Precut immature fruit slices showed a basal respiration rate of 160 mg carbon dioxide (CO_2) h^{-1} kg^{-1} fruit, while in the climacteric phase, the peak of respiration was 245 mg CO_2 h^{-1} kg^{-1} fruit (Selvaraj and Pal, 1989). The climacteric peak was attained after 3 days of ripening, and after 8 days of ripening, the respiration rate had dropped to a residual level of 60–70 mg CO_2 h^{-1} kg^{-1} fruit. Climacteric fruits such as jackfruit, peaches, plums, cantaloupe, bananas, pears and tomatoes continue to gain flavour and get sweeter by changing starch into sugar even after harvest (Kader, 2002, 2009). There are factors affecting the ripening process of jackfruit. Once opened and sliced, jackfruit turns brown and deteriorates rapidly after ripening. The ripening process depends on three factors: the maturity of the fruit, the ambient temperature of the surroundings and the relative humidity of the environment (Aseef *et al.*, 2017).

Ripening is influenced by starch content. Benkeblia (2019) noted that during ripening, the total and reducing sugars increased more significantly in ripe jackfruit. The starch content and total phenolic contents also increased significantly in ripe jackfruit. Jackfruit is well known as a starchy produce, and the starch content could be used as a maturity and ripening index.

Sweetness develops during ripening. Sugar accumulation during fruit ripening is caused by increased starch hydrolysis and decreased respiration-associated sugar breakdown (Li *et al.*, 2014). A likely scenario during fruit ripening of jackfruit is that glucose is first accumulated from starch hydrolysis, followed by glucose and fructose accumulation as a result of sucrose hydrolysis. This is supported by the fact that more glucose than fructose was measured at the ripening stage in jackfruit. A threefold increase in sucrose content was reported during ripening, while the concentration of glucose and fructose increased six- and fivefold, respectively (Table 13.1) (Selvaraj and Pal, 1989). When the fruits reached the fully ripe stage, starch hydrolysis is completed, and sucrose hydrolysis caused the molar ratio of glucose:fructose to be close to 1 (Li *et al.*, 2017). This biochemical process occurring in the fruit results in the delicious, sweet taste of ripe jackfruit.

Microscopic and chemical changes occurring during the ripening of both firm- and soft-fleshed forms of jackfruit was investigated by Rahman *et al.* (1995). They observed an increase in dry-matter content increased with maturity. Perianth from immature fruits had a high water-insoluble content (840–890 g kg^{-1} dry matter (DM)) consisting largely of cell-wall material (450–530 g kg^{-1} DM) and starch (approximately 330 g kg^{-1} DM). Microscopic examination of the fruit at this stage showed the perianth to contain thin-walled cells packed with starch granules, some organized into distinct clusters. In ripe fruits, the starch (20–110 g kg^{-1} DM) and cell-wall (170–200 g kg^{-1} DM) contents were substantially reduced, the extent of hydrolysis being greatest in the soft-fleshed form. Cell maceration and starch dissolution were evident in both forms of the fruit when examined by light microscopy but were more pronounced in the soft-fleshed form. Concomitant with the decrease in water-insoluble DM was a substantial increase of water-soluble material (660–790 g kg^{-1} DM), which included fructose (76–113 g kg^{-1} DM) and sucrose (approximately 95 g kg^{-1}

Table 13.1. Changes in biochemical constitutions in ripening jackfruit. Modified from Selvaraj and Pal (1989).

	Days after harvest			
Constituents	0	2	4	8
Flesh firmness (kg cm^{-2} pressure)	12.00	11.50	7.50	6.50
Flesh colour (OD at 448 nm)	0.103	0.143	0.181	0.203
Dry matter (%)	30.10	26.30	24.80	23.90
Alcohol insoluble solids (%)	8.10	4.10	3.90	3.40
Starch (%)	2.03	1.54	1.01	0.69
Sucrose (%)	2.90	7.20	7.30	8.70
Glucose (%)	0.80	4.30	4.80	4.90
Fructose (%)	1.30	5.90	5.70	6.60
Glucose:fructose ratio	0.64	0.73	0.84	0.75
Total acidity (as citric acid, %)	0.21	0.22	0.21	0.13
Sugar:acid ratio	23.30	78.80	85.50	153.20
Citric acid (mg%)	51.00	87.90	92.10	67.20
Malic acid (mg%)	68.50	85.20	83.50	39.70
Citric acid:malic acid ratio	0.74	1.03	1.10	1.69
Total soluble amino acid (mg%)[a]	396	386	333	210
Tannins (mg%)	6.40	3.10	2.80	2.30
Total lipids (%)	0.247	0.276	0.320	0.339
Unsaponifiables (%)	0.049	0.047	0.055	0.055
Phospholipids (mg%)	42.30	40.10	37.80	32.90
Free fatty acid as acid value	26.30	29.80	24.60	23.20
Total fatty acid (%)	0.139	0.135	0.175	0.169
Total sterols (mg%)	5.0	8.00	8.40	9.70
Squalene (mg%)	0.34	0.86	0.68	0.51
Fruit aroma[b]	–	+	+++	++++
Volatiles (%)	0	Trace	Trace	0.02

Values are the average of two independent determinations. OD, optical density.
[a]As isoleucine.
[b] –, No aroma; +, mid-aroma; +++, strong aroma; ++++, very strong aroma.

DM). Mannitol (22–68 g kg^{-1} DM) was also found in ripe fruits but not in immature fruits. The concentration of low-molecular-weight carbohydrate was greatest in the soft-fleshed forms. Polygalacturonase and pectin esterase activities were 12-fold and 40-fold higher in ripe fruits of the soft-fleshed form compared with those of the firm-fleshed form. As both forms of fruit demonstrate a common pattern of ripening, texture differences were evidently related to the extent of change, which was greatest in the soft-fleshed type. The firm-fleshed type of jackfruit may represent fruit in which cell-wall degradation is arrested or delayed during ripening and this is possibly related to a reduced capacity to produce pectic and other cell-wall degrading enzymes.

The ratio of fructose to glucose was reported to be different between soft- and hard-fleshed jackfruit. In fruits of soft-fleshed jackfruit, the activities of water-soluble invertase and amylase reached their peak at stage 3 of maturity and then decreased, while these enzymes changed little in fruits of hard-fleshed jackfruit. The activities of salt-soluble invertase and amylase changed differentially among genotypes, and no similarities were observed among genotypes of soft- or hard-fleshed jackfruit (Li *et al.*, 2014). During ripening of soft-fleshed jackfruit, the activities of water-soluble and salt-soluble galactosidase first increased and then decreased, while in hard-fleshed fruits, they rarely changed. The activity of water-soluble polygalacturonase increased during ripening in soft-fleshed jackfruit, but increased little during ripening in hard-fleshed fruits. These enzymatic changes are believed to be related to the different fruit texture developing into soft- or hard-fleshed jackfruit (Dong *et al.*, 2013).

The total titratable acidity in jackfruit is low (0.13% as citric acid) at the ripe stage and shows little change during ripening. Citric and maleic acids are the major non-volatile acids that have been identified. The concentration of malic acid decreased more, resulting in an increased citric acid:malic acid ratio in the ripe fruit (Table 13.1) (Selvaraj and Pal, 1989).

The protein content was found to range from 1.49% to 2.3% in edible pulp at the ripe stage (Bhatia *et al.*, 1955; Selvaraj and Pal, 1989). The total soluble amino acid concentration decreased during ripening. Sixteen amino acids were identified: α-aminobutyric acid, aspartic acid, asparagine, cysteine, glutamic acid, glutamine, glycine, isoleucine, leucine, lysine, methionine, phenylalanine, proline, serine, threonine and tyrosine, and their concentrations changed markedly during ripening. Asparagine, aspartic acid, glutamine, glutamic acid, glycine, α-aminobutyric acid, phenylalanine and proline are the major soluble amino acids in jackfruit (Selvaraj and Pal, 1989). Jackfruit contains very little tannin and their concentration decreases during ripening.

Total lipids, total fatty acids, unsaponifiables and total sterols increase during ripening, while phospholipids and fatty acids decrease (Table 13.1) (Selvaraj and Pal, 1989). The fatty acids that have been identified are lauric, myristic, palmitic, palmitoleic, celeic, linoleic and linoleric acids. Unsaturated fatty acid, namely palmitoleic, oleic, linoleic and linolenic acids, decrease during ripening.

Sensory evaluation of the aroma of ripening jackfruit has revealed that, at harvest maturity, the fruit has no aroma (–) (Table 13.1). The complex characteristic aroma was noticed at the half-ripe stage (+), and increased considerably (+++) at the three-quarters ripe stage, reaching the maximum concentration (++++) at the fully ripe stage. The initiation of aroma production coincides with respiration from the preclimacteric minimum to the climacteric peak (day 3 after harvest). However, the maximum production of aroma was during the postclimacteric decline (days 4–8 after harvest) in respiration (Selvaraj and Pal, 1989).

13.3 Maturity Indices

A ripe jackfruit is usually identified using a number of indicators. These include one or all of the following manifestations (Haq, 2006):

1. The jackfruit makes a hollow or empty-vessel sound when tapped.
2. The spines become flatter and wider, compared with the young and immature jackfruit.
3. The colour of fruit becomes paler, transforming from dark green to light green or light yellow.
4. A strong aroma develops that can be smelt even at a distance.
5. These changes happen 3–8 months after flowering.

Jackfruits are considered mature when the last leaf on the fruit stalk turns yellow, the rind spines become flattened and the fruit skin colour becomes pale (APAARI, 2012). Saha *et al.* (2016) stated that the rind colour changed with advancement of maturity. The rind was green at the earlier stage and changed to yellowish-green to yellowish-brown at the later stage. Thus, the changes of colour of the rind from green to yellow or greenish-yellow can be taken to indicate that the jackfruit is mature. The characteristic metallic sound was not heard from fruits at 80 and 90 days after fruit set (DAF), but was heard from fruits at 100 DAF at which point it was slightly to moderately detectable (Saha *et al.*, 2016). With advancing maturity, the metallic sound was increasingly detectable. It was moderately to clearly detectable at 120 DAF and clearly detectable at 130 DAF. This hollow metallic sound has been described as one of the most reliable maturity indexes of jackfruit in a number of studies in the literature. Saha *et al.* (2016) suggested harvesting jackfruit at an early stage when the metallic sound can be heard, particularly for commercial marketing. The flatness of the fruit stalk attachment could be used as a maturity indicator sign for jackfruit.

Changes in the physico-chemical properties of jackfruit including colour, texture, moisture content, total soluble solids (TSS), pH, titratable acidity, vitamin C, total starch and total sugar content at four maturity stages of jackfruit flesh were investigated by Ranasinghe and Marapana (2019). Their results clearly indicated a significant difference ($p < 0.05$) in physico-chemical traits of jackfruit flesh at the different stages. The colour parameters varied significantly during maturation, while hardness decreased with maturity. TSS increased with maturity from 3.4% to 19.6%, corresponding to the increase in total sugar content. The pH increased from the immature stage 1 (pH 5.27) to the mature stage (pH 6.25) and then decreased during ripening to pH 5.76, while variation in titratable acidity showed the opposite pattern, ranging between 0.17% and 0.29%. This explains why jackfruit, when overripe, tends to develop a sweetish-sour odour.

Internal changes occur during the advancement of maturity. Saha *et al.* (2016) described the qualitative changes observed in fruits harvested at 90 DAF and at later stages, and the quantitative measures of quality traits of the edible parts of jackfruits. The number of extractable bulbs obtained was more or less similar from the fruits harvested on different DAF. The weight of the bulb, flesh and seeds as well as the edible portion increased over time to maturity. The percentage of the edible portion was 50% for fruits harvested at 100 DAF. The ratio of the edible to non-edible portion was above 1 for the fruits harvested at 100 DAF and at later stages. It was noted that the expected criteria of jackfruit for using as fruits were larger bulbs with well-developed, yellow or orange-yellow flesh. Jackfruits harvested at 80 and 90 DAF had some of these traits, but their quality was not up to the expected level for excellent jackfruit flavour. The physico-chemical condition of the fruits at 80 DAF was the transition point from green (vegetable) jackfruit to being able to be used as a table fruit. The preference for utilizing the jackfruit as fruits increased with advancement of maturity, mostly because of the increasing flesh weight and expected colour development.

Interestingly, the vitamin C content increased with maturity of the fruit to a range of 2.18–8.17 mg 100 g^{-1} (Ranasinghe and Marapana, 2019). The total starch content also increased with maturity from 1.597% to 19.533% but then decreased with ripening to 6.237%. The variation in total phenolic content showed a decreasing pattern from the immature stage (46.969 mg gallic acid equivalents (GAE) g^{-1}) to the mature stage (3.754 mg GAE g^{-1}) and then increased slightly in the fully ripe stage (9.530 mg GAE g^{-1}). The physico-chemical properties of colour, texture, moisture content, pH, TSS, titratable acidity, vitamin C, total starch, total sugar, and total phenolic contents all changed significantly with maturity. TSS, pH, vitamin C content, total starch content and total sugar content increased with maturity, but titratable acidity and total phenolic content showed a decreasing pattern. Thus, fully matured jackfruit has a flavourful and excellent taste when eaten fresh.

The chemical composition (titratable acidity, moisture content, crude fibre, colour, pH, soluble solids, sugars and organic acids) and flavour changes during ripening of jackfruit were investigated by Ong *et al.* (2006). Significant ($p < 0.05$) changes in titratable acidity and colour occurred, but no significant changes were found in moisture and crude fibre contents during ripening. TSS and total sugars increased significantly ($p < 0.05$) throughout the ripening process. TSS in the top portion of the fruit were significantly ($p < 0.05$) higher than in the middle and bottom portions of the same fruit. A high content of malic acid was found in unripe fruit (day 1). In contrast, ripened fruit (days 5 and 6) contained a high amount of citric acid. Overall, the total amount of organic acids was found to decrease from the early to the later stages of ripening and a total of 23 volatile compounds were identified. The data obtained suggested that the ripening process of jackfruit was at its optimum at day 5 after harvest.

The time of harvest matters in terms of the sensory qualities of jackfruit. Chin and Nurshiwan (1989) found that sugar and Brix levels increased dramatically between days 2 and 3 of storage for fruit at both stages of maturity, and remained the same up to day 5. At the same time, the level of starch decreased markedly on days 2 and 3, remaining the same up to day 5. Increases in acidity and colour were observed. These changes account for the sensory observations when the fruit changes from raw, hard, powdery and light yellow to ripe, firm, sweet and deep yellow on day 3. Thus, initially, the flesh of the unripe fruit (at 15 and 16 weeks of maturity) is pale yellow and very hard, with a powdery raw taste and no aroma. When stored, the colour, aroma and taste begin to develop gradually with a significant change occurring on day 3. This is accompanied by a change from a raw, very hard flesh to a ripe, firm one. They further noted that on day 4, the aroma, taste and colour continued to develop but the fruit began to soften a little, and by day 5 there was a slightly fermented taste in the soft, watery fruit. This suggested that day 4 fruit were sweeter and more aromatic than day 3 fruit.

As the fruit quality of jackfruit cannot be improved after harvest, selecting the fruits at the correct stage of maturity is very important. The maturity of a fruit is an assessment of physiological development. It is usually described as two types: physiological maturity and commercial maturity (Shewfelt and Prusia, 1993). Physiological maturity is 'the stage of development when a plant or plant part will continue ontogeny even if detached', while commercial maturity is defined as 'the stage of development when a plant or plant part possess the prerequisites for utilization by the consumer for a particular purpose' (Watada *et al.*, 1984).

Maturity is an integral component of quality (Wills *et al.*, 1998). Maturity of a crop at harvest directly affects colour and size, and thus grade. The important quality characteristics such as texture, flavour and nutrient content, as well as perishability and susceptibility to adverse handling and storage conditions, are, in fact, a function of harvest maturity. Immature fruits do not have satisfactory sweetness and taste, while overmature fruits lose their attractiveness and freshness and became slimy in texture within a very short time (Pareek, 2001).

The time required to ripen, the ripened status and the storage life of jackfruit influences ripening. Jackfruits harvested at an early mature stage took more time to ripen than those harvested at later stages. Fruits harvested at 90 DAF did not ripe uniformly, while those harvested at 100, 110, 120 and 130 DAF were found to ripen well (Saha *et al.*, 2016). The time required to ripen, the storage life after ripening and the total storage life were highest (3.67, 2.33 and 6.00 days, respectively) for fruits harvested at 100 DAF and lowest (1.67, 1.33 and 3.00 days, respectively) for fruits harvested at 130 DAF. As the fruits harvested at 100 DAF ripened well and took longer (3.33 days) to ripen and retained a storage life (6.00 days) higher than fruits harvesed later, the fruits at this stage (100 DAF) could be selected for commercial harvesting.

In Malaysia, the variety 'J33' ('Tekam Yellow') is preferred by consumers due to its sweetness and crunchy texture. Depending on the experience of the growers, the recommended harvest maturity after bagging for 'J33' jackfruit is 90–95 days for the export market and 100 days for minimal processing. At 90–95 days, the fruit is considered to be physiologically matured (mature green) and firm to withstand postharvest handling. However, jackfruit has a relatively short storage life due to its susceptibility to fruit rot. *Lasiodiplodia theobromae* is one of the major fungal pathogens causing postharvest decay on many important horticultural fruits such as citrus, avocados, mangos, papayas, bananas and guavas, as well as jackfruits. In the early stages, fungal colonization by *L. theobromae* remains latent (quiescent) and does not cause any fruit decay; rather, infections tend to manifest after harvest, especially under conditions of high temperatures and relative humidity (Zhang, 2014).

13.3.1 Chemical traits of ripe jackfruit

In ripe jackfruit, the total sugar, reducing sugar and TSS were found to increase compared with freshly harvested fruits, but the vitamin C content and acidity decreased (Saha *et al.*, 2016). When only ripe fruits were considered, it was found that the total sugar, reducing sugar, non-reducing sugar and TSS were higher from late-harvested fruits. In harvested fruits allowed to ripen, the total sugar and TSS increased but acidity was found to decrease, as a result sugar:acid ratio and TSS:acid ratio increased in the ripe fruits. The vitamin C content was noted to be higher in ripe fruits harvested at 90 and 100 DAF (8.67 and 5.27 mg 100 g^{-1} flesh, respectively) compared with other harvesting dates. It was lowest (4.12 mg 100 g^{-1} flesh) in fruits harvested at 130 DAF. Physiological and commercial maturation of the fruit occurred at 100 DAF when the fruits should be harvested (Saha *et al.*, 2016).

13.3.2 Time of ripening

In Asia, jackfruit ripens primarily in March–June, April–September or June–August depending on the climatic region, with some off-season crops from September to December and sometimes a few fruits at other times of the year (Gunasena *et al.*, 1996; Azad, 2000). In Panruti, Kerala, India, jackfruit is harvested all year round (Padre, 2011). In Sri Lanka, the tree flowers in November–December and fruit ripens from March to June. At higher elevations, off-season fruits ripen during October–December (Gunasena *et al.*, 1996). In the West Indies, the fruits ripen in June, and in Florida, it is in late summer or autumn. In Darwin, Australia, the fruits ripen any time from June to April (Haq, 2006).

13.3.3 Ripening index

The ripening index is the ratio of TSS to titratable acidity, which indicates the maturity of fruit (Majidi *et al.*, 2011). A lower value of the ripening index is preferred as it is indicative of a longer shelf life and better flavour of the fruit. Changes in the ripening index for refrigerated storage and deep-freeze storage indicated that samples pretreated with citric acid were better at controlling the rise in the ripening index than samples pretreated with calcium chloride ($CaCl_2$) and control samples. Changes in the external appearance of the ripening fruit after harvesting (observed only on fruits at 15 and 16 weeks of maturity) were usually more marked, with colour and aroma being the main criteria. The increasing aroma and deepening of skin colour progressed through to day 5 when a slight alcoholic aroma was also detected and rust-like patches appeared on the skin.

13.4 Harvesting

Jackfruits are harvested at different stages of maturity depending on the intended use – immature for cooking and ripe for eating fresh. As the ripening proceeds normally even when the fruits are harvested a little earlier than at the optimum mature stage, it is not essential to wait until the fruits are fully mature before harvesting. There are, however, some differences in the taste from fruit to fruit, probably due to variations in maturity. Young fruits that are green to yellowish in colour and with well-expanded spines are harvested for use as vegetables during the first 2–3 months after fruit set, before the seeds harden, while mature fruits are harvested at ripening when they emit the jackfruit's characteristic aroma.

The fruits should be harvested without causing mechanical damage such as cuts or bruises. Harvesting is carried out by cutting the peduncle with a sharp knife and by traditional methods, such as the use of ropes and sickles for harvesting fruits located on high branches or at the top of the tree. It is desirable to maintain at least 2–3 cm of the fruit stalk to avoid latex staining on the surface of the fruit. The harvested fruit should be laid for some time with the stalks downwards to allow the latex to flow and coagulate.

13.5 Marketing

Jackfruit marketing involves three groups, namely producers, traders (middlemen) including wholesalers, and retailers (Haq, 2006). A stable marketing chain is not in existence for jackfruit in the Asia-Pacific Region. At the village level, collectors from outside as well as those from the village itself purchase the fruits and sell them in external markets, mainly at village fairs and road-side sheds (APAARI, 2012). Transport of jackfruit for the domestic market involves the use of unrefrigerated delivery systems. The fruits are transported to the local market (Fig. 13.1) in bulk, commonly by cycle-van (Fig. 13.2), bullock cart or truck (Fig. 13.3). Careless handling of the fruits often occurs during the loading and unloading processes. Protective measures against transport hazards, such as proper stacking and ventilation

Fig. 13.1. A local market. Photograph courtesy of Dr P.K. Pathak

and use of cushioning materials, are rarely adopted. The resulting damage inflicted during transport predisposes the fruits to high losses, which can be as much as 12–25%. Reddy *et al.* (2019) studied the marketing problems of jackfruit growers of the Cuddalore District of Tamil Nadu, India, and ranked the constraints of marketing in the following order: (i) cartel formation by wholesaler; (ii) lack of awareness and skills for value addition; (iii) lack of awareness on export procedures; (iv) lack of proper market information; (v) perishability; and (vi) lack of skill labour for harvesting. It is evident from the Garrett ranking analysis (Garrett and Woodworth, 1969) that farmers, although interested in participation in the export and value addition processes directly, are unable to proceed further due to a lack of awareness of the related technical procedure.

The production of jackfruit will expand mostly due to an expanding market for processed products. It is also estimated that the demand for fresh fruits will expand in countries such as Japan, Malaysia and the UK in addition to Singapore and Hong Kong and some Middle Eastern countries (Haq and Hughes, 2002). Jackfruit is gaining popularity even in the USA and in the United Arab Emirates due to emerging ethnic and mainstream marketing opportunities. Malaysia exports jackfruit to the Middle East, Europe, China, Australia, Hong Kong and Singapore, while Vietnam exports mostly to China and Guinea.

For long-distance transport, it is suggested that the fruits should be harvested at the colour turning stage (i.e. from green to yellowish-green) and when the portion of the stalk attached to the fruit is large enough to be used for handling (Kader, 2009). Uniform fruits of about 8–10 kg in weight are selected for special markets and for export, while extra-large and small fruits are sold locally.

The details of the distribution export channel of Malaysian jackfruit to Dubai were explained by Safari *et al.* (2019). Fruits with an 80% maturity index (full ripe) and an approximate weight of 10 kg each were selected. The fruits were harvested in the morning, cleaned

Fig. 13.2. Transportation of fruits by cycle-van. Photograph courtesy of Dr P.K. Pathak.

and transported to the exporters' warehouse by the evening. A normal truck was used for this purpose. The next day, in the warehouse, the fruits were washed in tap water and dried before export. The fruits were checked for quality before being packed in corrugated boxes. The fruits were stored in chillers at 10°C and exported by air fright in the evening of the same day (day 2 after harvesting). A non-stop flight from Kuala Lumpur, Malaysia, to Dubai, is around 7 h, while the fastest transit flight takes close to 10 h. The fruits arrived at the airport's warehouse on day 3 and were delivered to the buyer's warehouse on day 4. They were then expected to reach the retailers on the same day or the next day (day 4 or 5).

13.6 Storage

A considerable amount of jackfruit, especially that obtained in the glut season, goes waste due to lack of proper postharvest management. Proper postharvest technology for extending shelf life is therefore necessary. In addition, alternative ways of using jackfruits in the on-season play a significant role in reducing postharvest losses. Among these, processing is important. It adds diversified and attractive food items to the dietary menu, as well as contributing to the generation of income and employment (processing for product development is discussed in detail in Chapter 3, this volume). Some of the storage processes usually followed to extend the shelf life of jackfruit are discussed below.

13.6.1 Minimal processing

Jackfruits are minimally processed to increase their functionality without greatly changing their fresh properties. Minimal processing is usually done at the local level. Fresh jackfruit bulbs are stored in a covered container and then subjected to refrigeration temperatures. When kept cool and dry, a ripe jackfruit bulb can last up to 6 weeks. Jackfruit is bulky in nature, which makes it difficult to transport and store.

Fig. 13.3. Transportation of fruits by truck. Photograph courtesy of Dr W.Z.W. Sembok.

Although it is available almost all year round, there are peak seasons during which the fruit mainly rots away due to its perishable nature. Harvested ripe jackfruits may be stored for 4–5 days at ambient temperature and a relative humidity of 65–75%, while a low temperature of 13.3°C with relative humidity of 85–90% increase its shelf life up to 45 days (Thompson, 2014).

Minimally processed jackfruit has a short shelf life as it remains in a respiratory condition and physiologically active (Saxena *et al.*, 2009). Various methods to prolong the keeping quality of fresh-cut jackfruit have been proposed (Saxena *et al.*, 2012, 2013, 2015; Lakshmana *et al.*, 2013). The market for fresh-cut fruits has grown rapidly in recent years. Minimal processing is replacing more traditional preservation methods, as minimally processed products are better in term of sensorial quality and nutritional aspects (Ohlsson and Bengtsson, 2002). The International Fresh-cut Produce Association defines fresh-cut products as fruit or vegetables that have been trimmed and/or peeled and/or cut into 100% usable product that is bagged or prepackaged to offer consumers high nutrition, convenience and flavour while still maintaining its freshness (Lamikanra, 2001). However, the treatment used in minimally processed produce results in tissue wounding, which may have a deleterious effect on the produce (Siddiqui *et al.*, 2011).

Minimally processed fruits are generally evaluated based on colour, texture, physicochemical characteristics and sensory attributes (Rico *et al.*, 2007). In minimally processed products, the greatest hurdles are the changes in the physical attributes of the fruits (Soliva-Fortuny *et al.*, 2002; Toivonen and Brummell, 2008). Some of the methods currently employed in the fresh-cut industry include modified atmosphere packaging (Saxena *et al.*, 2008), mild heat treatment, edible coatings (Dávila-Aviña *et al.*, 2011) and calcium infusion (Gras *et al.*, 2003). As calcium has been associated with the maintenance of texture in various fruits, its inclusion in the bulbs is expected to retain the crispness of jackfruit bulbs.

Texture is a very important aspect in acceptance of fresh-cut fruits in general, and is one of the determining factors in the acceptability of fresh-cut jackfruit (Saxena *et al.*, 2013). With increasing consumer awareness about the health benefits of consuming more fruits, there has been an increase in the demand for fresh-cut fruit, thus creating an opportunity for minimally processed jackfruits. To maintain the desirable texture, infusions of calcium into the tissue of jackfruit pulps can be used to retain the textural integrity of the fruit up to at least 3 days by the immersion method and up to 10 days by the vacuum infusion method. The ascorbic acid content of jackfruits treated with calcium was comparable to that of untreated samples. Introduction of calcium by vacuum infusion to retain its textural characteristics is achievable, particularly in food service operations (Ramli *et al.*, 2017).

13.6.2 Extending the shelf life of minimally processed jackfruit bulbs

Marketing jackfruit in a ready-to-eat form offers several benefits to the consumers, as it reduces preparation time, provides a more uniform and consistent quality, requires less storage space, can easily be packed, and waste and handling are reduced. Fresh-cut products are highly perishable and demonstrate rapid postharvest quality degradation (Latifah *et al.*, 2018). The preparation of jackfruit for fresh-cut processing entails physical wounding of the tissue. This causes an increase in the rate of respiration and ethylene production, and also biochemical reactions responsible for changes in colour (including browning), flavour, texture and nutritional quality (Cantwell, 1996).

Storage under a modified atmosphere such as in polythene bags or packaging, or coating in polymeric films can prolong the shelf life of a wide variety of fruits (Sivakumar and Korsten, 2006; Dhalsamant *et al.*, 2017). MA treatments ultimately result in a build-up of CO_2 and depletion of O_2 in the internal atmosphere surrounding the fruit. In many cases, respiration, as well as ethylene production and perception by the fruit, are reduced, ripening is delayed and hence the shelf life of the fruit is extended markedly.

In wet markets, jackfruit bulbs are often sold fresh after washing in water and are packaged in polypropylene containers. This bulbs have a very short shelf life and need to be consumed shortly after purchase. Minimally processed jackfruit bulbs can be stored for longer under low temperatures (Fig. 13.4). Different methods such as (alone or combined) low temperatures, anti-browning agents, anti-respiratory substances, texturizing agents, preservatives and UV treatments are used to extend the shelf life of jackfruit bulbs (Anaya-Esparza *et al.*, 2018).

Storage temperature

Temperature management is key for preserving the freshness and quality of minimally processed jackfruit bulbs. At a low temperature (2°C), the bulbs can be stored for 7 days without much effect on the colour and freshness (Latifah *et al.*, 2018). Low-temperature storage decreases enzymatic activity and slows down biochemical changes (Anaya-Esparza *et al.*, 2018). Zeng *et al.* (2018), however, suggested storing fresh-cut jackfruit at 8–12°C. This temperature range delays the increase in cell membrane permeability and the accumulation of malondialdehyde in fruits, maintains the activity of superoxide dismutase, and improves the sensory quality of fresh-cut jackfruit. Low temperatures (0–4°C) affect the nutrient composition and taste, leading to loss of natural flavour.

Dipping treatment

Anti-browning agents or thermal blanching as a pretreatment can mitigate enzymatic browning (Saxena *et al.*, 2015; Xu *et al.*, 2015; Gayathri *et al.*, 2021). Tao *et al.* (2013) stated that decreasing (< 4.5) or increasing (>8) pH can inhibit enzymatic browning by inhibiting polyphenol oxidase activity in fresh-cut bulbs. Thermal blanching (immersion in water at 80°C for 1–2 min) before freeze drying inactivates over 95% polyphenol oxidase activity (Xu *et al.*, 2015).

Antimicrobial treatment is used extensively in the food industry, with chlorine being a common decontaminating agent, but its adverse health effects and ineffectiveness in maintaining

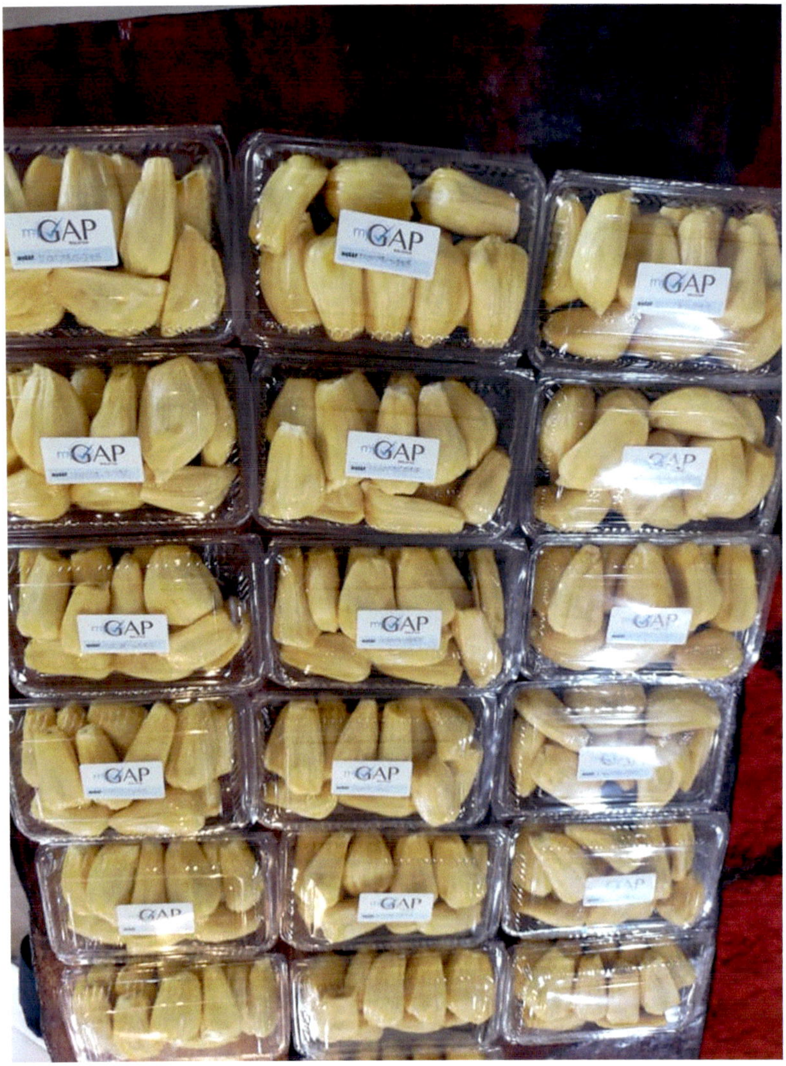

Fig. 13.4. Minimally processed packed bulbs for sale. Photograph courtesy of Dr W.Z.W. Sembok.

the sensory qualities, particularly taste, are a major concern. Ascorbic acid, citric acid, calcium lactate and potassium sorbate are examples of natural antimicrobials in fresh-cut jackfruit (Ulloa *et al.*, 2010; Acedo *et al.*, 2013). Acedo *et al.* (2013) explored the use of ascorbic acid and calcium lactate on microbial contamination of fresh-cut jackfruit. Microbial load was first determined following standard plating methods and was found to be higher in traditional wet markets than in supermarkets by about 0.7–1.0 log colony-forming units ml^{-1}. Jackfruit bulbs were dipped for 2 min in an aqueous solution of ascorbic acid (1.5%) or calcium lactate (2.5%) as the antimicrobial agent. A 2 min wash in 150 p.p.m. chlorine solution or distilled water and no wash served as control treatments. The treated bulbs were then packed in sterile, resealable 50 μm thick polyethylene bags and stored at 7–10°C for 5 days. They observed that calcium lactate treatment remarkably reduced aerobic bacteria and coliform bacteria and yeast and mould counts at both 3 h after treatment and after 5 days of storage by about 85–99.6%

Table 13.2. Percentage reduction in microbial counts on organic acid-treated and untreated minimally processed jackfruit at 3 h after treatment and after 5 days of storage at 7–10°C. Modified from Acedo *et al.* (2013) with permission from the International Society for Horticultural Science.

	Total aerobic count (% reduction)		Coliform count (% reduction)		Yeast and mould count (% reduction)
Treatment	After 3 h	After 5 days	After 3 h	After 5 days	After 5 days
Relative to no wash					
Ascorbic acid	82.60	NS	96.00	96.30	NS
Calcium lactate	85.20	99.40	91.10	99.60	97.90
Chlorine	91.90	99.70	90.00	NS	90.70
Relative to water wash					
Ascorbic acid	76.00	NS	89.50	NS	NS
Calcium lactate	79.70	96.60	NS	97.90	98.30
Chlorine	88.80	98.20	NS	NS	92.40

The percentage reduction was calculated using the actual microbial count. NS, microbial count not significantly different from that of no wash or water wash.

(Table 13.2) relative to that of no wash. These effects were similar to the effect of chlorine, except that chlorine did not affect the coliform load after 5 days of storage. Ascorbic acid was as effective as calcium lactate and chlorine at 3 h after treatment, but had no marked effect after 5 days of storage. Colour, soluble solids and sensory quality showed no significant difference among treatments. Probiotic treatment with *Lactobacillus plantarum* at 3% was reported to be more effective than ascorbic acid and chlorine treatment for advancing storage periods and to have no adverse effects on the physico-chemical and sensory qualities of fresh-cut jackfruit (Troyo and Acedo Jr., 2015).

Calcium has been associated with the maintenance of texture, and its infusion into the bulbs is expected to retain the crispness (Ramli *et al.*, 2017). Infusion of 1% $CaCl_2$ in fresh-cut jackfruit bulbs, packaged in aluminium foil trays wrapped with cling film and stored at 5–6°C extended the shelf life (5 days versus 3 days for the control), maintained a high carotenoid and vitamin C content along with a low phenol content, thus making it nutritionally superior (Gayathri *et al.*, 2021). Better firmness and colour retention of fresh-cut bulbs was noted by pretreatment dipping of bulbs in 1% $CaCl_2$ and 0.02% ascorbic acid for 30 min and storage at 4°C (Saxena *et al.*, 2012). Vacuum infusion of 1% calcium for 15 min into the tissue of jackfruit bulbs enabled the textured integrity of the bulb to be maintained for up to 10 days at 8°C (Ramli *et al.*, 2017).

Edible coatings

Edible films and coatings are also used to minimally process fruits to increase their shelf life. An edible coating acts as a barrier against O_2 and CO_2, reduces the respiration rate and ethylene production, and prevents decay and colour changes during storage (Anaya-Esparza *et al.*, 2018). Chitosan-coated fresh-cut jackfruit could be stored for 15 days at 3 ± 1°C (Jiang *et al.*, 2012). Teja *et al.* (2016) evaluated the effects of different coatings (pectin and *Aloe vera* gel) compared with a commercial gelling agent on postharvest quality of minimally processed jackfruit bulbs during 7 days of storage at 6°C. Coating with *Aloe vera* gel was found to be very effective in preventing moisture loss, weight loss and better retention of quality. Bulbs dipped in potassium sorbate and $CaCl_2$, treated with 1-methylcyclopropene (1-MCP) and coated with edible coatings (xanthan gum, sodium alginate or gelatin gum) showed a decrease in ripening rate, and thus weight loss and respiration rate were lower and the higher-quality values such as colour, TSS and titratable acidity were maintained, as well as pH, for 12 days of storage at 5°C (Vargas-Torres *et al.*, 2017).

UV radiation treatments

UV radiation is used in the industry, including postharvest processing, and has minimal impact on the original flavour and texture, thus ensuring maintenance of food quality (Guerrero-Beltr·n and Barbosa-C·novas, 2004). The high-frequency, electromagnetic UV waves can extend shelf life by stress-induced defence responses that reduce infection and cellular damage (Charles and Arul, 2007). Jackfruit bulbs were minimally pretreated with UV radiation (240 nm, 5 min) and packaged in polyethylene bags. The treatment significantly reduced the population of mesophilic aerobes and total coliforms, yeasts and moulds. The bulbs could be stored for 14 days at 5°C, without significant changes in colour or other sensory attributes (Bizura Hasida *et al.*, 2013).

Vasudeva *et al.* (2020) reported that the shelf life of fresh-cut jackfruit bulbs could be extended by pretreatment of bulbs combined with modified atmosphere packaging and cold storage. Fresh-cut jackfruit bulbs were pretreated with 1% $CaCl_2$ and 0.25% ascorbic acid packed in transparent polypropylene covers with different gas compositions (2% O_2 + 3% CO_2 + 95% nitrogen (N_2); 2% O_2 + 5% CO_2 + 93% N_2; 3% O_2 + 3% CO_2 + 94% N_2; 3% O_2 + 5% CO_2 + 92% N_2; 4% O_2 + 3% CO_2 + 93% N_2; 4% O_2 + 5% CO_2 + 91% N_2) along with the control (passive gas composition) and stored in cold storage (6±2°C and relative humidity 85±5%). Among the different gas compositions, the bulbs packed with 3% O_2 + 5% CO_2+ 92% N_2 showed an increase in storage life up to 24 days by reducing the respiration rate (3.96 mg CO_2 kg^{-1} h^{-1}) and minimizing the physiological loss in weight (0.23%), and retained higher firmness (4.05 kg cm^{-2}) and maintained TSS (28.70%) and higher titratable acidity (0.52%) compared with 12 days of storage life in the control.

Bulbs colour may be affected by the process involved in modified atmosphere packaging. Changes in the physico-chemical properties of fresh-cut tender jackfruit during storage can be related to change in its colour. Rana *et al.* (2018) using image analysis, observed that the browning of fresh-cut tender jackfruit slices occurred more rapidly in controls and normally packed groups than in modified atmosphere-packed and vacuum-packed groups. The change in colour values for the control and normally packed groups were higher for every single investigative technique, while for the vacuum-packed group, the change in colour values was moderate. Image analysis can be utilized to obtain colour data from the whole surface of samples, and is able to express different colour values and limit errors, which is not possible by manual inspection. Additionally, the study distinguished the changes in colour values by the degree of brightness. Rana *et al.* (2018) also found that by using vacuum packaging, the rate of enzymatic browning in fresh-cut tender jackfruit slices can be reduced. The shelf life of fresh-cut tender jackfruit slices increased from 2 days to 10 days when vacuum packed in 0.05 mm polyethylene film.

Hypoxia delays fruit ripening and senescence as it decreases ethylene production and tissue sensitivity to ethylene (Kader *et al.*, 1989). Jackfruits held in a sealed, low-density polyethylene bag (LDPE) failed to ripen normally compared with other treatments (Ying *et al.*, 2020).

The physical properties of frozen confections (overrun, melting resistance and hardness), using refrigerated, vacuum-packed jackfruit bulbs (of different weeks) maintained their frozen confectionery quality (Johari *et al.*, 2020). Jackfruit frozen confectionery products made with frozen, vacuum-packed bulbs also showed similar results. All jackfruit frozen confections had lower overrun compared with the controlled frozen confection sample (without jackfruit addition). The refrigerated and frozen vacuum-packed jackfruit bulbs could be stored for 1–3 weeks and 1–6 weeks, respectively, and showed no significant effects on the physical properties. Thus, refrigeration for 1–3 weeks was used in the factory to store vacuum-packed jackfruit bulbs. This work contributed to solving the surplus issue faced by jackfruit plantations and processing factories by converting 'Mastura' jackfruit into frozen confectionery products (Johari *et al.*, 2020).

Controlled atmosphere storage

One way of maintaining quality in the fruit is to use controlled atmosphere storage. The bulbs are pretreated with $CaCl_2$ (1%), ascorbic acid (0.02%), citric acid (1%) or sodium benzoate

(0.045%) for 30 min and packaged in polyethylene bags with 3 kPa O_2 and 5 kPa CO_2, and can be stored for 35 days at 4°C. This atmosphere decreased the respiration rate and ethylene production, preserved firmness and colour, and minimized the microbial load (Saxena *et al.*, 2008). These authors further reported that precut jackfruit bulbs kept under a controlled atmosphere (3 kPa O_2 + 6 kPa CO_2) in synergy with pretreatment involving anti-respiratory, anti-browning and antimicrobial agents showed beneficial effects in terms of higher retention of phenolic and ascorbic acid. The interaction of controlled atmosphere pretreatment and chitosan coating was found to render microbial and physiological stability to the precut produce. Fresh-cut bulbs pretreated with 1% $CaCl_2$ and 0.25% ascorbic acid and packed in a polypropylene cover with a gas composition of 3 kPa O_2 and 5 kPa CO_2 showed an increase in storage life up to 24 days by reducing the respiration rate (3.96 mg CO_2 kg^{-1} h^{-1}), minimum physiological loss in weight (0.23%) and retention of higher firmness (4.05 kg cm^{-2}), and maintained quality (Vasudeva *et al.*, 2020).

13.6.3 Storage of tender jackfruit

Green jackfruit, as a cooked vegetable, is a dish that is enjoyed by the Asian community. Therefore, presentation of jackfruit as a fresh-cut product would be convenient to reduce preparation time in the kitchen and is a product that would be welcomed by consumers. However, enzymatic browning, microbial contamination and excessive tissue softening were found to be the major limiting factors in fresh-cut jackfruit commercialization (Navindra *et al.*, 2009; Marak *et al.*, 2019).

Tender green jackfruit cubes of 25 mm treated with 1% citric acid and 2% ascorbic acid and vacuum packed (550 mbar atmospheric pressure) in 80 μm LDPE vacuum pouches can be stored for 15 days at 2–4°C. This treatment inhibits browning and deterioration of fresh-cut green jackfruit (Navindra *et al.*, 2009). Sudheer *et al.* (2019) developed a protocol for increasing the shelf life of fresh-cut tender jackfruit. Tender fruits (60–70 DAF) were sliced and steam blanched for 3 min. The blanched slices were then dipped in 1% sodium chloride and 0.05% sodium benzoate and stored under refrigerated conditions. The tender jackfruit slices retained their quality for 9 days in storage under these conditions. Marak *et al.* (2019) reported that tender jackfruit can be brined at 8% salt in glass jars or in standing pouches of 75 μm for a period of at least 6 months. The brined tender jackfruit can be used as a vegetable in curries during the off-season and can easily be transported to places where it is not available. Vacuum packaging reduces the rate of enzymatic browning in fresh-cut tender jackfruit slices.

13.6.4 Storage of ripe fruits

Temperature storage

Freshly harvested ripe fruits can be stored for 4–5 days at 23–35°C or for 2–6 weeks at 11–14°C at a relative humidity of 85–95% (ICUC, 2005; Kader, 2009). Fruits stored at temperatures below 12°C before transfer to higher temperature show chilling injury symptoms including dark brown discoloration of the skin, pulp browning and a deterioration in flavour and increased susceptibility to decay (ICUC, 2005). As discussed earlier, jackfruit is susceptible to postharvest fruit rot caused by *L. theobromae*, which remains latent in the early stages but manifests after harvest, especially under conditions of high temperature and relative humidity. In jackfruit, *L. theobromae* tends to develop rapidly after the onset of the ripening process during cold storage or at ambient temperature. Ripe fruits packed in a perforated LDPE bag and kept at 12°C with 85–90% relative humidity could be stored for 2 weeks, had lower scores of disease incidence, had a better appearance and maintained quality (Ying *et al.*, 2020).

1-MCP treatment

For fruits treated with 1-MCP at 0.5 mg l^{-1}, the storage life was extended by 6–10 days at ambient temperature (25±1°C). 1-MCP treatment inhibited ethylene production, delayed the initiation of starch degradation by 2 days and markedly reduced the sucrose content. 1-MCP treatment also inhibited the decrease in starch, pectin and cellulose and the increase in total soluble sugars, reducing sugar and soluble pectin (Wang *et al.*, 2014).

13.7 Conclusion

Very little research information is available on the postharvest management of jackfruit. Optimal harvesting of the fruits depends on the intended use. Immature fruits are harvested at 3–6 months after fruit set for use as a vegetable. When eaten as a fresh fruit, the fruits are harvested at maturity. Ripeness brings out the best flavour, texture, taste and smell. Harvesting is carried out by cutting the peduncle with a sharp knife and by traditional methods. A stable marketing chain is not in existence for jackfruit in the Asia-Pacific Region. The fruits are transported to the local market in bulk by cycle-vans or trucks in an unrefrigerated delivery system.

Jackfruit is a climacteric fruit. The fresh jackfruit can be stored at 12–16°C and 85–90% relative humidity for 4–6 weeks compared with only 4–5 days at room temperature (25–27°C). Minimally processed jackfruit bulbs can be stored at low temperatures (2–8°C) and treatments such as anti-browning agents, anti-respiratory substances, texturizing agents, preservatives, UV treatment and a controlled atmosphere are used to increase their shelf life.

References

Acedo, J.Z., Varron, D.A.C., Emnace, I.C., Lauzon, R.D. and Acedo Jr., A.L. (2013) Bioprotective potential of lactic acid bacteria in fresh-cut jackfruit (*Artocarpus heterophyllus* Lam.). *Acta Horticulturae* 989, 209–217. DOI: 10.17660/ActaHortic.2013.989.26.

Anaya-Esparza, L.M., González-Aguilar, G.A., Domínguez-Ávila, J.A., Olmos-Cornejo, J.E., Pérez-Larios, A. *et al.* (2018) Effects of minimal processing technologies on jackfruit (*Artocarpus heterophyllus* Lam.) quality parameters. *Food and Bioprocess Technology* 11(9), 1761–1774. DOI: 10.1007/s11947-018-2136-z.

Aseef, R.M., Manikandan, K., Kavino, M., Vijaykumar, R.M. and Ganeshan, N.M. (2017) Biochemical evaluation of local genotypes of jackfruit (*Artocarpus heterophyllus* Lam.) in Pudukkottai district. *Journal of Pharmacognosy and Phytochemistry* 6, 2533–2536.

Asia-Pacific Association of Agricultural Research Institutions (2012) *Jackfruit Improvement in the Asia-Pacific Region – A Status Report*. APAARI, Bangkok.

Azad, A.K. (2000) Genetic diversity of jackfruit in Bangladesh and development of propagation methods. PhD thesis, University of Southampton, UK.

Benkeblia, N. (2019) Variation of reducing and total sugars starch, total phenolic contents in unripe and ripe jackfruit (*Artocarpus heterophyllus*) and West Indian locust (*Hymenaea courbaril*) fruits. *International Journal of Plant Science and Horticulture* 1, 49–55. DOI: 10.36811/ijpsh.2019.110005.

Bhatia, B.S., Siddappa, G.S. and Lal, G. (1955). Composition and nutritive value of jackfruit. *Indian Journal of Agricultural Science* 25, 303–305.

Bizura Hasida, M.R., Nur Aida, M.P., Zaipun, M.Z. and Hairiyah, M. (2013) Microbiological changes of minimally processed jackfruit after UV-C treatment. *Acta Horticulturae* 1012, 1025–1030. DOI: 10.17660/ActaHortic.2013.1012.138.

Cantwell, M. (1996) Fresh-Cut Products: Maintaining Quality and Safety, Postharvest Horticulture Series 10. University of California, Davis.

Charles, M.T. and Arul, J. (2007) UV treatment of fresh fruits and vegetables for improved quality: a status report. *Stewart Postharvest Review* 3, 1–8. DOI: 10.2212/spr.2007.3.6.

Chin, A.H. and Nurshiwan, Z. (1989) Jackfruit quality for canning in syrup. *MARDI Research Journal* 17, 266–274.

Dávila-Aviña, J.E. de J., Villa-Rodriguez, J., Cruz-Valenzuela, R., Rodriguez-Armenta, M., Espino-Díaz, M. *et al.* (2011) Effect of edible coatings, storage time and maturity stage on overall quality of tomato fruits. *American Journal of Agricultural and Biological Sciences* 6(1), 162–171. DOI: 10.3844/ajabssp.2011.162.171.

Dhalsamant, K., Mangaraj, S. and Bal, L.M. (2017) Modified atmosphere packaging for mango and tomato: an appraisal to improve shelf life. *Journal of Packaging Technology and Research* 1(3), 127–133. DOI: 10.1007/s41783-017-0021-2.

Dong, L., Wang, Y., Li, Y., Li, H., Zhang, J. and Ye, C. (2013) Activity of glycosidase and polygalacturonase in jackfruit fruit. *Journal of Southern Agriculture* 44, 924–929.

Garrett, H.E. and Woodworth, R.S. (1969) *Statistics in Psychology and Education*. Vakils, Feffer and Simons Pvt, Bombay, India.

Gayathri, G.S., Mini, C. and Anith, K.N. (2021) Effect of pre-storage treatments on shelf life and biochemical quality parameters of minimally processed jackfruit (*Artocarpus heterophyllus* L.) bulbs. *Environment and Ecology* 39, 482–488.

Gras, M.L., Vidal, D., Betoret, N., Chiralt, A. and Fito, P. (2003) Calcium fortification of vegetables by vacuum impregnation: interactions with cellular matrix. *Journal of Food Engineering* 56, 279–284.

Guerrero-Beltr·n, J.A. and Barbosa-C·novas, G.V. (2004) Advantages and limitations on processing foods by UV light. *Food Science and Technology International* 10(3), 137–147. DOI: 10.1177/1082013204044359.

Gunasena, H.P.M., Ariyadasa, K.P., Wikramasinghe, A., Hearth, H.M.W., Wikramasinghe, P. *et al.* (1996) *Manual of Jack Cultivation in Sri Lanka*, Vol. 48. Forestry Information Service, Forest Department, Sri Lanka.

Haq, N. (2006) *Jackfruit* (*Artocarpus heterophyllus*). Southampton Centre for Underutilized Crops, University of Southampton, Southampton, UK.

Haq, N. and Hughes, A. (2002) *Fruits for the Future in Asia*. International Centre for Underutilized Crops, University of Southampton, UK.

ICUC (2005) *Training Manual on Processing and Small Business Development*. International Centre for Underutlized Crops, University of Southampton, UK.

Jagadeesh, S.L., Reddy, B.S., Swamy, G.S.K., Gorbal, K., Hegde, L. *et al.* (2007) Chemical composition of jackfruit (*Artocarpus heterophyllus* Lam.) selections of western ghats of India. *Food Chemistry* 102(1), 361–365. DOI: 10.1016/j.foodchem.2006.05.027.

Jiang, Q.-H., Ye, S.-Q., Ye, C-.H., Lv, Q.-F. and Han, R. (2012) Study on preservation of fresh-cut jackfruit by chitosan film. *Modern Food and Technology* 28, 14–17.

Johari, A.M., Rahman, N.A.A., Baharuddin, A.S., Basha, R.K. and Wakisaka, M. (2020) Effect of low temperature storage on Mastura (J37) jackfruit bulbs on the physical quality of jackfruit confection. *Journal of Agriculture and Food Engineering* 2: 0016.

Kader, A.A. (2002) Postharvest biology and technology: an overview. In: Kader, A.A. (ed.) *Postharvest Technology of Horticultural Crops*. Publication No. 3311, University of California, Oakland, California, pp. 15–20.

Kader, A.A. (2009) *Jackfruit: Recommendations for Maintaining Postharvest Quality. Produce Facts*. Department of Plant Science, University of California, Oakland, California.

Kader, A.A., Zagory, D. and Kerbel, E.L. (1989) Modified atmosphere packaging of fruits and vegetables. *Critical Reviews in Food Science and Nutrition* 28(1), 1–30. DOI: 10.1080/10408398909527490.

Lakshmana, J.H., Jayaprahash, C.L.J., Kumar, R., Kumaraswamy, M.R., Kathiravan, T. *et al.* (2013) Development and evaluation of shelf stable retort pouch processed ready to- eat tender jackfruit (*Artocarpus heterophyllus*) curry. *Journal of Food Processing & Technology* 04(10), 1–6. DOI: 10.4172/2157-7110.1000274.

Lamikanra, O. (2001) *Fresh-Cut Fruits and Vegetables: Science, Technology and Market*. CRC Press, Boca Raton, Florida. DOI: 10.1201/9781420031874.

Latifah, M.N., Shukor, A.R.A., Aziz, I.A., Habsah, M., Talib, Y., Rahman, K.M. and Jabir, H. (2018) Quality of minimally processed jackfruit stored at different temperatures. *Journal of Tropical Agriculture and Food Science* 28, 87–94.

Li, Y., Deng, L., Wang, Y., Li, H., Zhang, J. *et al.* (2014) Characterization of sugar metabolism of soft and hard flesh jackfruit during ripening of fruit. *Journal of Fruit Science* 31, 85–90.

Li, Y., Duan, X., Liu, S., Li, Y., Zhang, X. *et al.* (2017) Changes in soluble sugar accumulation and activities of sucrose-metabolizing enzymes during fruit ripening of jackfruit. *Journal of Agricultural Science* 9(8), 155. DOI: 10.5539/jas.v9n8p155.

Majidi, H., Minaei, S., Almasi, M. and Mustofi, Y. (2011) Total soluble solids, titratable acidity and ripening index of tomato in various storage conditions. *Australian Journal of Basic and Applied Science* 5, 173–176.

Marak, N.R., Nganthoibi, R.K. and Momin, C.W. (2019) Process development for brining of tender jackfruit. *International Journal of Current Microbiology and Applied Sciences* 8(4), 2408–2414. DOI: 10.20546/ijcmas.2019.804.280.

Navindra, B., Arvind, R. and Hassina, B.B. (2009) Effects of antibrowning agents on the shelf life of fresh-cut green jackfruit (*Artocarpus heterophyllus* Lam.). *Journal of Applied Horticulture* 11(1), 35–40. DOI: 10.37855/jah.2009.v11i01.06.

Ohlsson, T. and Bengtsson, N. (2002) Minimal processing technologies of foods with non-thermal methods. In: Ohlsson, T. and Bengtsson, N. (eds) *Minimal Processing Technologies in the Food Industry*. Woodhead Publishing, Cambridge, UK, pp. 34–60.

Ong, B.T., Nazimah, S.A.H., Osman, A., Quek, S.Y., Voon, Y.Y. *et al.* (2006) Chemical and flavour changes in jackfruit (*Artocarpus heterophyllus* Lam.) cultivar J3 during ripening. *Postharvest Biology and Technology* 40(3), 279–286. DOI: 10.1016/j.postharvbio.2006.01.015.

Padre, S. (2011) *Forgotten Kalpavriksha*. National Jackfruit Fest, Thiruvananthapuram, Kerala, India.

Pareek, O.P. (ed.) (2001) *Ber*. International Centre for Underutilized Crops, University of Southampton, UK.

Rahman, A.K.M.M., Haq, E., Mian, A.J. and Chesson, A. (1995) Microscopic and chemical changes occurring during the ripening of two forms of jackfruit (*Artocarpus heterophyllus* L.). *Food Chemistry* 52(4), 405–410. DOI: 10.1016/0308-8146(95)93290-8.

Ramli, R.A., Azmi, A., Johari, N.R. and Noor, S.M. (2017) Minimally processed jackfruit: opportunity for the food service industry. *Journal of Food Research* 6(1), 1. DOI: 10.5539/jfr.v6n1p1.

Rana, S.S., Pradhan, R.C. and Mishra, S. (2018) Optimization of chemical treatments on fresh cut tender jackfruit slices for prevention of browning by using response surface methodology. *International Food Research Journal* 25, 196–203.

Ranasinghe, R.A. and Marapana, R. (2019) Effect of maturity stage on physicochemical properties of jackfruit (*Artocarpus heterophyllus* Lam.) flesh. *World Journal of Dairy and Food Sciences* 14, 17–25. DOI: 10.1155/2019/4327183.

Reddy, I.C., Prabakar, C., Devi, K.S., Ponnarasi, T. and Peter, Y.S. (2019) An economic analysis of jackfruit production and marketing in Cuddalore district of Tamil Nadu, India. *Plant Archives* 19, 2901–2809.

Rico, D., Martín-Diana, A.B., Barat, J.M. and Barry-Ryan, C. (2007) Extending and measuring the quality of fresh-cut fruit and vegetables: a review. *Trends in Food Science & Technology* 18(7), 373–386. DOI: 10.1016/j.tifs.2007.03.011.

Safari, S., Razali, N.A. and Mustafa, R. (2019) Distribution channel assessment: a case study in exporting Malaysian jackfruit to United Arab Emiratis (UAE) by air shipment. *International Journal of Agriculture, Forestry and Plantation* 8, 75–85.

Saha, M.G., Islam, M.N. and Molla, M.M. (2016) Determination of harvest maturity of jackfruit. *Bangladesh Horticulture* 2, 23–26.

Saxena, A., Bawa, A.S. and Srinivas Raju, P. (2008) Use of modified atmosphere packaging to extend shelf-life of minimally processed jackfruit (*Artocarpus heterophyllus* L.) bulbs. *Journal of Food Engineering* 87(4), 455–466. DOI: 10.1016/j.jfoodeng.2007.12.020.

Saxena, A., Bawa, A.S. and Raju, P.S. (2009) Optimization of a multitarget preservation technique for jackfruit (*Artocarpus heterophyllus* L.) bulbs. *Journal of Food Engineering* 91(1), 18–28. DOI: 10.1016/j.jfoodeng.2008.08.006.

Saxena, A., Bawa, A.S. and Raju, P.S. (2012) Effect of minimal processing on quality of jackfruit (*Artocarpus heterophyllus* L.) bulbs using response surface methodology. *Food and Bioprocess Technology* 5(1), 348–358. DOI: 10.1007/s11947-009-0276-x.

Saxena, A., Saxena, T.M., Raju, P.S. and Bawa, A.S. (2013) Effect of controlled atmosphere storage and chitosan coating on quality of fresh-cut jackfruit bulbs. *Food and Bioprocess Technology* 6(8), 2182–2189. DOI: 10.1007/s11947-011-0761-x.

Saxena, A., Maity, T., Raju, P.S. and Bawa, A.S. (2015) Optimization of pretreatment and evaluation of quality of jackfruit (*Artocarpus heterophyllus*) bulb crisps developed using combination drying. *Food and Bioproducts Processing* 95, 106–117. DOI: 10.1016/j.fbp.2015.04.005.

Selvaraj, Y. and Pal, D.K. (1989) Biochemical changes during the ripening of jackfruit (*Artocarpus heterophyllus* L). *Journal of Food Science and Technology* 26, 304–307.

Shewfelt, R.L. and Prusia, S.E. (1993) *Postharvest Handling – A System Approach*. Academic Press, California.

Siddiqui, M.W., Chakraborty, I., Ayala-Zavala, J.F. and Dhua, R.S. (2011) Advances in minimal processing of fruits and vegetables: a review. *Journal of Scientific and Industrial Research* 70, 823–834.

Sivakumar, D. and Korsten, L. (2006) Evaluation of the integrated application of two types of modified atmosphere packaging and hot water treatments on quality retention in the litchi cultivar 'McLean's Red.' *The Journal of Horticultural Science and Biotechnology* 81(4), 639–644. DOI: 10.1080/14620316.2006.11512117.

Soliva-Fortuny, R.C., Grigelmo-Miguel, N., Hernando, I., Lluch, M.Á. and Martín-Belloso, O. (2002) Effect of minimal processing on the textural and structural properties of fresh-cut pears. *Journal of the Science of Food and Agriculture* 82(14), 1682–1688. DOI: 10.1002/jsfa.1248.

Sudheer, K.P., Sarabya, S. and Sankalpa, K.P. (2019) Process prorocal standardization and shelf life study of minimal processed fresh-cut tender jackfruit. *Pharma Innovation Journal* 8, 164–169.

Tao, Y.M., Yao, L.Y., Qin, Q.Y. and Shen, W. (2013) Purification and characterization of polyphenol oxidase from jackfruit (*Artocarpus heterophyllus*) bulbs. *Journal of Agricultural and Food Chemistry* 61(51), 12662–12669. DOI: 10.1021/jf403828e.

Teja, T.R., Santhi, K.K. and Narsingharao, A. (2016) Edible coating of fresh jackfruit. *International Journal of Science, Environment and Technology* 5, 1658–1668.

Thompson, A.K. (2014) *Fruit and Vegetables: Harvesting Handling and Storage, Vol. 2*. Blackwell Publishing, Chichester, UK. DOI: 10.1002/9781118653975.

Toivonen, P.M.A. and Brummell, D.A. (2008) Biochemical bases of appearance and texture changes in fresh-cut fruit and vegetables. *Postharvest Biology and Technology* 48(1), 1–14. DOI: 10.1016/j.postharvbio.2007.09.004.

Troyo, R.D. and Acedo Jr., A.L. (2015) Optimizing the application of probiotic lactic acid bacteria for increased antimicrobial efficacy on fresh-cut jackfruit (*Artocarpus heterophyllus*). *Acta Horticulturae* 1088, 419–423. DOI: 10.17660/ActaHortic.2015.1088.73.

Ulloa, J.A., Aguilar, J.R., Rosas-Ulloa, P., Galavíz-Ortíz, K.M. del C. and Ulloa-Rangel, B.E. (2010) Effect of soaking conditions with citric acid, ascorbic acid and potassium sorbate on the phesicocmical and microbiological quality of minimally processed jackfruit. *CyTA – Journal of Food* 8, 193–199.

Vargas-Torres, A., Becerra-Loza, A.S., Sayago-Ayerdi, S.G., Palma-Rodríguez, H.M., García-Magaña, M. de L. *et al.* (2017) Combined effect of the application of 1-MCP and different edible coatings on the fruit quality of jackfruit bulbs (*Artocarpus heterophyllus* Lam) during cold storage. *Scientia Horticulturae* 214, 221–227. DOI: 10.1016/j.scienta.2016.11.045.

Vasudeva, K.R., Krishna, H.C., Suresha, G.J. and Swamy, G.S.K. (2020) Effect of modified atmosphere packaging on quality of fresh-cut jackfruit bulbs (*Artocarpus heterophyllus* L.). *Acta Horticulturae* 1299, 409–414. DOI: 10.17660/ActaHortic.2020.1299.61.

Wang, J.N., Chen, J.P., Gong, D.A.Q., Feng, F., Lyu, Q.F. *et al.* (2014) Effect of 1-MCP treatment on post-harvest physiology of jackfruit (*Artocarpus heterophyllus* Lam). *Acta Horticulturae* 36, 56–61.

Watada, A.E., Herner, R.C., Kader, A.A., Romani, R.J. and Staby, G. (1984) Terminology for the description of developmental stages of horticultural crops. *HortScience* 19, 20–21.

Wills, R., McGlasson, B., Graham, D. and Joyce, D. (1998) *Postharvest: An Introduction to the Physiology and Handling of Fruit, Vegetables and Ornamentals*. CAB International, Wallingford, UK.

Xu, F., He, S.Z., Chu, Z., Zhang, Y.J. and Tan, L.H. (2015) Effects of heat treatment on polyphenol oxidase activity and textural properties of jackfruit bulb. *Journal of Food Processing and Preservation* 40(5), 943–949. DOI: 10.1111/jfpp.12673.

Ying, J.C., Jaffar, N.S., Hussin, W., Hamid, M. and Muhsin, N.I. (2020) Maintaining postharvest quality of jackfruit during cold storage. *International Journal of Agriculture, Forestry and Plantation* 10, 282–286.

Zeng, L.P., Fan, H.Y., Zhang, M.X., Li, X.G. and Yan, S. (2018) Effect of different storage temperature on biochemical quality of fresh-cut jackfruit. *Acta Agriculturae Jiangxi* 30, 28–32.

Zhang, J.X. (2014) *Lasiodiplodia theobromae* in citrus fruit. In: Bautista-Banos, S. (ed.) *Postharvest Decay Control Strategies*. Academic Press, USA, pp. 309–335.

Index

Page numbers in *italic* type refer to figures.

CABI – who we are and what we do